## About the Author

MICHAEL WHITE is the former science editor of British *GQ*, as well as the former director of scientific studies and science lecturer at D'Overbroeck's College, Oxford. In addition to being the author of hundreds of articles covering the cutting edge of science, he is a musician and formerly a member of the successful eighties band the Thompson Twins. He was a consultant for the Discovery Channel series *The Science of the Impossible* and is the author of more than twenty books, including the international bestselling biographies of Stephen Hawking (with John Gribbin) and Isaac Newton. White lives with his wife and children in Australia.

## Also by Michael White

# Acid Tongues

## and Tranquil Dreamers

### Eight Scientific Rivalries
### That Changed the World

## Michael White

## Perennial
*An Imprint of* HarperCollins*Publishers*

First Perennial edition published 2002.

*Designed by Nicola Ferguson*

The Library of Congress has catalogued the hardcover edition as follows:

White, Michael.
Acid tongues and tranquil dreamers : tales of bitter rivalry that fueled the advancement of science and technology / Michael White.—1st ed.
p.   cm.
Includes bibliographical references and index.
ISBN 0-380-97754-0
1. Science—History.   2. Technology—History.
3. Competition (Psychology).
Q125.W32 2001
509—dc21          00-046309

ISBN 0-380-80613-4 (pbk.)

02 03 04 05 06  ❖/RRD  10 9 8 7 6 5 4 3 2 1

*For Kevin,*
*a great friend,*
*from keyboard to keyboard*

*The reasonable man adapts himself to the world: the unreasonable one persists in trying to adapt the world to himself. Therefore all progress depends on the unreasonable man.*

—George Bernard Shaw, *Maxims for Revolutionists*

# CONTENTS

# ACKNOWLEDGMENTS

I would like to extend my special thanks to David Milner, who came up with the idea of writing about scientific rivalry and was good enough to ask me to attempt this book. From there the idea grew to involve many others. Huge thanks go to Jennifer Brehl and all those at Morrow/Avon who became excited by my proposal, especially Lou Aronica, Jennifer Hershey, Lisa Queen, and Peter Schneider.

More thank-yous to Andy Brice; Marcus Chown; Kevin Davies; Russ Galen; Bill Gates; Geoff Mulligan; the archivists at the Lyndon B. Johnson Space Center, Houston; Cambridge Local History Library; Microsoft PR (Text 100), London; Peter Robinson; and Malcolm Wheatley.

Last, but certainly not least: Thank you, Lisa, India, George, and Noah; you've been very patient.

# Acid Tongues

## and Tranquil Dreamers

# INTRODUCTION

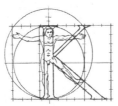

## The Long Road to Reason

In science you don't need to be polite, you only have to be right.
—*Winston Churchill*

A FORM OF SCIENCE EMERGED FROM the sun-blanched olive groves and sandy promontories of Greece. Twenty-five hundred years ago Anaximander, Pythagoras, and Anaxagoras walked this land, and perhaps the slap of sandals on rock or the beating of the waves on the seashore inspired thoughts of rhythm, of cadence, of harmony and symmetry, for they had taken vague ideas from what was known of Babylonian and Egyptian culture and had begun the process of quantifying, interpreting, imagining.

A century and a half after these pioneers, Aristotle's polymathic view stood as quintessential Classical culture, and a tiny group of people centered in Athens delved into many aspects of the world they witnessed. They employed deductive reasoning, syllogistic logic to draw conclusion C from ideas A and B, and this is why they practiced only "a form of science": the Greeks shunned the concept of experiment but excelled with their thoughts of the fundamentals. They wondered what the universe might be made from, whether it might be a mere abstraction. Could the universe have been formed around a conceivable pattern? Were numbers repeated in its obvious glory?

It is intriguing to imagine the great philosophers of Classical times squabbling over priority, angry words exchanged in the amphitheater, a young radical expelled from the academy perhaps, but we can trace little of their thought, and any rivalry or bitter exchanges there may once have been are forgotten, the tongues of conflict long stilled. However, we may at least study faint echoes of the way intellectuals fought over cerebral property.

Aristotle (384–322 B.C.) created a vast mosaic, an interpretation based upon what he saw and what he imagined, and through prominence and serendipity his philosophy lingered long, astonishingly long. For two thousand years Aristotelianism was taught in the rain of Oxford and Paris and lauded at the Frankfurt Book Fair just as it had been celebrated and honored in sunny Macedonia, where Aristotle had instructed his most illustrious pupil, Alexander (later, the Great).

But Aristotle's was not the only voice, and the best remembered is not necessarily the best. The ideas of Democritus (c. 460–c. 370 B.C.), brought to us in the writings of Lucretius (c. 95–55 B.C.), describe a mechanical universe, a physical realm in which atoms form the most fundamental conglomerates and from their collisions create all movement and dynamism.* Democritus and his followers applied atomism to every aspect of the observed world and even tried to explain human behavior as a consequence of atomic collisions.

"This fright, this night of the mind, must be dispelled not by the rays of the sun, nor day's bright spears," Lucretius wrote two thousand years ago, "but by the face of nature and her laws. And this is her first, from which we take our start: nothing was ever by miracle made from nothing."[1]

Two generations after Democritus died, Plato (c. 428–347 B.C.) destroyed this vision with his semimystical interpretation of the universe. Within sci-

---

* Lucretius wrote his accounts some three centuries after the death of Democritus.

entific development, Democritus had taken a leap forward and Plato moved the world two steps back. "Plato was a great philosopher," one historian has declared, "but in the history of experimental science he must be counted a disaster."[2] And with Plato's greatest pupil, Aristotle, the world took another step back.

But none of these men feuded. Democritus was dead before Plato did his work; instead, their battleground lay some twenty centuries in the future within the vaulted halls of the great universities and upon the crisp pages of newly published books written by the founders of the Enlightenment: René Descartes (1596–1650), Pierre Gassendi (1592–1655), Isaac Newton (1642–1727), and Robert Boyle (1627–1691).

In the fourth century B.C. Aristotle's words had found immediate resonance, his comforting description of the four elements—fire, earth, air, water—his cozy ideologies and reassuring connections that placed humans immutably at the center of things, all this gave the living a sense of meaning. And as Aristotle's ideas—that we see because our eyes project particles that bounce off viewed objects; that an apple falls because it is trying to establish its rightful place in the universe; that an object moves through the air because as it does so the displaced air in front of it flows behind it instantaneously and pushes it on—these took priority, molding the thinking of one hundred generations. Democritus was forgotten.

And so, for Western civilization, darkness.

Some of the earliest human ideas that had flowed west to Greece as civilization dawned now dispersed across the world, greatly modified. The Arabic culture that nurtured so many ingenious natural philosophers and mathematicians expanded the canon enormously: Ibn-al-Haitam (965–1020), Rhazes (860–930), and the alchemists of Persia who passed west through Alexandria, leaving their inimitable mark. In Europe memory faded and the ideas of the Classical tradition were kept alive only in the monasteries.

But in the monasteries all was not well. To maintain spiritual peace and earthly power, the theologians and monks needed to find an amalgam of natural philosophy (the teachings of the heathen Greeks) and Christianity, a marriage of Aristotle and the Gospels. And perhaps here, in this dark era, we had the potential for the first grand rivalry, a clash born out of obvious inconsistencies between observation and faith. But somehow it was avoided. The intellectual fathers, the holy men who looked at the world and wondered, men like Thomas Aquinas (1225–1274) and Albertus Magnus (c. 1200–1280), facilitated a strange, short-loved compromise, a meld of Aristotelianism and Christianity they called Scholasticism.

But not all were entirely deceived. Roger Bacon (c. 1220–1292) was bigger than the myopic age in which he lived. Devout and pious to a fault, he was also a critical observer and did not always take the word of Aristotle as second only to that of God. As he watched, learned, and gathered information, as he dipped a toe into forbidden alchemy and dared to question, he tried to loosen the overtaut bond between inherited wisdom and inherited faith. In a triumvirate of visionary books, *Opus Majus, Opus Minor,* and *Opus Tertium,* he argued against elements of the Aristotelian creed (for creed is what it had become by this time). Foolishly perhaps, he arranged for exquisitely bound editions of his books to be presented to the pope, Nicolas IV, who, far from appreciating Bacon's genius, had him arrested for heresy and imprisoned for life.

Long before Roger Bacon it must have become clear that the only way to discover Truth, to probe deeper than the word of Aristotle and the word of God, was to keep silent outwardly but to shout loudly inside. And so, for a millennium, from the foundations laid by the Arabs of the seventh century until the analytical blending of alchemy and science that sparked Isaac Newton's genius, the alchemists who worked in secret laboratories and fled prosecution across continents carried the torch of investigation.

Alchemists—men who studied an early form of chemistry with philosophical and magical associations, the chief aims of which were to transform the base metals into gold and to discover the elixir of perpetual youth— were misguided fantasists certainly. Nonetheless, they were intrepid individualists who consciously took a path less well traveled. They were not true scientists, but they had imagination and determination and they did not accept the givens so cherished by their rivals, the orthodox philosophers and theologians. The rivalry that was later to burst into the open with the attempted intellectual assassinations of Copernicus, Bruno, Galileo, Hobbes, and Darwin had its origins within the scantly recorded tension between the dogmatists of the Church and the naïve experimenters who spent their lives at the caldron risking the noose.

On a philosophical level, the alchemists were unified; they shunned what they had been told to believe. But we must not forget that concurrently they were in agreement about nothing else. Each alchemist pursued his own path, created his own rules, and this is one of the reasons they should not be considered scientists. Alchemists were rivals par excellence. They had common goals—the elucidation of the twin pillars of the Hermetic tradition, the elixir of eternal life and the philosophers' stone—but each had his own agenda. Alchemists recorded their findings in coded form, and in order to ennoble and at the same time obfuscate their findings, they drew upon the rich culture of mysticism, from the cabala to the New Testament. This was not simply to keep their ideas from the eyes of the Church and the State; it sealed discovery in a cocoon, removed it from interpretation, extrapolation, and, most crucially, theft.

And so these men were pulled in two directions simultaneously. The enemies of orthodoxy, driven to learn more than Aristotle or the theologians could offer them, risked everything to unveil what they perceived as Truth. But while struggling to prize secrets from Nature's jealous clasp, they missed the chance to develop any truly profound understanding of the universe because they could never share their philosophies, never communicate their findings, never begin to build the edifice of science as later generations would do. Torn between the effects of positive and negative rivalry, they achieved little of lasting value.

LEONARDO DA VINCI (1452–1519) was no alchemist; rather, he was the first scientist, yet, sadly, he too suffered the alchemist's failing and could never bring himself to communicate his revelations. Leonardo was paranoid, but in part for good reason. There were those who wanted to plagiarize his ideas, rivals who planted assistants in his workshop to steal, and there were spies from Rome waiting for him to slip up. To counter these agents Leonardo filled thirteen thousand pages of notebooks using mirror writing (as a defense against the casual glance over his shoulder perhaps), and he hid his discoveries from everyone save his most trusted disciples.

But defending himself against the prying eyes of the Holy See was far from easy. Leonardo spent most of his life in Milan and Florence during a period in the history of those cities when holy repression was at its least effective. But as soon as he left these regions he needed to be constantly aware of censure. Famed throughout Italy and revered by the powerful,

Leonardo still had to tread carefully in pursuit of his more risqué interests. One of his assistants, he said, "hindered me in anatomy, denouncing it before the Pope."[3] Leonardo offended few with his ideas for flying machines, and he delighted his patrons with ingenious designs for weaponry, but the Church did object to his nocturnal habits: For his deceitful assistant, horrified by the sight of his master up to the elbows in human viscera, such work was a violation; and it was no less so to the cardinals. Such censorship forced Leonardo into a peripatetic life, keeping him barely one step ahead of his enemies until he was given protection by the young Francis I at Cloux, where, during his final years, he could do as he pleased.

As a scientist, Leonardo had as his greatest rival the Catholic Church, and ideological conflict both spurred him to new heights of ingenuity and caused immense damage. Leonardo was a free spirit who refused to be nailed to any cross. While he offered Rome every outward sign of respect, his only vision of God was a pantheistic one, and, most unusually for an intellectual of the time, he made almost no mention of the divine in anything he wrote. Facing the wrath of the Church, he was defensive but proud; any attempt at prohibition made him work harder, dig deeper, and find more that would shock and infuriate, if only he had been able to speak out.

But Leonardo published nothing (his *Treatise on Painting* is his only complete work, finally published in 1651), for as much as he wanted to reveal Truth, claiming: "There is nothing more deceptive than to rely on your own opinion, without any other proof, as experience always proves to be the enemy of the alchemists, necromancers, and other ingenious simpletons," he simply could not.[4] Before he died, he entrusted his notes to his closest companion, Francesco Melzi, who spent the rest of his life attempting to catalog and clarify the thousands of pages Leonardo left the world. When Melzi died, his son, Orazio, a man with no interest in Leonardo, filed the papers away in an attic room on the family estate. There they remained for almost two centuries, lost to the world and the evolution of science. The Church may not have stopped Leonardo working, but it effectively cut out his tongue.

LEONARDO WAS NOT the only hero. Three others from the pre-Newtonian era figure hugely in the struggle for reason. Each fought in his own way and was victimized by papal power; each became an enemy of ignorance.

Nicolaus Copernicus (1473–1543) was one of the Church's own, a Polish canon with medical training and a fascination with astronomy. He knew well the power of his greatest rival, the agonies it could inflict, and he could not face them. Working in secret, he gathered astronomical observations and scribbled for thirty years before submitting his thoughts for publication: but only when he knew he was dying. Copernicus had no close family, no one Rome could persecute after he had gone, and he must have felt a swell of satisfaction as the first copy of his treatise was placed beside him on the bed.

The year was 1543, and although Copernicus could not have known it, science had won. For sure, the victory celebrations would lie some time in the future, and others would suffer, even die, for their knowledge before that glad day, but as *De revolutionibus orbium coelestium* (*On the Revolutions of the Heavenly Spheres*)—the first great scientific work of the new age of printing and one of the most important—emerged from the press, Rome, if only it had realized the true importance of the book, would have shivered.

The cardinals had not yet noticed anything was wrong for two reasons. First, Copernicus's publisher, a Lutheran minister, had, without the author's consent, included a preface to *Revolutions* in which he declared that the treatise was merely an aid to calculation of planetary movement and not a statement of reality. Second, Copernicus had either deliberately or unintentionally confused his own message.

He had noticed that the stars and the planets moved in such a way that the earth could not possibly lie at the center of the universe or "the heavenly spheres," but he had retained many Classical ideas in the way he explained these observations. Furthermore, he began his treatise boldly by asserting that the Sun lies at the center of the universe but then appeared to change his mind. After the first few pages, he complicated his theory more and more with unnecessary refinements, finally placing the Sun slightly off-center. This prevarication made the entire work almost unreadable and frequently contradictory. Although the whole runs to 212 sheets in small folio, the heart of *Revolutions* may be found in the first twenty pages. As a result of these confusions, *Revolutions* did not have the immediate scientific impact it should have and went unnoticed by the Church for half a century, finding its way onto the Index Librorum Prohibitorum, the Index of Forbidden Books, only in 1616.*

Even so, Copernicus had good reason for maintaining secrecy. In his treatise he had rejected the words that had for so long massaged the egos of men, the geocentric model taught since ancient times—the very essence of Aristotelianism. "In the midst of all dwells the sun," he wrote in those crucial first twenty pages. "Sitting on the royal throne, he rules the family of planets which turn around him. . . . We thus find in this arrangement an admirable harmony of the world."[5]

Absolute anathema to the enemy: When this heliocentric vision was eventually understood, it was of course immediately branded as heretical. But it was too late. The Church may have been slow to understand the radical nature of what had happened, but some intellectuals of the period had worked through the muddle that was *De revolutionibus orbium coelestium* and from it had drawn their own conclusions.

*F*EW CHOOSE TO DIE in a way that changes all of history. Jesus Christ is one who had that opportunity and made the most of it. Giordano Bruno (1548–1600) was another and became the first and only martyr to science.

Late in January 1600, Bruno stood in chains before the Inquisition court of the Roman Church in the Vatican and was condemned to death by His Holiness Pope Clement VIII. Bruno's crime had been to publish heretical works—*The Ash Wednesday Supper*, *De immenso et innumerabilibus seu de universo et mundis*, and *De vinculis in genere*—based upon *Revolutions* and blended with his own idiosyncratic vision of natural philosophy.

Bruno had been persecuted for decades, his books banned, his ideas repressed, but like Leonardo da Vinci a century before him, he had always succeeded in keeping one step ahead of the Church and spent most of his life in liberal or Protestant states, in England and Germany. But in 1591 he received an offer to teach a Venetian nobleman named Giovanni Mocenigo and made the strange decision to return to his native Italy.

It was a trap, and Bruno was led to the slaughter. Mocenigo was work-

---

*It was removed in 1835.

ing for the Inquisition, betrayal was in his blood. Bruno faced trial first in Venice and then in Rome, where he was incarcerated in a tiny cell for seven years, tortured and humiliated, then burned alive. Bruno was everything the Church despised and feared, a rival who offered an alternative vision of the universe. He was not burned over some pedantic detail of Catholic doctrine or transient political view but because he possessed the power of communication—people listened to him, and they read his inflammatory words. Three quarters of century earlier, Martin Luther had shaken the roots of Catholicism by attacking the structure of the Church and lambasting the pope for his decadence. But Bruno, like Leonardo, like Copernicus, like Kepler, like all Truth-seekers, attacked the givens, the philosophical foundations. These men offered no mere alternative form of worship, as Lutherans and Calvinists did, but a full-blown rival ideology.

The cardinals had tried to silence Giordano Bruno with edicts, proclamations, excommunications, but in the end they were forced to seal him in a room six feet square before piercing his tongue with a metal spike to stop him from spreading his subversion to the crowds who flocked to see him burn in Campo di Fiori, the Field of Flowers.

Bruno's downfall stemmed from a refusal to accept orthodoxy, to attempt an amalgam of the science of Copernicus (which the Church persecutors still did not understand fully) and a belief in a Catholic God. To the Holy See, Bruno was an arch-heretic, but he had never lost his faith in the divine, and in some ways he was a traditional Catholic. Unfortunately for him, the world was not ready for a man who spoke of life on other planets, a God that was more pantheistic than biblical, and a science that discarded almost everything Aristotle had taught. As early as the 1580s Giordano Bruno had become a champion of Democritus and the atomists and questioned what had been cast in stone, asking: What is matter? What is energy? How could an infinite universe exist? And if it did, what did it mean?

Bruno offered a poetic vision of these things; he, like Leonardo da Vinci, had no math. Only now, in a world explained by quantum mechanics and guided by relativistic insights, can Bruno's worldview be appreciated. His model of a universe in which all things are interconnected on an atomic level bears comparison with ideas that sprang from such exotica as superstring theory during the 1990s.

As visionary as his ideas were, and as otherworldly, the Church knew Bruno as a deadly rival, one it had to murder. His vision was warm and personal, it touched both the poet and the analyst; one day it would influence men like Werner Heisenberg and Albert Einstein, and in the sixteenth century it sent a cold chill along the spines of the God-fearing cardinals.

$\mathcal{B}$UT IT DID NOT END with a burning; how could it? A generation after Bruno's fire burned low, the Church was to find another enemy in its midst, a man from a very different mold than Bruno, but a natural philosopher who was both appreciative and fearful of Giordano's legacy.

By the time Galileo Galilei (1564–1642) came into conflict with the Mother Church, he was already the most famous scientist in the world, a close friend of the pope, Urban VIII, and respected by both nobility and clergy far beyond his native Tuscany. Galileo had walked a slender tightrope throughout his career. Wary of all that had gone before him, he was conscious of the need for delicacy, and outwardly he had been a good Catholic. But by 1632, when he published his *Dialogue on the Two Great World Systems, Ptolemaic and Copernican*, he had come to believe he could portray Copernican astronomy in a way that would not be considered heresy and that, in so doing, he might open up a tiny, seemingly innocuous crevice into which reason might flow.

He was wrong. His friend the pope, a very intelligent man but one who had grown hard and impatient in direct proportion to his burgeoning power, realized Galileo's attempted deception, and within months of publication of the *Dialogue* he had its author hauled before a committee of ten cardinals to face charges of "vehement suspicion of heresy." While preparations for the trial began and Galileo was kept under house arrest, a furious pope had the Tuscan ambassador to Rome brought before him to hear: "Your Galileo has ventured to meddle with things that he ought not to, and with the most important and dangerous subjects that can be stirred up these days."[6]

Like that of Bruno thirty-three years earlier, Galileo's crime was to present Copernicus's heliocentric model as fact, thereby threatening the supposed infallibility of Aristotle and, by association, ecclesiastical dogma. But Galileo was no wandering philosopher, no excommunicated cleric; he was treated with enormous respect, held in custody in palatial quarters in the Vatican, and given the consideration due his stature. Still, such respect went only so far. After a trial lasting two months, he was found guilty of propagating heretical ideas in his published works. Galileo was not made of the same stuff as Bruno, but then not many are. He was certainly wiser and more self-protecting, yet even this demoralized and humiliated sixty-nine-year-old who had been ordered to his knees to declare before the court "with sincere heart and unfeigned faith" that Copernicus had been wrong could not resist a final swipe at his enemies. Legend has it that as he left the courtroom and the cardinals' belligerent stares he whispered: "Nevertheless it does indeed move."

Luckily for Galileo, few heard the comment. He was spared the stake and sentenced instead to house arrest in a small villa, Il Gioiello (the Jewel) near Florence. He died there nine years later.

Galileo was an altogether different sort of "scientist" than Bruno. Empirical, an analytical experimenter, he was the first "mathematical scientist," whose technique was an inspiration for Descartes, Newton, Boyle, and others. And as Bruno was forgotten, only to reemerge as an influential force during the twentieth century, Galileo's version of science would lead directly to the *Principia*, the steam engine, the car, and the space shuttle. The Church wanted none of this. Under pain of rearrest and certain execution, Galileo was banned from publishing his thoughts, and the *Dialogue* was added to the Index Librorum Prohibitorum.

But the Church's day was passing, its tyranny becoming increasingly ghettoized, for Galileo was not alone. Within months of his arrest, helpers had managed to smuggle his writings from the villa and to find very willing publishers in the free states of northern Europe. In this way one of Galileo's most influential works, *Discourses and Mathematical Demonstrations Concerning Two New Sciences* (written during his final days at Il Gioiello), reached a wide audience far from Italy.

Just weeks before Galileo's death, the great atheist thinker Thomas Hobbes (1588–1679) managed to gain an audience with him and imparted the extraordinary news that the *Dialogue* had been translated into English, a rare thing indeed in 1642.

With the persecution of Galileo, the rivalry between the Church and natural philosophy had reached a great turning point. After Galileo, nothing would be the same again: Experimental science had begun, and its value had been passed on to a burgeoning audience. Galileo had planted seeds that would bring a rich harvest, and although a long road still lay ahead, reason could no longer be everywhere tempered by the demands of Rome.

Even so, it was a torturously slow process. When Newton went up to Cambridge in 1661, the curriculum was little different from the Hellenic model taught there since the foundation of the university at the beginning of the thirteenth century. But the words of Galileo, Descartes, and Boyle, and the musings of the alchemists, were there to save him. Even in Darwin's student days, a century and a half after Newton, the great universities were retrogressive in the extreme and every student was required to swear an oath to follow the Thirty-nine Articles of the Anglican Church. Darwin demurred, but, as we will see, by the end of his life he was to have the final say on the relevance of religion.*

This battle between science and the Church has been discussed and dissected since it began, and we have only to consider how Galileo was treated, how ideas were suppressed, and how Bruno died to see this schism was certainly a genuine one; indeed, the Index Librorum Prohibitorum says it all. Even so, recently it has become fashionable for some scholars to attempt to show that the Church was actually more enlightened than had been believed previously and that it did not try to smother reason and innovation. One important piece of modern research in this field attempts to demonstrate that many Catholic scholars between the sixteenth and eighteenth centuries were encouraged by the Vatican to conduct serious astronomical observations.[7]

This is true, the Church did sanction certain astronomers and allow observatories to be built, paid for from the coffers of the Vatican. Giovanni Cassini (1625–1712) was the most famous astronomer to benefit from this patronage, and he conducted many observations of the Sun in Church-funded observatories during the 1650s. The official purpose of these observations was to produce a more accurate calendar so that Church officials

---

*Galileo's findings were officially accepted by the Roman Church in 1992, twenty-three years after humans first walked on the Moon, and in 2000 Pope John Paul II asked God's forgiveness for the wrongs committed by Rome in past centuries. The Church has yet to admit fault for its treatment of Bruno.

could establish the date for Easter each year, which the Council of Nicaea in A.D. 325 had set at "the first full moon after the vernal equinox."*

But this is not the end of the story, for in allowing such things the Church had clear ulterior motives. Keen to find any form of evidence to support its anti-Copernican convictions, the Church created its own team of astronomers to scan the heavens in search of proof with which it could fight science on its own terms. This was naturally a self-fulfilling exercise, because powerful figures in the Vatican suppressed evidence that might appear to support Copernicus and advertised anything that could be used to refute it. The world would know of the results only if they were favorable to the Church.

However, it has been claimed that some observations made by Church astronomers did indeed contradict Aristotle. An oft-cited example is the story that Jesuit astronomers noticed dark patches on the surface of the Sun and were the first to describe the phenomenon (later known as sunspots), even though this observation went against the teachings of Aristotle, who had declared that all heavenly bodies were flawless spheres.

Perhaps such discoveries were made and accounts written, but these workers were forbidden to publish their findings or to include news of such discoveries in their lectures. In other words, they were neutered, they became intellectual eunuchs. If nothing else, this alone demonstrates that the Church was a closed system, its senior figures afraid of rival doctrines. It was an authority that operated as a purely oppressive force. It should not be forgotten that the Church tolerated those who argued against the geocentric model but only so long as (like Copernicus's cautious publisher) they left their speculations as just that—mathematical curios—and decidedly not competing worldviews.

WHEN WE CONSIDER the Herculean figures on the long road to reason and the contributions they have made to our modern worldview, it is only natural to wonder what drove them, what led them to discover. Indeed, such consideration makes us question the very meaning of discovery. In so many ways scientists and the natural philosophers who preceded them have

---

*When the hours of light and darkness are equal.

much in common with artists, musicians, writers. Indeed, many who have excelled as scientists or artists have shown twinned talents. Sir Edward Elgar was trained as a chemist, Aleksandr Borodin was a distinguished professor of chemistry, Albert Einstein was considered a capable amateur violinist, and Leonardo was of course recognized first as a consummate artist and only later as a scientist. Perhaps such people hold the key to understanding how discovery happens, for scientific innovation and artistic creation bear remarkable similarities and seem to stem from an identical impulse.

Many people perceive the scientist as a dry individual, a caricature, a man with wispy hair in a white lab coat. This is of course a stereotype beloved of the media and as false an image as the artist in a beret and paint-daubed smock. Such hype is not to be trusted. Furthermore, the notion that science is a dry pursuit or, worst still, purely a "useful" endeavor must be ignored. Certainly scientists are people of their time, they share the concerns of us all and work within a limited cultural framework, but in a direct parallel with that of the artist, the work of the scientist transcends time and place. Science is indeed "useful," but so are words, paints, violin strings.

"The scientist does not study nature because it is useful to do so," France's greatest theoretical scientist, Henri Poincaré, wrote toward the end of the nineteenth century. "He studies it because he takes pleasure in it, and he takes pleasure in it because it is beautiful. If nature were not beautiful it would not be worth knowing, and life would not be worth living. I am not speaking of course of that beauty which strikes the senses, of the beauty of qualities and appearances. I am far from despising this, but it has nothing to do with science. What I mean is that more intimate beauty which comes from the harmonious order of its parts, and which a pure intelligence can grasp. . . . Intellectual beauty, on the contrary, is self-sufficing and it is for it, more perhaps than the future good of humanity, that the scientist condemns himself to long and painful labours."[8]

In other words, science is transcendent, and this transcendency holds a key to its meaning. Transcendence imparts power and creative energy, it fuels dreams and ambition, and, most crucially, it makes science universal. The physicists working on the Manhattan Project to build a bomb to destroy whole cities knew this. They knew their bomb would be "useful," and they were terrified the enemy, their rivals in Germany, would acquire it first. But they also knew what would come from this work would be universal, a science that would eventually benefit humanity and offer insights into the deeper meaning of existence. Crick and Watson were chasing a Nobel Prize, seeking fame and glory, but as they worked to unveil Nature's paradigm, they were excited still more because they were revealing a universal truth, they were swimming in the current of life.

During the middle of the Napoleonic Wars, Bonaparte was outraged to learn that the Institut de France he had founded in 1795 had awarded its first scientific prize to an Englishman, Humphry Davy. But Napoleon's scholars knew that science was more important than nations and would outlast mere empires.

Transcendence holds one key, but the here and now holds the other. Ego is married to creativity, ambition and aggression are wed to the drive to discover. The creative mind is possessed by angels and devils. The angels offer transcendence, they capture the inspirational moment and make diamonds from dust; the devils look across the laboratory bench and whisper of ambition, offer challenges, and spur rivalry; each group plays its part.

Scientific discovery is based upon the excitement of argument. The scientist argues with Nature. The scientist says: This is what I think, now let's see if it's true. If the idea is shown to be false, then the scientist rethinks, reworks, and tries again; experiment follows experiment until ideas and observation coalesce.

Argument and the threat of challenge are certainly not drives that motivate only the scientist, and it is not only science that has advanced through rivalry. We need only consider Shakespeare spurred on to greater creative heights when faced with Christopher Marlowe's genius. Let us recall Antonio Salieri draining his soul of music to overshadow Mozart, Leonardo and Michelangelo attempting to outdo each other while painting murals on opposite walls of the Council Chamber of the Palazzo della Signoria in Florence. But argument is ingrained in science, and the scientist can operate only by challenging the rules, arguing with what is already known and attempting to find new, better answers to fundamental questions.

"You'd think that scientists would have a degree of saintliness that would be almost unbearable," the Nobel Prize–winning physicist Dr. Leon Lederman said recently. "It doesn't work that way. The competition goes on at all levels—the international, the national, the institutional, and finally the guy across the hall."[9]

The scientist lives in a dynamic world in which ideas are changing and developing all the time. Scientists need to compete so their ideas may evolve; standing still means annihilation, a fact known long before Darwin quantified it. And, considering the often esoteric nature of some science, it is remarkable how often discoveries and ideas come to more than one thinker contemporaneously. In some cases, it is almost as though a concept is there in the ether crying out to be interpreted and two or more great minds rush in and seize it; and, of course, priority disputes usually follow.

Many of the conflicts described in this book are of this type. Newton and Leibniz, Lavoisier and Priestley, Crick and Watson racing against Rosalind Franklin and Maurice Wilkins are just a few examples. But there are many others. The arguments over who first discovered electromagnetic induction raged for years, and today Michael Faraday rather than the American Joseph Henry is seen as the victor. John Couch Adams calculated the orbit of an unknown planet later christened Neptune only months before a French astronomer, Urbain Leverrier, published almost identical findings. Who should be considered the true discoverer of antibiotics, Alexander Fleming or Howard Florey? In more recent times, Richard Leakey and Donald Johanson have fought over who discovered the oldest fossil to be identified as a species of prehuman ancestral to other known forms of human. And we should not forget the very public clash between Robert Gallo and Luc Montagnier, each investigating the nature of AIDS, whose opposing theories appeared simultaneously.

But priority disputes are not the only sources of conflict in science. In some cases what appears to be a scientific clash stems from unspoken but nevertheless deep-rooted animosities and hidden agenda. As we shall see, Newton felt threatened by the very presence of Leibniz, who was working with equal ingenuity in areas of science Newton considered his domain. Lavoisier and Priestley held different religious and political views as well as fundamentally incompatible scientific opinions. The English mathematician John Wallis (1616–1703) battled viciously with the philosopher and mathematician Thomas Hobbes and helped destroy the latter's scientific reputation not only because he disagreed with his mathematics but because Wallis, a devout Christian, felt compelled to quash the atheistic opinions

Hobbes had expressed in his masterpiece *Leviathan*. Wallis wrote of Hobbes, "Mr. Hobbes loves to call that chalk which others call cheese,"[10] an expression that has lasted well.

A century before Wallis and Hobbes's exchanges, the Danish astronomer Tycho Brahe (1546–1601) engaged in a merciless campaign to discredit a rival that eventually led to a lawsuit. Brahe, whom Galileo loathed, had proposed a radical, anti-Copernican theory for planetary movement in which the Sun revolved around the earth and all the other planets orbited the Sun. A few months after he established this idea, a little-known astronomer named Ursus independently postulated an almost identical hypothesis. Tycho was outraged and immediately branded Ursus a thief. But underlying this dispute was the fact that Tycho Brahe was of noble birth and renowned for his elitism, while Ursus, astronomer to the court of Rudolf II in Prague, had come from a peasant family.

If egos, personal vendetta, and scientific conviction combine, the blend can be heady indeed; spite becomes almost boundless. At the height of a dispute based upon the way each viewed a detail of reproductive biology, Voltaire (1694–1778) suggested his rival John Needham (1713–1781) was a homosexual. "What I shout, a Jesuit transfigured among us, a teacher of young men! This is dangerous in every way!" Voltaire exclaimed.[11]

To his face, Darwin's great supporter Thomas Henry Huxley called his reviled opponent, Richard Owen, "a liar"; and because Robert Oppenheimer thwarted his plans to push the Los Alamos scientists into his preferred research path, Edward Teller testified that Oppenheimer had Communist sympathies and encouraged Senator Joseph McCarthy to strip him of his security rating and dismiss him from all government positions.

In 2000 a French archaeologist, Emilia Masson, told police investigators that her rival, Henry de Lumley, had deliberately destroyed evidence at a dig because it supported her theory over his.

And scientific dispute is sometimes not the sole preserve of individual scientists. The Manhattan Project and the German effort to construct an atomic bomb were central to a race involving large teams of scientists, each funded by warring nations. In a similar way, the space race was a gargantuan scientific effort propped up by the Cold War.

During ancient times, the results of rivalries took huge expanses of time to alter perceptions. During the Middle Ages and the Renaissance, scientists (natural philosophers) were almost entirely ignorant of each other's existence and there was very little communication between thinkers; no mechanism had yet been found for open dialogue, or to fall out. In these remote times, rationalists and visionaries had a single common enemy, the Church. Later, learned societies formed in which the earliest experimenters could discuss ideas with mathematicians and engineers, philosophers and medics. One of the first was the Pinelli Circle, established during the 1590s in Padua by the wealthy nobleman Gianvincenzo Pinelli, friend and supporter of Galileo. With the invention of movable type and the establishment of the first printing presses, ideas could flow between nations, and the societies that were created on the Pinelli model—the Royal Society in London, the Académie des Sciences in Paris, and later the Berlin Academy, the St. Petersburg Academy, and others—became centers for debate, experiment, and, of course, conflict.

Many of these societies published their own journals. Most influential was the *Philosophical Transactions* of the Royal Society, in which scientists could present their findings for open discussion with other members and subscribers abroad. This method of communication eventually led to the establishment of learned journals, science magazines such as *Nature*, which was first published in November 1869. This new forum did much both to publicize science and to act as arbiter in scientific disputes.

Beyond the tales of personal bitterness and the painful effects of some scientific rivalry, we must consider what comes from these battlefields. And it seems the overwhelming effect has been to propel science forward. The late Karl Popper observed that the longer two intelligent people argue, the better their arguments become; each is being all the time improved by the other's attacks. So whatever lies at the heart of a dispute, be it a priority race, nationalistic fervor, personal hatred, or any combination of these, strife and competition have done much to advance our understanding of the universe.[12] This has happened in many ways. Most significant, competition has driven individual scientists, research teams, institutions, and nations toward greater effort. In some cases such competition has nurtured, within just a few years, significant scientific advance that would otherwise have required decades.

Beyond this, disputes that reach the gaze of others within the scientific community spur interest in the subject of the quarrel. An example of this is the ongoing disagreement between the supporters of the late astronomer royal Sir Martin Ryle and their determined opponent Fred Hoyle over the nature of the early universe. Ryle postulated what has become known as the big bang theory, which states that the universe began with a singularity of infinite density.* His opponent advocates the theory of the steady state universe. The dispute has been documented in dozens of books and hundreds of research papers and has drawn into the debate many scientists who would otherwise have shown little interest in trying to find evidence to support either party. As a result, enormous progress has been made in efforts to solve the big question of how the universe began.

Taking this forward another stage, scientific disputes have also heightened public awareness of science. Rivalry and vituperative exchange make great headlines, and a race to achieve a scientific goal adds a certain glamour to subjects that many people find difficult to understand. Darwin's bold assertions, expressed through his mouthpiece Thomas Huxley, heralded an age in which the media and the public could begin to follow learned exchanges between scientific rivals. This continued into the age of electricity and the debacle over AC and DC involving the world-famous "Wizard of Menlo Park," Thomas Edison. The world was stunned by the race for the structure of DNA, in which Crick and Watson were triumphant, and as children many of us watched fuzzy black-and-white TV pictures showing the latest developments in the space race. I was certainly not the only young person who was inspired by these events and encouraged to take up science because of them.

But not every dispute has led to a positive outcome. Isaac Newton fought with many of his contemporaries, and his earliest clash, with the then curator of experiments at the Royal Society, Robert Hooke, led to an outcome that was decidedly detrimental to scientific progress. When Hooke dismissed a paper of Newton's (*Theory of Light and Colour*) in which he described his optical experiments, the hypersensitive Newton took his

---

*"Big bang" was created by Hoyle as a term of derision and came from a BBC Radio interview in 1950, when he described Ryle's theory as "a Big Bang." He then said that the idea was "about as elegant as a party girl jumping out of a cake."

work away from the Royal Society and refused to allow its publication for thirty years, thereby slowing our understanding of optics because no other scientist could learn from the work.

In a different field and a century later, the row between Voltaire and John Needham over how the mechanism of reproduction works on a fundamental level (about which both were actually wrong) is believed to have slowed progress because it distracted others in the field from pursuing more fruitful lines of investigation. The historian of science George Sarton has said of this sorry event: "Thus was the fine observational tradition of the seventeenth century interrupted, or at any rate slowed down for more than a century, by discussions which were irrelevant."[13]

More recently, Crick and Watson's triumphant elucidation of DNA may have produced different headlines if their greatest rival, Rosalind Franklin, had been able to work with her colleagues at King's College in London rather than allow what we could call negative rivalry to stymie her best efforts.

$\mathcal{T}$HE EIGHT DISPUTES covered here constitute a sample taken from the long history of scientific rivalry and were chosen in an attempt to cover the period from Newton's era to the present day, and to encompass as diverse a collection of disciplines as possible. As well as this, the choices were made to illustrate the various forms of rivalry I have touched upon in this introduction—personal, national, and industrial.

Rivalry is a reflection of humanity, and as human culture has changed, so has the guise of rivalry. In simpler times scientists pitted their wits to reveal Nature and sometimes also to expose what they believed to be falsity in their competitors. As science became public property, it was used by the governments that represented entire nations. Today, and perhaps only temporarily, perspectives have changed again, the balance of power has, at least in part, shifted from politics to commerce, and with it there has been a resultant shift in the way scientists compete. The best example of this is considered in the final dispute, between the "techno-moguls" Bill Gates and Larry Ellison.

But whatever form it takes and however it may be transmogrified by the society in which it operates, rivalry is there in every lab, in every corner of the world, and in every age, blending itself with the "search for beauty," the transcendence, and the need to defrock mystery that motivates all scientific endeavor. Science of all ages and all disciplines is an organic thing, a very human thing. *Vive la concurrence!*

# ONE

$$y = \int_{0}^{\infty} (2x-1)^{3} . \delta x$$

## Second Inventors Count for Nothing

*Great men are like women, who never give up their lovers
except with the utmost chagrin and mortal anger. And
that, gentlemen, is where your opinions have got you.*
—*Caroline of Ansbach to Gottfried Leibniz*

*London, October 1711*

THE ASTRONOMER ROYAL, JOHN FLAMSTEED, would have
liked to run up the stairs of the Royal Society to punch Newton's nose, but
gout prevented him. Instead, assisted by his servant, he slowly ascended the
grand staircase and entered the meeting room where Newton was waiting
for him.

In 1711, Sir Isaac Newton was the most famous scientist in the world.
Knighted six years earlier, he was both scientific megalith and public servant, president of the Royal Society and master of the Royal Mint. The

society he presided over was growing in power and influence, and only two months earlier it had moved into its first independent home, a beautiful town house in Crane Court, in the heart of the City of London.

With a public credo that served as one of the earliest models for cooperation between natural philosophers,* it was nevertheless a society of wealthy gentlemen who liked to argue. The Royal Society dedicated itself to: "the advancement of the knowledge of natural things and useful arts by experiments, to the glory of God the creator and for application to the good of mankind." In this it served science well, but many of the individuals who contributed to the life of the Royal Society despised one another, and some could barely bring themselves to sit in the same room as their scientific colleagues.

Newton and Flamsteed were just such individuals. They rarely spoke, and by 1711 each went out of his way to hinder the work of the other and to embarrass and humiliate the other whenever possible. Newton had summoned Flamsteed to Crane Court on this October afternoon as though the astronomer royal was an errant schoolboy, and Flamsteed had taken his time to appear there, deliberately antagonizing the president. Newton was charging Flamsteed with withholding data from his government-funded observatory and claimed that the astronomer should willingly share his findings with the scientific community. But this was merely the tail end of a dispute that had dragged on for almost two decades. Flamsteed knew that Newton merely wanted the data to include in his new edition of the *Principia mathematica*. Furthermore, the astronomer royal felt used, and believed he had been paid a pittance for a lifetime of dedicated effort, thanks in part to the powerful influence Newton exerted over decisions made throughout the scientific establishment.

In the grand first-floor Council Chamber, with its tall windows offering a view of the narrow courtyard beyond, Flamsteed settled himself carefully into a chair to face the president. He eyed him defiantly.

Newton immediately asked him if he had at last brought with him the required data. Flamsteed merely smiled and said he had not. Then, raising his voice, he added bitterly: "I was robbed of the fruits of my labours."

"We are then robbers of your labours?" Newton shouted back.[1] Flam-

---

* The term *scientist* was not coined until 1840 by the scholar William Whewell.

steed, seething from years of what he believed to be ill treatment, could not restrain himself, and there followed one of the most furious rows the Royal Society had ever seen. As lesser figures huddled outside the doors to eavesdrop and the late October sun set in the west, Newton called Flamsteed "a puppy" and the astronomer royal shouted back that the sixty-nine-year-old Sir Isaac was an "abominable thief." They argued for hours. Flamsteed claimed Newton was trying to acquire data from his observatory by illicit means, and Newton retaliated that he had royal permission to obtain such information. After the two had exhausted themselves, Flamsteed stormed out to his personal kingdom at the Royal Observatory in Greenwich, defiant and unbowed, and Newton turned to searching for new schemes to bend the astronomer's will to his own.

Although this was one of the more dramatic episodes in Newton's long succession of clashes with other scientists of his day, it was neither the most heartfelt nor the most protracted. This dubious honor goes to the three-decade-long dispute between Britain's greatest scientist and his European counterpart, the "Continental Newton," Gottfried Wilhelm von Leibniz, over who could claim priority for the invention of a mathematical technique called the calculus. But to understand how this conflict arose, as well as its significance for the evolution of science, we must first look at the characters of the two men at the heart of the battle and the social climate that led to their clash.

ISAAC NEWTON WAS an only child born on Christmas Day 1642 in the tiny village of Woolsthorpe in Lincolnshire. His father had died before Isaac's birth, and his mother was left to run the family's small farm. They were not a wealthy family but lived well for the time, approximately a middle-income, home-owning family today. In fact, young Isaac's life could have been idyllic except that when he was three years old his mother, Hannah, decided to remarry, and the boy was left in the care of his elderly grandparents in the Woolsthorpe manor house that had been the family home for four generations.

This traumatized Newton. Years later he would write of his hatred for his stepfather, the Reverend Barnabas Smith, who at sixty-three was more than thirty years older than Newton's mother. Psychologists and historians have long pondered Newton's writings in his private notebooks and pointed to evidence that, as a young man at Cambridge University, he still cherished dreams of killing Smith and his mother for what they had done to him. He spewed bile onto the page, bile infected with contempt and hatred for his half siblings, the three children Hannah bore Smith before his death eight years after their marriage.

In a confessional account now known as the Fitzwilliam Notebook and started soon after Newton began his degree, Newton listed his sins, which included "threatening my father and mother Smith to burne them and the house over them" and "wishing death and hoping it to some."[2]

Newton's distress is understandable. Smith showed little interest in his stepson and insisted his wife devote her energies to her new family. From Isaac's perspective, his mother had simply been taken from him. The boy must have assumed he had done something terribly wrong to precipitate her unexpected departure, Isaac grew confused further by her short-lived and infrequent visits and unannounced disappearances. But what is both striking and pertinent to his later behavior toward his contemporaries is that time seems to have exacerbated his pain. These notebooks prove that, rather than gain emotional distance from the events of his childhood, over the years he merely calcified his bitterness.

Of Newton's later epoch-making genius little could be seen from his earliest years. Among his contemporaries at the king's school in the largest town of the area, Grantham, he was renowned for making models of windmills and lanterns, for kites and sundials, but for the most part he was viewed as a morose character who did not excel academically until he was about thirteen or fourteen years old. The story goes that one day on the way to school, the class bully punched Isaac in the stomach. Enraged, Newton challenged him to a fight after school. Although the other boy was much larger, Newton won the fight, apparently dragging the boy along the ground before scraping his rival's face roughly across a stone wall. Significantly, the one who had started the trouble was a place higher than Newton in the school academic ratings. As the boy lay nursing his bloodied nose, Newton leaned into his face and declared that he would not rest until he had not only overtaken him but become the best pupil in the school. And he stuck to his word. Within a year Newton was the star pupil and the apple of Headmaster Henry Stokes's eye.

How much of this is elaborated and how much is truth is not so important as the fact that this event seems to have marked a turning point in Newton's life. From this time on he relished study and could not be drawn away from his books. This greatly angered his mother, who had returned to manage the farm after Smith's death, when Newton was eleven. She saw nothing of value in academic matters and wanted her eldest son to learn the ways of the farmer so he could someday take over the estate.

Newton ignored her completely and, with the help of Hannah's Cambridge-educated brother William Ayscough, and Henry Stokes offering moral support, he went his own way. He already had at his disposal some useful resources with which to improve his mind and to begin an autodidactic course. The doting Stokes supplied him with books, and his stepfather had owned a respectable, if largely untouched, library. Newton also lodged in Grantham for a while, close to the king's school, and stayed with a family named Clark. Mr. Clark was an apothecary whose brother, Dr. Joseph Clark, had been a scholar at Cambridge. When Joseph Clark died, the apothecary allowed his inquisitive young lodger to study his brother's collection of texts whenever he liked, and in this way Newton gained an introduction to Descartes, Galileo, Aristotle, Plato, and many other classic texts during the years before he went up to Cambridge.

As an undergraduate, Newton continued to be an outsider. He appears to have had only one friend, his roommate, John Wickins, and he formed a productive working relationship with the first Lucasian professor of mathematics, Isaac Barrow, then in his early thirties.* Newton was a natural misanthrope, and he devoted all his time and energies to study and shunned the usual student activities of drinking, whoring, and gambling. When he was young he was a conventional puritan and often found himself ridiculed mercilessly by his freewheeling contemporaries, a group composed primarily of the sons of wealthy landowners and nobility who had no interest in learning. Later, he was to develop unorthodox religious views that led him to the heretical beliefs of the Arians, the only religious group (other than Roman

---

*The Lucasian professorship of mathematics was established in 1664 by Henry Lucas, M.P., and received a royal warrant from King Charles II. It is today held by Stephen Hawking.

Catholics) deliberately, consciously, and conspicuously omitted from the Toleration Act of 1689.*

Newton also became an alchemist and spent more time studying the ancient art than he did upon purely scientific research. From his earliest days as Lucasian professor, he was gripped by the possibilities of alchemy and wrote almost a million words on the subject: It was said after his death that he had owned the finest collection of Hermetic texts of the era. Most crucially, his greatest scientific achievements, the elucidation of the law of gravitation, his three laws of motion, and his development of optics, came from a distillation of mathematics, astronomical observation, and musings derived from his alchemical experiments.[3]

Although Newton's religious leanings and his fascination with alchemy have no direct connection with the central theme of our story, it is important to recognize that, both as a scientist and in the way he chose to worship, Newton always took an unconventional path.

Newton was also an unlovable character with little natural charisma. When in his fifties he became master of the Royal Mint, he made sure he attended the execution of every forger and clipper he and his assistants had tracked down, even though this was certainly not a requirement of the position. He is said to have laughed only once, when a nonscientist fellow at Trinity asked him seriously what possible use the world could have for Euclid.

A man marked by internal anguish and emotional turmoil as a child, from his earliest days as an undergraduate at Cambridge, Newton retreated into science and the arcane investigation of the Hermetic tradition in a deliberate attempt to isolate himself from the world. He is thought to have enjoyed only two close personal relationships throughout his adult life. The first was with his roommate John Wickins; the second developed during the early 1690s, when he shared an intimate companionship with a much younger mathematician named Nicholas Fatio de Duillier. Newton and Fatio conducted alchemical experiments together and talked of one day

---

*This was an act of Parliament passed by King William in an early attempt to settle the powerful undercurrent of religious conflict in England. It allowed a much greater degree of religious freedom, but, as noted, it stopped short of sanctioning either Catholicism or Arianism.

sharing rooms.[4] Newton was secretive by nature, but he also kept many of his ideas to himself for good reason. Apart from his deep interest in alchemy, which would have attracted the opprobrium of many of his colleagues, who could have used it against him had it been known, his homosexual leanings might also have endangered him. Coupled with this, Newton's deeply held but unorthodox religious views rejected the traditional teaching of the Anglican Church, the church that dominated English society and controlled the university curriculums. As an Arian, Newton could not accept the doctrine of the Holy Trinity; he believed that Jesus and God are not of one substance but that Christ was the first created being. If his faith had been made public, he would have been expelled from the university and possibly all scientific and educational institutions, effectively ostracizing him from society.

But in spite of his character flaws, his unorthodox beliefs, and his misanthropic nature, as a brilliant young scholar Newton progressed rapidly through the hierarchy of Trinity College, Cambridge, because he took great care to impress those who would ensure his ascendancy. Yet even when he became the youngest professor of mathematics at Cambridge, taking over Isaac Barrow's Lucasian chair in 1669, he was almost entirely unrecognized outside the university.

Newton's need for secrecy and his sense of isolation inevitably led to a degree of paranoia, and even as a young scholar with much to prove he was diffident about publishing anything he discovered. He was pushed into the spotlight only after he had learned that one of his mathematical ideas had been published by another mathematician, the Dane Nicolaus Mercator, who, naturally unaware of Newton's unpublished thoughts, stole his thunder with a work titled *Logarithmotechnia*, a treatise on the use of logarithms. This impetus, coupled with the persuasion of Barrow, eventually pushed Newton to the mouthpiece of the scientific community, the newly formed Royal Society in London, where he was initiated as a fellow in January 1672.

This forum encouraged debate and discussion, and for the first time Newton could communicate with other scientists, mathematicians, and philosophers. However, although for most thinkers this was an enlivening and stimulating environment, it merely enriched some of the darker aspects of Newton's character. The inevitable result was a series of clashes with other academics and thinkers with similarly large egos and reputations to protect.

The first of these conflicts occurred only a few months after Newton accepted his fellowship, when he crossed the path of the society's curator of experiments, Robert Hooke.

Like Newton, Hooke was a man with an extremely high opinion of himself. He was capable and hardworking but insisted that he had invented or discovered almost anything worth consideration long before any of his contemporaries, even if, as was almost always the case, he could offer no proof of priority. For Hooke, there was always an excuse for why he could not show drawings, models, or explanations for the things he was supposed to have achieved effortlessly long before any other claimant, and although he commanded respect from his peers, they were fully aware of this annoying habit.

With egos such as Newton's and Hooke's coexistence in the same scientific "club" was an impossibility. The first point of conflict came when Newton offered the Royal Society a reflecting telescope (only six inches long), the first of its kind ever built. Having failed singularly to find instrument makers capable of constructing this tiny wonder, Newton had not only designed and fashioned the instrument with his own hands but made the tools he needed to do it. But although he appreciated its true value, he hadn't wanted to show the telescope to his peers. Writing to an unidentified friend, he described the device as being able to magnify "about 40 times in diameter which is more than a 6 foot tube [in other words, a refracting telescope] can do, I believe with distinctness. . . . I have seen with it Jupiter distinctly round and his satellites, and Venus horned."[5]

Hooke of course immediately declared that he had built a very similar instrument some years earlier and that he could fit his into the fob of his watch. Newton, unused to Hooke's bravura and easily angered, immediately took exception to these claims, and it took the great powers of reconciliation of the president of the Royal Society, Henry Oldenburg, to pacify the Lucasian professor and to silence Hooke's bluster.

However, a much more serious clash occurred a few weeks later, when Newton, after a great deal of persuading from Oldenburg, Barrow, and other esteemed members of the scientific community, finally agreed to submit to the Royal Society his new *Theory of Light and Colour*. The paper was read in his absence, and as curator of experiments, Hooke was asked to write a report on Newton's claims and offer it to the society for analysis.

At first Hooke prevaricated. When finally he brought himself to write the appraisal, he simply dismissed Newton's findings out of hand and even boasted that he had spent only three hours reviewing Newton's work, assuming it to be flawed and therefore unworthy of his precious time. In his report he declared: "I have perused the excellent discourse of Mr. Newton . . . and I was not a little pleased with the niceness and curiosity of his observations. But although I wholly agree with him as to the truth of those he has alleged, as having by many hundreds of trials found them so, yet as to his hypothesis of solving the phenomenon of colours thereby I confess I cannot yet see any undeniable argument to convince me of the certainty thereof. For all the experiments and observations I have hitherto made, nay and even those very experiments which he alleged, do seem to me to prove that light is nothing but a pulse or motion propagated through an homogeneous, uniform, and transparent medium."[6] The reason for this deliberate aggression was that Newton's theory of light supporting a corpuscular theory (that light is made up of invisible particles) was at complete variance with Hooke's own ideas, that light must be thought of as a wave.

Ever captious, Newton was outraged, and there followed an angry exchange of letters between the two scientists that was stopped by the further intervention of Oldenburg. Newton's anger was quelled only when Hooke was forced to reexamine the Lucasian professor's paper and come up with a detailed report. This he did but again came down heavily against Newton. But this time, sensing something was wrong, Oldenburg gave Newton the opportunity to argue against Hooke's judgment. Newton wrote a detailed discourse that brilliantly demolished Hooke's arguments, and the society extracted a public apology from the curator of experiments. It was only then that Newton calmed down somewhat.

However, from that point on, the two men loathed each other. This clash was but the first in a long succession of public quarrels that lasted the remainder of Hooke's life (he died in 1703). Along the way, one row led Newton to threaten resignation from the Royal Society. Even after he retracted this, he fell silent on the society for the best part of two years and refused to allow it to publish his optical theories, a move that slowed the development of this field for at least three decades, until Newton was himself president of the Royal Society and published one of his great masterpieces, *Opticks*, in 1704. In this we may see a clear example of how rivalry might both elevate and enervate.

In 1677 Hooke became secretary of the Royal Society and Newton fell into a threnetic sulk, flatly refusing to attend a single meeting until the mid-1680s. The only person who was able to draw him out of his shell was Edmund Halley, who in 1684 visited Newton in Cambridge and, employing great powers of persuasion, influenced his fellow scientist to begin writing the book that changed the path of scientific development radically, the *Principia mathematica*.

Halley was indeed one of very few of Newton's contemporaries who enjoyed a warm professional relationship with the Lucasian professor (Christopher Wren was another). This was because Newton saw immediately that Halley understood his work and appreciated his scientific method—his strict inductive approach supported by a strong mathematical foundation. It is also important to remember that not only was Halley a social climber and a politically adept scientist, extremely diplomatic but, unlike those of others at the Royal Society, his own ideas never challenged Newton's.

Although Newton and Hooke despised each other, they kept up a false gentlemanly exchange in their letters and their meetings at the Royal Society in London, but each laced his comments with poisoned thorns. The most famous of these is Newton's seemingly innocent comment directed to Hooke: "If I have seen further it is by standing on the shoulders of Giants."[7] Hooke was a physically deformed dwarf.

By the time Hooke died in 1703, some quarter century after the two men first clashed, Newton's life had changed utterly. From being a respected but obscure professor at Cambridge, Newton had gained huge recognition for the *Principia*, which established him as the greatest living scientist, a man respected through most of Europe. But as he had been working on this great book, he had also been experimenting with alchemy, and his long-term goal had been to find what we would today call a grand unified theory—a theory that could explain how the universe functions with the broad strokes of cosmology and within the microcosmic world of subatomic particles. Newton lived during a time when relatively little was known of cosmological laws and even less about the subatomic realm; indeed it was not until the end of the eighteenth century that the idea of atoms reemerged as a sound scientific principle upon which the later ideas of subatomic mechanics and quantum theory would develop.

Newton's search for universal laws through the heretical practices of alchemy, combined with his inability to come to terms with his sexual feelings for Nicholas Fatio de Duillier, led him in 1693 to a nervous breakdown. Less than three years later he had moved from Cambridge to London, given up practicing both science and alchemy, and accepted a senior position at the Royal Mint. By 1705 he had been promoted to master of the Mint, had served as an M.P., was knighted, and was made president of the Royal Society.

But none of these achievements and this huge recognition slowed his inner anger and outward bitterness. Newton and Hooke rarely spoke during the secretary's final years, but by the early 1690s Newton had found a new enemy in, the astronomer royal, John Flamsteed.

The early lives of Isaac Newton and John Flamsteed bear striking similarities. A biographer of Flamsteed called them "two troubled creatures, one the son of a tradesman, the other the son of a yeoman, now great officers of the Crown and rivals for world fame."[8]

And many of Flamsteed's personal and religious views were akin to Newton's approach to life. In his autobiography, Flamsteed declared: "God suffers not man to be idle, although he swim in the midst of delights; for when he had placed His own image (Adam) in a paradise so replenished (of his goodness) with varieties of all things, conducing as well to his pleasure as sustenance, that the earth produced of itself things convenient for both. He yet (to keep him out of idleness) commands him to till, prune, and dress his pleasant verdant habitation; and to add (if it might be) some lustre, grace, or convenience to that place which, as well as he, derived its original from his Creator."[9]

From the same social background as Newton, Flamsteed was the younger by four years. He was puritanical, but unlike Newton he maintained an orthodox faith throughout his life. Serious and often cantankerous as an old man, he worked hard to be thorough and meticulous and was quite brilliant in his own narrow field of research, but he did not share Newton's amazing intellectual precision.

Flamsteed's mother had died when he was three years old, and this left him emotionally scarred, just as Newton had been by the circumstances of his childhood. When he was fifteen Flamsteed developed a rheumatic fever and was bedridden for months. The disease left him crippled but also gave him the opportunity to discover the joys of learning.

From this time astronomy became for him an obsession. He sailed through his career at Cambridge, where he may have encountered the slightly older Newton, and under the aegis of the then secretary of the Royal Society, Christian Oldenburg, he was offered the newly created position of astronomical observer for King Charles II, a position that soon became known as the office of astronomer royal. It came with a very modest salary of one hundred pounds per year, but he had free use of a state-of-the-art observatory beautifully designed by Wren and ideally located on a hill overlooking the village of Greenwich, to the southeast of London.

Flamsteed and Newton had been colleagues at the Royal Society for many years before they clashed. The astronomer royal had been critical of the *Principia* and seemed to think that Newton should have given him some credit for using his published observational data, claiming this had helped support Newton's theories. "Little notice taken of Her Majesty's observatory (with very slight acknowledgements of what he had received from the observatory)," he said of it.[10]

But Newton may not have heard about Flamsteed's feelings, because the first bad blood between them arose when Newton wanted to gather data for a proposed new edition of the *Principia*. In an exchange of forty-two letters between September 1694 and January 1696, what began as a cordial sharing of thoughts and ideas turned into a succession of increasingly vitriolic missives, by the end of which Newton and Flamsteed had become bitter enemies.

The trouble had started when Flamsteed offered Newton some calculations in addition to the requested data. Unfortunately, Flamsteed had made some silly errors in the calculations. Instead of politely ignoring these or gently pointing out the mistakes, Newton lashed out, telling the astronomer that he wanted not his conclusions but his information. "I want not your calculations but your observations only. . . . If you like this proposal, then pray send me first your observations for the years [sic] 1692 and I will get them calculated and send you a copy of the calculated places. But if you like it not, then I desire you would propose some other practicable method of supplying me with observations, or else let me know plainly that I must be content to lose all the time and pains I have hitherto taken about the Moon's theory."[11]

This prompted Flamsteed to retaliate. Angered and hurt, he declared: "I never took anything of any for communicating of my skill or pains, except of those who forced themselves upon me to devour my time . . . pray therefore lay by any prejudicial thoughts of me, which may have crept into you by malicious suggestions."[12] Clearly Flamsteed was not seeking an argument, and a conflict could have been avoided, the argument calmed; but Newton, thinking Flamsteed was being either lazy or slapdash, seemed determined to cause maximum upset. He offered Flamsteed money for his data, a move guaranteed to insult any honest and committed scientist of the time. In disgust, Flamsteed replied: "All the return I can allow or ever expected from such persons with whom I corresponded is only to have the result of their studies imparted as freely as I afford them the effect of mine or my pains."[13]

By now Newton had come to see Flamsteed as little more than a technician, a mere gatherer of data for his free use. At the same time, he regarded himself as the genius who translated raw observed facts (gathered by those he considered semieducated) into all-consuming theories capable of changing the way people thought. To Newton, it was Flamsteed's duty as a public servant to offer his observational data to another natural philosopher.

But this never happened. In his private notes, Flamsteed called Newton "hasty, artificial, unkind, arrogant."[14] And from the 1690s until Flamsteed's death in 1719, the two men could not bring themselves to be civil to each other. Their mutual hatred grew out of all proportion to the original cause of their dispute and was fueled by Newton's rudeness and insensitivity. From this point on Flamsteed did everything in his power to thwart Newton and utterly refused to impart to him anything he was not forced to relinquish.

Throughout most of this dispute Newton maintained a position of immense power within the scientific community, and he could call upon some of the most influential figures in the land to help him pummel Flamsteed. Allies included Sir Christopher Wren; the royal physician, Dr. John Arbuthnot; Queen Anne's husband, the imbecilic Prince George, and even, for a while, the queen herself. Ultimately, Flamsteed was forced by Newton's trickery to give him almost everything he wanted. Using his power and influence, Newton had persuaded the royal family to sponsor a catalog of the heavens, *Historia coelestis*, written by Flamsteed in their honor. In it were to be found all the data the president of the Royal Society sought for his researches. Unable to refuse a royal command, Flamsteed could only protest about this underhandedness to Newton, who merely turned the tables and accused him of treason, snapping: "The Queen would be obeyed."[15]

Flamsteed's only semblance of revenge came with the posthumous publication of an alternative version of his catalog, to which he had added a preface that contained a scathing attack upon his hated enemy: "His design was . . . *to make me come under him*," he wrote, "force me to comply with his humours, and flatter him, and cry him up as Dr. Gregory and Dr. Halley did . . . He thought to work me to his ends by putting me to extraordinary charges . . . *Those that have begun to do ill things, never blush to do worse to secure themselves.* Sly Newton had still more to do, and was ready at coining new excuses and pretences to cover his disingenuous and malicious practises. I had none but very honest and honourable designs in my mind: I met his cunning forecasts with sincere and honest answers, and thereby frustrated not a few of his malicious designs. . . . I would not court him. . . . For, honest Sir Isaac Newton (to use his own words) would *have all things in his own power*, to spoil or sink them; that he might force me to second his designs and applaud him, which no honest man would do nor could do; and, God be thanked, I lay under no necessity of doing."[16]

But, for all Flamsteed's words, throughout this battle Newton had maintained the overwhelmingly dominant position. Flamsteed was a thorn in Newton's side and had hindered the development of his scientific work, but he never presented a threat, nor did he at any point succeed in undermining Newton's reputation. Even so, as this row developed from a misunderstanding to open hostility, Newton had to simultaneously fight another, far more important battle, a dispute that grew out of deep-rooted prejudice and anxiety and did threaten his position as the world's leading thinker. As well as challenging the reputation and achievement he had spent a lifetime acquiring, this fight became the most important of Newton's career, and its outcome was both profound and influential on the future progress of mathematics and much of science.

$\mathcal{T}$HE GERMAN MATHEMATICIAN and natural philosopher Gottfried Leibniz was born in Leipzig in 1646. His father, Friedrich Leibnütz, was a professor of moral philosophy at the University in Leipzig, and Gottfried grew up in an environment of strict Lutheran piety, leading him to a traditional religious outlook he maintained his entire life. His father died when Gottfried was six, and although he enjoyed a thorough and progressive school education, like Flamsteed, Newton, and many intellectuals of the period, he followed an autodidactic course before taking up his official studies.

In the spring of 1660, just before entering his university, Leibniz spent two months at the University of Jena, in what is now central Germany, and through the strong mathematics fraternity there he became interested in the ideas of the Neo-Pythagoreans, who believed that the structure of the universe was pinned to a simple mathematical code. These ideas were never to leave him and informed many of his later work in mathematics.

He entered the University of Leipzig at Easter 1661, just a few months before Newton went up to Cambridge, and was enrolled in the law program. He was so successful he qualified for his doctorate in law by the age of twenty, but according to the rules of the university, he was too young to receive his degree. Leibniz took this very badly, and although he never explained his reasoning for this assertion, he claimed it was due to a conspiracy precipitated by the dean's wife.[17] Although this tells us nothing of what might really have happened, it helps us understand Leibniz's character a little more, serving as it does to show that even as a young man he was no stranger to intrigue and academic politics.

Whatever the root of this schism with his university, Leibniz was deeply embittered and left Leipzig for Nuremberg, where in 1666 he submitted a paper titled *De casibus perplexis* (*On Perplexing Cases*), which was so brilliant it not only gained him his doctorate but brought an offer of a law professorship. This he declined, choosing instead to pursue his scientific interests.

By this time Leibniz was well read in the classic works of Galileo, Kepler, and Descartes, and, in a parallel with Newton, throughout his orthodox degree course he had continued with his private researches. His keenest interest had been in the field of logic, and before his twentieth birthday he had composed a paper called *De arte combinatoria* (*On the Art of Combination*), which is now seen by some scholars as an early theoretical model for the computer. He was also interested in alchemy, and, almost exactly contemporaneously with Newton, he expended great effort pursuing the philosophers' stone. Leibniz is considered by many to have been Newton's equal; one historian has gone as far as to say that he was "one of the greatest polymaths in history."[18] Another described Leibniz and Newton as "two of the greatest geniuses of the European world, not only of their own time but of its whole long history."[19]

Like his English rival, Leibniz was a multifaceted character with a range of talents. He was a good administrator and was employed for a short time as diplomat and lawyer for Elector Johann Philipp von Schönborn. He was manually dextrous and constructed a functioning calculating machine, a device that was noticed by Oldenburg and other leading lights of the Royal Society and gained his admittance as a foreign fellow when he first visited London in 1673. But above all, Leibniz was a master of logic and pure mathematics.

The conflict between Gottfried Leibniz and Isaac Newton was very different from the battles the Lucasian professor fought with Hooke and Flamsteed. Hooke's fights with Newton were based upon jealousy and petty rivalry; Flamsteed and Newton argued over the rights to use scientific material. In spite of the fact that Hooke and Flamsteed held higher official positions within the international scientific community than Leibniz, they were quite mismatched against Newton's intellect. Leibniz was a world-class intellectual who had blossomed from an early age. And, indeed, it may be argued that there was insufficient room for two such geniuses living simultaneously, in view of which a conflict was quite inevitable.

After his visit to England, Leibniz returned home to Paris to learn his employer, the elector of Mainz, had died. This proved to be a great turning point in Leibniz's career, for instead of returning immediately to Mainz or Leipzig to seek fresh employment, he decided he would give himself time to follow through intellectual pursuits and to dedicate himself to mathematics, reasoning that it would be better to live on the verge of poverty in Paris than to immediately take a menial post serving a local governor or regional dignitary.

Thus began his own *anni mirabiles*. During a two-year period between 1673 and 1675, working in almost complete isolation, Leibniz produced what he thought was a revolutionary canon of higher mathematics, including a technique called the infinite series and, most crucially, a version of the calculus. But, because he had come from a relatively isolated intellectual background, where many of the latest ideas about science and mathematics were unknown, he actually had an inflated opinion of his achievement. In fact he had simply mirrored Newton's own development of a decade earlier. This misunderstanding lay at the root of the conflict between Leibniz, the genius amateur, and Newton, the equally brilliant establishment figure.

To understand the essence of the dispute and to appreciate not only its significance for the mathematics and science of the day but its ramifications for us living in the twenty-first century, we should take a close look at what exactly is meant by the calculus, the subject that took over the lives of Leibniz and Newton and their supporters.

$I$N THE SEVENTEENTH century, most educated people found simple addition and subtraction difficult, and only a very few knew how to multiply or divide numbers. Indeed, Samuel Pepys had to learn multiplication after he became a senior naval administrator. But there were and had long been a rare breed who had developed mathematics, even though little connection was then made between pure mathematics and its application to the material world.

Primitive cultures had employed simple mathematics to build monoliths such as Stonehenge and the pyramids, and the Babylonians developed a form of astrology that applied basic mathematical principles. But the first sophisticated mathematical ideas came from Greece. Archimedes, Pythagoras, Euclid, and Plato were dedicated to the belief that mathematical patterns lay at the heart of the universe. Plato even had a sign over the door of his famous academy in Athens that read: LET NO MAN ENTER WHO KNOWS NO GEOMETRY.

Ptolemy, during the first century A.D., and the Arabs of the sixth century, each developed mathematical notation and expanded the canon of arithmetic, devised new aspects of algebra, and enlarged our understanding of geometry. These ideas were developed by the scholars of the Middle Ages and the Renaissance, men such as Roger Bacon of Oxford and Luca Pacioli, a friend of Leonardo da Vinci.

A hundred years before the birth of Newton, geometry had evolved into the best understood mathematical discipline. Mathematicians could calculate areas and volumes of many shapes, and they understood the geometric and trigonometric relationships between triangles and other geometric figures. Algebra was less well defined, but frequent use was made of powers and roots, and both architects and astronomers used quite sophisticated manipulations of formulae in their work. The mathematician François Viète demonstrated the value of symbols by using plus and minus signs for operations, and letters to represent unknowns. This form of notation helped make possible the great mathematical advances of the seventeenth century. In Italy during the early sixteenth century, Niccolò Tartaglia and Scipione Ferro had discovered general solutions for what are called cubic equations (where one of the terms is in the third power or cubed such as $x^3 - 2x = 4$, and these were published in 1545 in a book called *Ars magna* by another mathematician, Girolamo Cardano.

In 1614 John Napier invented logarithms, which were used up to recent times, when calculators took over the laborious process of long-winded multiplication and division, and by the 1630s, a decade before Newton's birth, the Flemish mathematician Simon Stevin of Bruges had introduced decimal fractions.

However, before Newton and Leibniz arrived on the scene, the greatest mathematician of the seventeenth century was the Frenchman René Descartes. His most profound mathematical finding was that an equation was not the only way in which mathematical terms could be related. During the 1630s he devised the idea of constructing coordinates to represent pairs of numbers relating to algebraic terms (usually $x$ and $y$). These came to be known as Cartesian coordinates and opened up the vast range of possibilities offered by the drawing of graphs (lines and curves bordered by axes). The technical name for this branch of mathematics is analytical geometry, and it appeared in an appendix called *La géométrie*, tacked onto the end of Descartes's *Discourse on Method*, first published in 1637.

Descartes's technique transformed the world of mathematics, and within a few years of publication the *Discourse* had influenced the work of mathematicians and astronomers throughout Europe. Isaac Barrow, Newton's predecessor at Cambridge, John Wallis at Oxford, and David Gregory, along with Descartes's countryman Pierre de Fermat, all used findings explained in the *Discourse* as a springboard for their own efforts. Of particular interest was the study of curves drawn between the axes of the graph using Cartesian coordinates.

A simple example is the graph produced by representing the distance traveled by an object dropped from a high point (such as a tower) against the time for which it has fallen. Galileo had shown that the speed of the ball increases with time. If after one second the ball has fallen 16 feet, after two seconds, 64 feet, and after three seconds, 144 feet, it is obviously accelerating. Descartes then reasoned that by plotting these values on a graph with distance on the $y$-axis and time on the $x$-axis, a curve would be produced.

Now, it is comparatively easy to calculate the properties of straight lines represented graphically. For example, the area under a straight line can be calculated by simple geometry known to the Babylonians, and the gradient (or steepness) of a straight line may be found by dividing the change in the values along the $y$-axis by those on the $x$-axis.

So, if our graph was a straight line, the gradient or steepness of the line (the change in distance with time) would have given us the speed of the ball. But how could properties represented by curves be calculated in a similar way? In other words, what if the distance was not changing in proportion to the time as it did in a simple case in which a straight line graph was produced?

It was soon realized that the only possible way to determine properties of curves such as the one in our problem was to imagine them as constantly shifting straight lines, and that if a straight line was drawn next to a curve and touched it at a particular point, this line could approximate the curve. Mathematicians called this straight line a tangent and found they could then treat a tangent like any other straight line. In other words, they could, for example, find its gradient and work out a value for the speed of the ball at a particular point. But this was still an approximation and a very rough one at that.

The only way to achieve a precise answer to a problem such as finding the exact acceleration of an object was to use an algebraic technique. A geometric process would always be an approximation, no matter how fine the tangent or how accurate the graph. An algebraic solution would come from a manipulation of equations rather than a mechanical process.

Suppose we know the equation of a curve is $d = t^2 - 2t + 1$, where $d$ is distance in feet and $t$ is time in seconds. Newton realized that the gradient of this curve could be calculated by applying a technique called differentia-

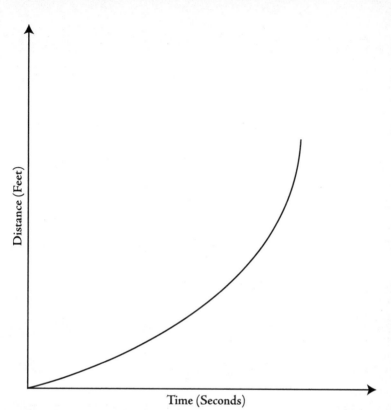

Figure 1. The curve produced by plotting distance against time for a falling ball

tion. In differentiation the coefficient of each term in an equation, that is, the number before the letters (the letter $t$ in our example), is multiplied by the power assigned to the letter, and the power is then reduced by 1. This means if we differentiate the equation $d = t^2 - 2t + 1$, we get a gradient equal to $(2 \times 1)t - (2 \times 1)$, or $2t - 2$. So, if we want to find the gradient when time equals, say, 3 seconds, we put $t$ equal to 3 and find the gradient (or the speed in this case) at that moment. In our example this is $(2 \times 3) - 2$, which equals 4 feet per second.

Differentiation has a vast range of applications. It is used to find the rate of change of any quantity at any instant as it changes in relation to another quantity. These include ways to describe the motion of pendulums, meth-

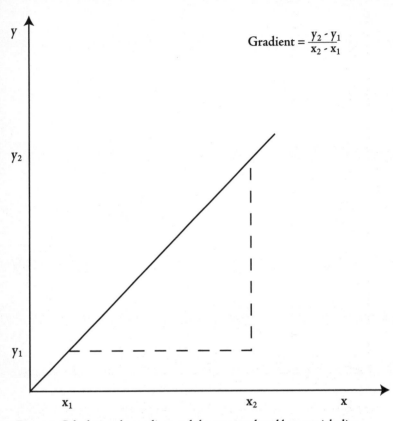

$$\text{Gradient} = \frac{y_2 - y_1}{x_2 - x_1}$$

Figure 2. Calculating the gradient and the area produced by a straight line

ods of calculating loads and stresses for beams in architectural designs, calculating the speed and acceleration of any moving object from a flywheel to the space shuttle, as well as finding maximum and minimum values in a variety of mechanical processes.

Differentiation is the simplest form of calculus. A more complex technique is integration, which is the reverse of differentiation. With integration, we begin with a simple equation and take it "back" to the more complex form. In the example just discussed, integrating the (linear) equation gradient = $2t - 2$ gives us the original (quadratic) equation: $y$ (distance) = $t^2 - 2t + 1$.

Integration is used by scientists to find things such as the center of grav-

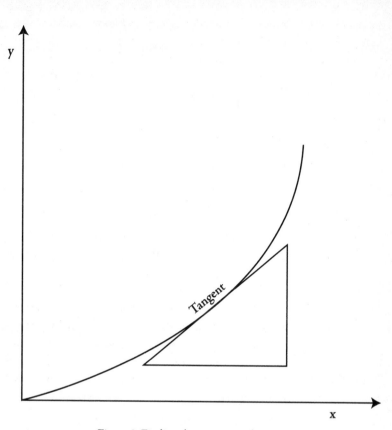

Figure 3. Finding the tangent to the curve

ity of an object, moments of inertia of revolving objects, and what are called volumes of rotation, all of which are essential to the day-to-day work of structural engineers, space scientists, and theoretical physicists.

THE CONFLICT BETWEEN Leibniz and Newton began soon after Leibniz managed to ingratiate himself into the European scientific community and to emerge from his years of research in a Paris garret. During his visit of 1673, Leibniz was introduced to Oldenburg at the Royal Society, and through him he had begun to correspond with the collector and publisher

John Collins. Collins was in sporadic contact with Newton, and for many years he had wanted to publish the Lucasian professor's work. As a prominent publisher, he was a focal point for the scientific scene of the time, and he introduced Leibniz to many European mathematicians and philosophers through a dense international network centered in the Royal Society. As the ideas flowed, Leibniz sent them to Collins, who understood enough (with Oldenburg's help) to see that the young German was making significant progress.

Collins encouraged him, sending on the latest ideas circulating within the Royal Society. This is a crucial point; because Collins was untrained in mathematics, his letters consisted of mere gossip and occasional vague references to the work of many mathematicians.[20]

Newton and Leibniz had not met in London during the latter's visit, and the Lucasian professor probably did not learn about Leibniz until around 1675. But within twelve months of Leibniz's visit both Collins and Oldenburg were beginning to see the potential for conflict between the two men. They were both aware of Newton's mathematical work, both knew he had developed a method of infinite series and a version of the calculus during the 1660s, and although Leibniz's notation was quite different, his technique was fundamentally the same. They also knew from the conflict with Robert Hooke just how self-protecting Newton could be. He had threatened to resign from the Royal Society because of arguments with Hooke only two years earlier, and Oldenburg was quite aware of the man's volatile nature. Convinced Leibniz was about to publish his discoveries, Oldenburg and Collins tried to press Newton into publishing first.

But Collins had painted himself into a corner. He could not tell Newton he had passed on information to Leibniz, even if it was of no real significance. An admission that he was communicating the professor's thoughts in any form to another mathematician without his permission would certainly have precipitated Newton's outrage and the breaking of all ties with the publisher. Yet he also saw that this recent turn of events could be a godsent opportunity because, if Newton could be persuaded to publish without knowing about the correspondence with Leibniz, the work would almost certainly go to him.

Unfortunately for Collins, this intrigue was unfolding contemporaneously with Newton's rows with Hooke over his *Theory of Light and Colours*.

Consequently, the prickly Newton would not have contemplated having his mathematical work published by Collins or anyone else, and any request for permission would have been either ignored or met with a hostile rejection.

Some five years earlier, after failing to persuade Newton to publish what was then his newly conceived calculus, an exasperated Collins had written to James Gregory declaring: "I desist and do not trouble him [Newton] any more."[21] And, in September 1675 he repeated himself, telling Gregory: "I have not written to or seen Newton this 11 or 12 months, not troubling him as being intent upon chemical studies and practices, and both he and Dr. Barrow are beginning to think mathematical speculations grow at least nice and dry, if not somewhat barren."[22]

Collins then had little influence with which to attempt to head off a clash, and it was up to Oldenburg to persuade Newton to write directly to Leibniz under the pretext that the German had a number of mathematical queries only he was qualified to answer with any surety. Newton finally and reluctantly complied.

But Newton being Newton, he could not write a simple letter to Leibniz; instead he produced two elaborate discourses. These were later known as the *Epistolia prior* (the first letter), written in June 1676 and running to eleven pages, and the *Epistolia posterior* (later letter), composed in October, which extended to nineteen pages of closely packed lettering. Together they summarized Newton's mathematical discoveries and were designed to show Leibniz that he had arrived at a version of the infinite series and other breakthroughs many years earlier. But even then, nervous that others would steal his ideas, Newton pointedly left out any mention of the calculus. Instead, he added an encrypted version of the material in the form of an anagram, which was meant to act as a type of patent.

"I cannot proceed with the explanation of the fluxions [the calculus] now," he declared. "I have preferred to conceal it thus: 6accdae13eff7i3i9-n4o4qrr4s8t12vx." Translated, this has been interpreted as: "given any equation involving any number of fluent quantities, to find the fluxions, and vice versa," an encoded message that gave the meaning of the calculus, techniques we now call differentiation and integration.[23]

In a covering letter to Oldenburg, Newton then tried irritably to draw a veil over the correspondence, commenting: "I hope this will so far satisfy M. Leibniz that it will not be necessary for me to write any more. . . . For having other things in my head, it proves an unwelcome interruption to me to be at this time put upon considering these things."[24] Still unaware of what Leibniz had achieved and with a typically wary glance over his shoulder, Newton wrote again to Oldenburg two days later and insisted: "Pray let none of my mathematical papers be printed without my special license."[25]

Communications with the Continent were bad, and the *Epistolia posterior* did not reach Leibniz until June 1677, a full eight months after it had been sent. In the interim he had acquired an official position at the court of Johann Friedrich, duke of Brunswick Lüneburg in Hannover and had again visited London, where he had met Collins for the first time.

Remarkably, during the visit, the publisher had allowed his guest free access to his collection of papers and correspondence, and this added fuel to Newton's later claims that Leibniz was a common thief and that Collins had been his accomplice. Yet Leibniz did not find in the collection much he did not already know, and it is significant that he completely ignored papers on the calculus, which adds evidence to the theory that Leibniz had already devised his own method sometime earlier.[26]

Oldenburg died later that year, and for a time communication between Newton and Leibniz came to a halt. As the Lucasian professor concentrated on alchemy and began to work on the *Principia*, the question of the calculus and the obscure German mathematician Oldenburg had made so much of apparently melted into the background of Newton's mind. But then, in October 1684, some two months after Halley's visit to Cambridge, during which he had regenerated Newton's interest in celestial mechanics, Leibniz delivered his first paper on the calculus, published in *Acta eruditorum*, a learned journal produced by Leibniz's former academic home, the University of Leipzig.

We can imagine how Newton received this news. He must have been deeply shocked because later he claimed to have immediately concluded Leibniz's work was pure and unmitigated plagiarism. When Leibniz's paper arrived, Newton had been working night and day on the *Principia*, and such preoccupation may have deflected some of his initial fury. Even so, it was a pivotal moment in the relationship between Newton and Leibniz; after it, the two men never maintained any form of civilized communication.

Newton knew that he had now to make up for lost time and that it was imperative he establish his priority to the calculus. His first move was to add a passage to the manuscript of the *Principia* he was then finishing. This became *The Scholium to Book II, Section II, Proposition VII*, and it reads: "In letters which went between me and that most excellent geometer, G. W. Leibniz, ten years ago, when I signified that I was in the knowledge of a method of determining maxima and minima, of drawing tangents, and the like, and when I concealed it in transposed letters involving this sentence [the encryption from the *Epistolia posterior*] that most distinguished man wrote back that he had also fallen upon a method of the same kind, and communicated his method, which hardly differed from mine, except in his forms of words and symbols."[27] Newton's second move was to attempt to make the scientific community realize that he had arrived at the calculus first by carefully planting elements of his methods, letting his priority become known to other scientists and mathematicians who could be trusted to pass on the word.

Oblivious to the furor his paper had produced in Cambridge, Leibniz had no reason to think he should not promote his own priority, and it was not until six months after the appearance of his work in *Acta eruditorum* that he learned something was amiss. In the spring of 1685, as Newton worked on with the *Principia*, Leibniz received a letter from a friend, Otto Mencke, the professor of philosophy at Leipzig, warning him that the invention of the calculus would be attributed to Professor Newton at Cambridge.[28] This did not seem to trouble Leibniz, who felt sure it would be possible to announce the work as a parallel creation. "As far as Mr. Newton is concerned, I have letters from him and the late Mr. Oldenburg in which they do not dispute my quadrature with me, but grant it. Nor do I believe that Mr. Newton will ascribe it to himself, but only some inventions about infinite series which he in part also applies to the circle. Mr. Mercator, a German, first came upon this and Mr. Newton developed it further, but I arrived at it by another way. Meanwhile, I acknowledge that Mr. Newton already had the principles from which he could well have derived the quadrature, but one does not come upon all the results at one time: one man makes one contribution, another man another."[29]

Unfortunately, the simple idea that "one man makes one contribution, another man another" was quite alien to Isaac Newton. In fact, it is because he believed the very opposite that the dispute between him and his German rival escalated into the intellectual earthquake it did. At first glance the row appeared to originate from Leibniz's desire to proselytize and Newton's apparent distrust of publicity. But the problem between the two men was actually based upon far more personal drives and primal urges, at least on Newton's part.

Seventeen years earlier, in 1668, Newton had been infuriated by the mathematician Nicolaus Mercator's publication of *Logarithmotechnia* because it contained material he had already researched and explained. And this had not been an isolated reaction. Many years later he referred grudgingly to Christiaan Huygens's work on centrifugal forces with the comment "What Mr. Huygens has published . . . about centrifugal forces I suppose he had before me."[30]

Newton maintained an obsessive conviction that there could be only one genuine interpreter of divine knowledge in the world at any one time and that he was this unique being. The historian Frank E. Manuel has described Newton as having "a fierce independence, a conviction that he was the chosen intermediary between the Creator and mankind, a sense of omnipotence."[31] Indeed, one of Newton's most vocal supporters, the mathematician David Gregory, reported that Newton became "offended" and "angry" when anything appeared in print that seemed to contradict his doctrine.[32] The idea that anyone else could acquire independently the same insights and breakthroughs he had was, for Newton, simply unimaginable. Because of this, in Newton's eyes Leibniz was a worthless, contemptible thief who had intentionally stolen the ideas he had derived. And to make things worse, the German had gone on to loudly declare to the world that he alone had arrived at his conclusions.

As soon as Newton learned of Leibniz's work, he reached the conclusion that the material he had shown Oldenburg and Collins over a decade earlier had been passed on to his rival. However, it is apparent the letters that passed between Collins and Leibniz never contained anything more helpful than nontechnical summaries of recent mathematical debates, in particular those conducted at the Royal Society in London. An example is a catalog Collins sent Leibniz in April 1675, which contained current developments in the concept of the infinite series by Newton, Gregory, and others. It did not contain a single demonstration and arrived long after Leibniz had developed his own method.

Newton could not accept this. The possibility that Leibniz might have arrived at the same revelations independently could not be entertained, even for a moment. Newton's reaction was to attack, to destroy his enemy, to drive him into the ground without a trace of mercy. And there were many who were ready to help him do this. Without ever truly understanding the deeper reasons for their master's reaction to Leibniz, Newton's supporters were as keen to fight his battles as they were to attempt concurrently to bend poor John Flamsteed to Newton's will. What began as a personal contest between two scientists dragged into a four-decade-long battle, which did not even end with Leibniz's death in 1716. The battle lines between Newton's defenders in London and their rivals in Europe developed into a war of clashing ideologies, a schism in philosophical thought as well as mathematical practice, which led to repercussions that would last generations.

After the *Principia* was published and Newton's reputation blossomed, the issue of the calculus became a cause célèbre throughout Europe. British mathematicians such as John Wallis stoked the embers by claiming Newton's clear priority. In 1693 he published the first installment of his three-volume *Mathematical Works* and made only scant reference to the issue of the calculus. But then, after hearing fresh rumors from the Continent concerning the dispute, he regretted not having said more and wrote to Newton asking permission to publish both the *Epistolia posterior* and the *Epistolia prior* in the preface to volume 2. "I had intimation from Holland," he commented in the letter, "that your notions of fluxions pass there with great applause, by the name of *Leibniz's Calculus Differentalis*."[33]

Naturally, Newton was all for Wallis's defense and agreed immediately, but this only angered the mathematicians on the Continent who were already beginning to see Leibniz as their hero and Newton as the villain of the piece. Leibniz's friend and supporter the mathematician Johann Bernoulli even had the temerity to write to Leibniz suggesting that Newton had plagiarized *his* work.[34] And because he was growing increasingly angry over the entire issue, Leibniz himself began gradually to think in this way too. In his reply to Bernoulli, he wrote: "I could easily believe that Newton possessed some very remarkable knowledge at that time [1676], which, in his usual way, he had greatly polished up in the subsequent period."[35]

But Newton's supporters grew just as agitated. In 1699, while struggling to maintain a reputation as a philosopher and existing on the fringes of the scientific establishment, Fatio de Duillier took his turn to publicly attack Leibniz's claim. Over a dozen years earlier the German mathematician had dismissed Fatio's rather obtuse and unworkable version of the calculus, and de Duillier decided the time was right to take his revenge. In a mathematical treatise titled *Lineae brevissimi descensus investigatio geometrica duplex* (*A Two-fold Geometrical Investigation of the Line of Briefest Descent*), de Duillier pronounced: "But I now recognise, based upon the factual evidence that Newton is the first inventor of this calculus, and the earliest by many years; whether Leibniz, the second inventor, may have borrowed anything from him, I should rather leave to the judgment of those who have seen the letters of Newton and his other manuscripts. Neither the silence of the more modest Newton, nor the unremitting exertions of Leibniz to claim on every occasion the invention of the calculus for himself, will deceive anyone who examines these records as I have."[36]

Some questions have been raised concerning the extent to which Newton may have influenced de Duillier's attack, but it seems unlikely Newton had any direct impact on the mathematician's decision to go on the offensive over the issue. By this time Newton and de Duillier were no longer intimate; their close relationship had ended some seven years earlier. Their only contact would have been when they happened to meet at the Royal Society. No correspondence between them from this time survives, and indications are that rather than being encouraged or pleased by this passage, Newton was actually rather embarrassed by it. Although Fatio was merely voicing the private thoughts of many, not least Newton himself, at this stage of the dispute the attack was simply too forthright.

However, such attacks were not without precedent. During the 1650s the philosopher Thomas Hobbes, a man widely hated and ostracized by his fellow academics, poured out his feelings for the then Savillian professors at Oxford, John Wallis and Seth Ward, in an addendum to his book *De corpore*: "So go your ways, you Uncivil Ecclesiastics, Inhuman Divines, Dedoctors of morality," he wrote, "unasinous colleagues, egregious pair of Issachers, most wretched Vindices and Indices Academiarum."[37]

Yet for the most part, surviving correspondence from the era shows how frequently bitter enemies expressed hypocritical politeness throughout their personal letters. Until their battles became public, Newton and his rivals Hooke, Flamsteed, and Leibniz unfailingly referred to each other as "honourable colleague" or "esteemed friend," and littered their correspondence with compliments such as "I value your friendship very highly."[38] But just beneath the surface emotions seethed, and these same rivals fought vigorously over their reputations and the way they wished to be perceived by future generations as well as their own contemporaries. Voltaire later referred to such hypocrisy when he wrote in a letter to a relative that he was compiling what he called a "Dictionary for the Use of Kings," in which, he said: " 'My dear friend' means 'you are absolutely nothing to me.' By 'I will make you happy,' understand 'I will bear you as long as I have need of you.'"[39]

In this way both Newton and Leibniz followed convention. When they were in direct communication with an enemy, they invariably used subtle insults rather than direct verbal attacks, shunning outright assault for insinuation buried beneath grandiloquence. For this reason alone, de Duillier's diatribe was unwelcome and embarrassing. Like Hobbes's rejoinder from a generation earlier, the attack had been included in one of his published works and so did not require traditional courtesies, but nevertheless it went too far too soon. Newton would not have endorsed it, nor even wanted it if he had been approached first.

Fatio may have been Newton's most aggressive champion, but his most effective supporter was a mathematician named John Keill. As a young lecturer at Oxford taught by David Gregory, Keill had long been a dedicated Newtonian who later became the first lecturer anywhere in Britain to teach experimental philosophy. In his lectures he delivered the latest ideas on hydrostatics, dynamics, and optical phenomena based entirely upon Newton's two great works, the *Principia* and *Opticks*. Soon after Gregory's death in 1708, Keill came to Newton's attention via a paper containing a direct refutal of Leibniz, which he had offered for publication in the journal of the Royal Society, the *Transactions*. "All of these [laws] follow from the now highly celebrated arithmetic of fluxions which Mr. Newton, without any doubt, first invented, as anyone who reads his letters published by Wallis can readily determine," he declared. "Yet the same arithmetic, under a different name and method of notation, was afterwards published by Mr. Leibniz in the *Acta Eruditorum*."[40]

Although this too was an aggressive declaration that Newton was the first to the calculus, it fell short of claims of being in any sense vulgar or extravagant because Keill refrained from implying that Leibniz was a thief.

Also, by this time, almost a decade after Fatio's slander, such thinly veiled attacks were seen in some quarters as fair game because by then Leibniz had opened himself to attack by writing an "anonymous" review of Newton's *Opticks* in the January 1705 edition of *Acta eruditorum*. In this review he had taken the dangerous step of accusing Newton of plagiarism, writing: "Instead of the Leibnizian differences [Leibniz's version of the calculus] then, Mr. Newton employs, and has always employed fluxions. He has made elegant use of these both in his *Mathematical Principles of Nature* and in other publications since, just as Honoré Fabri in his *Synopsis geometrica* substituted the progress of motions for the method of Cavalieri."[41]

Most observers knew Leibniz had written the review, but where he pushed the argument between himself and Newton into new, uncharted territory was with his reference to Honoré Fabri, a notorious plagiarist of the time. This naturally infuriated Newton and his supporters, but Leibniz merely dismissed their anger by calling their reference to his comment "the malicious interpretation of a man who was looking for a quarrel."[42] This time though Newton was right to be angry, just as Leibniz had been over de Duillier's bluster.

Although it had been published in the October 1708 issue of the *Transactions*, Leibniz did not see a copy of Keill's missive until late in 1710, but when he did he was no less angry about it than Newton had been about the review of his *Opticks*. He immediately drafted an angry letter to the Royal Society, in which he demanded an apology. His grievances were then aired at a meeting of the society on April 5, 1711, during which his letter was read to the gathering. As a result, Keill was asked to write an apology.

By this stage it is likely Newton began to see that Keill's enthusiasm could be employed to his further advantage. The apology sent to Leibniz six weeks after the meeting certainly bears the master's imprimatur, and for his part Keill, like many other young followers of Newton, was more than keen for preferment. He had been recently passed over as Gregory's successor to the Savillian professorship at Oxford and was clearly willing to be manipulated by the great Isaac Newton. "I suggest only this," Keill wrote in his response to Leibniz, "that Mr. Newton was the first discoverer of the Arithmetic of Fluxions or Differential Calculus; however as he had in two letters written to Oldenburg (which the latter transmitted to Leibniz) given pretty plain indications to that man of most perceptive intelligence, whence Leibniz derived the principles of that calculus or at least could have derived them; but as that illustrious man did not need for his reasoning the form of speaking and notation which Newton used, he imposed his own."[43]

But for all Keill's smooth writing and word wrangling, this was no apology at all, and Newton's vindictiveness clearly burns through the gossamer. Here he makes it apparent that it would have been somehow impossible for Leibniz to have arrived at the calculus independently and that he had simply been fed information by the now long-dead Oldenburg (who of course could not contradict the charge). Newton makes clear his assertion that all Leibniz had done was to rearrange the master's work, "imposing his own" notation.

Not surprisingly, Leibniz instantly dismissed this ungracious response, and early in 1712 he replied to the Royal Society with matching aggression. This letter was read before a general meeting on January 31. "What Mr. John Keill wrote," he declared, "attacks my sincerity more openly than before; no fair-minded or sensible person will think it right that I, at my age and with such a full testimony of life, should state an apologetic case for it, appearing like a suitor before a court of law."[44]

Now, at long last, the lines of demarcation had been drawn, the time for gentlemanly insinuation had passed.

Given the opportunity to bring things to a head, Newton immediately stretched his privilege as president to the absolute limit, and perhaps influenced by Leibniz's comment about his "appearing like a suitor before a court of law," Newton decided to set about creating an official committee to investigate the dispute.

From the start, Leibniz had little chance of even holding his own in a contest with Newton. Beneath the thin veneer of officialdom and impartiality, Newton guided his servants with the experience of a lifetime of intrigue and self-protection. He called his judges "a numerous committee of gentlemen of several nations."[45] Eleven in number, at least six were Newton devotees, including Edmund Halley, Dr. Arbuthnot, and Abraham De Moivre.[46] Newton oversaw every detail of the investigation and wrote the committee's report, the *Commercium epistolicum*, published only six weeks after formation of the committee and, tellingly, a mere seven days after the last three members had been appointed. In fact, this committee and its report were such a sham the members did not even sign it, and their identities were unknown until revealing documents were unearthed in the society's archive during the nineteenth century.[47]

After detailing the reasons for Newton's claim and the shaky premise of Leibniz's contribution, the report concluded: "For which reasons we reckon Mr. Newton the first inventor and are of the opinion that Mr. Keill in asserting the same has been in no way injurious to Mr. Leibniz."[48]

Some years later Newton went on record explicitly denying any role in the committee that judged Leibniz's priority claim. Writing to Abbé Varignon in 1719, he remarked: "I was so far from printing the *Commercium Epistolicum* myself that I did not so much as produce the letters in my custody . . . lest I should seem to make myself a witness in my own cause."[49]

Perhaps he had deluded himself into believing this was the case, yet for Newton, usually the most pious of men, this was a very rare example of a barefaced lie; normally when the need arose, he was merely "economical with the truth." For evidence of this lie we need look no further than his own work: *Account of the Commercium epistolicum*, published anonymously a few years after the event and taking up all but three pages of the *Transactions* for January and February 1715. Even a cursory look shows this is definitely by Newton because it contains numerous technical references and details of his own mathematical insights only he could have written. One scholar has said of this document, "His identity [as the author] is unmistakable."[50] To accompany this, there is a draft of the committee's report, written in Newton's hand, thus proving his culpability.[51]

More than any other document written by Newton, his *Account of the Commercium epistolicum* reveals the extent of his determination and the lengths he would go to in order to destroy anyone who crossed him. Leibniz was his most despised opponent because his claims cut the deepest, endangering Newton's self-image and personal esteem. And, if the *Commercium epistolicum* itself was an unalloyed attack upon Leibniz's intellect, the *Account* is a character assassination in which Newton adds accusation to accusation and employs every scrap of evidence at his disposal. He seems to have become so obsessed and consumed with hatred for Leibniz that he slips into unadulterated hypocrisy, which reaches its most extreme when he writes of his enemy: "But no man is a witness to his own cause."[52] At no point does Newton seem aware of how far he is going with his writing nor how blindly he is passing judgment upon another with accusations that could be far better leveled at him.

Even after Leibniz was long dead Newton could not leave him alone. He went on to add comments to the drafts of the *Account* and commentary to the margins of his notes such as "Second inventors count for nothing."[53] Some of these documents run to a dozen drafts and fill no fewer than five hundred folios. The *Commercium epistolicum* and the *Account of the Commercium epistolicum* were bludgeons with which Newton attempted to pummel Leibniz's reputation. Although he never succeeded in breaking the man's will and failed completely to rewrite history, Newton at least managed to convince himself he had. In the final months of his life, he boasted to his doctor, Samuel Clarke, that he "had broken Leibniz's heart with his reply to him."[54]

But actually, he had failed in this too. Contrary to Newton's expectations, news of the *Commercium epistolicum* was received stoically by Leibniz. The document was published in January 1713 and distributed to academic centers throughout Britain and Europe. Leibniz first heard about it two months later, when his friend Johann Bernoulli, whose son Nikolaus had picked up a copy in Paris, wrote to him detailing its contents: "You are at once accused before a tribunal consisting, as it seems, of the participants and witnesses themselves," he roared, "then documents against you are produced, sentence is passed: you lose the case, you are condemned."[55]

Rather than this document having the effect of at least silencing Leibniz, now offended and wounded by what he saw as a betrayal by the Royal Society, it slowly turned his anger to outrage. "I have not yet seen the little English book directed against me . . . those idiotic arguments which (as I gather from your letter) they have brought forward deserve to be lashed by satirical wit," he replied to Bernoulli. Then, holding nothing back, he tore into Newton's own claim to the calculus: "He knew fluxions, but not the calculus of fluxions which he put together at a later stage after our own was already published. Thus I have myself done him more than justice, and this is the price I pay for my kindness."[56]

Within three months Leibniz and Bernoulli had struck back at the *Commercium epistolicum* (although Bernoulli insisted he remained incognito). They composed a pamphlet, the *Charta volans* (published on July 29, 1713), that followed the Royal Society pamphlet to the academic centers of

Europe. Although it was anonymous and contained no indication of where it had been published, its source was unmistakable. Leibniz had gained something of a reputation for "anonymous" compositions, and the conceit of referring to himself in the third person throughout the pamphlet fooled no one, least of all the president of the Royal Society.

"Newton took to himself the honour due to another," the author of the *Charta volans* declared, "he was too much influenced by flatterers ignorant of the earlier course of events and by a desire for renown. . . . Of this Hooke too has complained, in relation to the hypothesis of the planets, and Flamsteed because of the use of his observations."[57]

One can imagine Newton reading his copy and with each attack growing steadily more incensed. But actually, if Newton was not blinded by his own fury he might have known that Leibniz was fading fast, falling rapidly into fatal illness, and neglected by his employers. He was a match for Newton intellectually but completely outclassed and outgunned in all other respects.

Newton was president of the most important scientific institution in the world, the Royal Society, where he had been dubbed the Perpetual Dictator by his opponents.[58] He had the devoted attentions of a large body of supporters and disciples who would lay their careers at his feet. In stark contrast, Leibniz was neglected in his position as archivist for the elector of Hannover (the future King George I of England). He too had presided over an academic society, but it had been in the then cultural backwater of Berlin. He too had his supporters, but even they were in awe of Newton and would have been in fear for their careers if they crossed him too obviously. For this reason even Leibniz's most determined supporter, Johann Bernoulli, preferred his attacks to remain anonymous and went in for snide remarks behind Newton's back rather than head-on conflicts.

Leibniz fell into decline at the Hannoverian court. He was forced to earn a meager salary by writing a history of the House of Brunswick-Hannover, and in spite of his great talents, he was denied superior postings and academic preferment. Even when his patron, the great-grandson of James I, ascended the English throne in 1714, Leibniz's attempts to move to England to become royal historian were thwarted, and he spent his last days in what had become an almost deserted court. "We dwell here in a kind of solitude since our court has gone to England," he wrote to a friend.[59]

There were many sincere attempts to reach an amicable solution to this now famous dispute between Europe's two most respected scientists. A conscientious effort to find a settlement came from a young cleric and keen philosopher, the noble-born Abbé Antonio-Schinella Conti. He visited England as part of a delegation of European philosophers to witness the solar eclipse of 1714 and stayed on in London, where he gradually ingratiated himself with Newton and others at the Royal Society.

Although a charming and well-connected man (he was a friend of both Leibniz and King George I's daughter-in-law Princess Caroline), Conti exaggerated his claims to have had influence over the cantankerous Newton. He succeeded in gaining a degree of Newton's trust, but his efforts at bringing together the two sides and acquiring kudos as peace broker failed utterly. In fact, the dispute was never truly settled; the most that can be said for the final years of the conflict is that the public show of animosity faded as the rivals grew older and Leibniz's health began to fail.

As I have mentioned, one of the most striking aspects of Newton's character was that his anger increased rather than diminished as time passed. He was almost totally unforgiving, particularly over a matter so close to his heart as a priority dispute. But, given the fact that Leibniz's patron was now king of England, the dispute could not last long under public scrutiny and a semblance of resolution came in 1716, when King George instructed Newton to write a letter of reconciliation to Leibniz via Conti.

The king could command his servants to communicate, but he could not make them say what they did not believe to be true, so the resulting missive, sent by Newton in February 1716, was little more than a reiteration of the facts put in slightly milder tones than those offered in the *Commercium epistolicum*: "But as Leibniz has lately attacked me with an accusation which amounts to plagiarism: if he goes on to accuse me, it lies upon him by the laws of all nations to prove his accusation on pain of being accounted guilty of calumny," he declared. And, still unrelenting in his insistence that the Royal Society acted fairly, he attacked Leibniz's claim to the contrary: "The *Commercium Epistolicum* contains the ancient letters and papers . . . collected and published by a numerous committee of gentlemen of several nations appointed by the Royal Society."[60]

Ill and now almost totally alone at the Hannoverian court, Leibniz had little energy to comment further and merely dubbed the letter a *cartel de défi*—an act of defiance.

And so the personal vendetta gradually drained away and the debate was left to fester beyond the confines of documents, pamphlets, reviews, and letters. Leibniz died in November 1716, his funeral attended by the single servant who had remained with him at the Hannoverian court. As one of his friends later noted in a memoir: "Leibniz was buried more like a robber than what he really was, the ornament of his country."[61]

But despite the neglect, his later supporters, no less patriotic and determined than Newton's, ensured their master's work survived him. For generations European scholars refused to accept that Newton was the inventor of the calculus, and a settlement was only reached during the nineteenth century. Since then it has been accepted that the two great rivals who fought out a priority battle spanning half their lives had achieved independently the mathematical breakthroughs each had claimed as his own.

Isaac Newton died in 1727 and was given a funeral worthy of a monarch. He was buried in Westminster Abbey, where a grand marble monument was erected four years later. The monument depicts the many facets of Newton's long and illustrious career (but conspicuously omits any reference to his greatest preoccupation, alchemy). It dominates a region of the abbey now known as Scientists' Corner, where other British scientific luminaries—Faraday, Maxwell, and Darwin—are buried. Greatly honored and his work cherished during his lifetime, Newton could not have ended his mortal existence more differently than Gottfried Leibniz.

But that was certainly not the end of the story. From our viewpoint, the forty-year feud between Newton and Leibniz ended in a draw. Newton is properly seen as the first man to have devised the calculus, and Leibniz is no longer seen as a plagiarist; his notation is used throughout the world. But Newton never stopped believing he had been wronged; to have been first was not enough.

Newton never came close to acknowledging Leibniz's contribution, not even in private, and never forgave Leibniz for what he saw as a criminal attempt to intrude upon his sacred domain, his divine mission. As Newton crushed his English rivals, Hooke and Flamsteed, he had unceremoniously deleted their names from later editions of his masterpiece the *Principia*, even though he had used ideas, concepts, and data original to each of these men. For the third edition of the *Principia*, which appeared in 1726, a full ten years after Leibniz's death, Newton took great pains to have every single mention of the German's name deleted. As far as Newton was concerned, Leibniz had never existed.

$\mathcal{I}$GNORING FOR A MOMENT the enormous impact this dispute had upon the history of science, it is often forgotten that the battle between Leibniz and Newton established our modern method of publishing scientific ideas. Today a paper is submitted to an established journal such as *Nature* or *Science*. It is then sent to a set of referees, scientific peers who judge the content and either support it or criticize it. If a paper gains the acceptance of these referees, it may then be published, along with exhaustive references linking the work with its antecedents. In this way the scientific community can see clearly what an author of a paper is claiming to have discovered and plagiarism and priority disputes may be avoided.

The first arena for this method was the publication of the Royal Society, the *Transactions*, and the technique has evolved into a fine method that helps define scientific discovery and greatly enhances the communication of science. Today any breakthrough deemed significant enough by the press reaches the media of the world almost as soon as it is published in a scientific journal.

Newton would probably not have approved of this development. To him the important aspect of science was not communicating it but doing it. Indeed, the entire dispute with Leibniz hinged upon the fact that Newton had done the work but had told almost no one. According to Newton, the work was completed when the concept had been fully elucidated; the idea that others should know about it seemed irrelevant to him. Leibniz was every bit as talented as Newton and his work equally profound, but he understood that an element of its value came from peer acceptance, through which science itself could evolve. And in spite of the fact that Leibniz was less celebrated during his lifetime, it was his version of the calculus that had the greater immediate impact upon the future of mathematics and science.

The principal difference between Newton's notation and Leibniz's was ease of use. Newton, we should remember, worked in complete isolation. He cared little for a notation that would be easy to communicate and instead created a system to suit himself. In a letter to an associate he crowed that he had deliberately made his masterpiece, the *Principia mathematica*, as unreadable as possible, "to avoid being baited by little smatterers in mathematics."[62]

In stark contrast, Leibniz communicated frequently and openly with other mathematicians and natural philosophers and wanted to express his mathematical ideas in a universally understandable form. Because of this, it is his notation that has become part of standard mathematical language and is used universally today. During Newton's lifetime British mathematicians led the world, but their refusal to accept Leibnizian notation out of misplaced loyalty to Newton meant they lost this advantage during the subsequent fifty years. Leibniz even sensed this beginning to happen during his own final years and wrote of his creation: "The inventor [himself] and the very learned men who employed his invention, have published beautiful things which they have produced with it; whereas the followers of Mr. Newton have not effected anything in particular, having hardly done more than copy the others, or wherever they wanted to pursue the matter they have tumbled into false conclusions. . . . Hence it can be seen that what Mr. Newton had found was attributed more to his own genius than to the advantage of the invention."[63]

Ironically the advance of science on the Continent came through a powerful combination of Leibnizian calculus and Newtonian physics. Of particular importance were the extraordinarily talented Bernoulli brothers, Nikolaus and Daniel, sons of Leibniz's ally Johann Bernoulli. At the same time, their friend the renowned mathematician Leonhard Euler did much to amalgamate Newtonian mechanics and Leibnizian mathematics. Euler's greatest work, *Mechanica*, published in 1736 (just nine years after Newton's death) laid the theoretical framework that bridged Newton's principles with the practical applications of the engineers who initiated the Industrial Revolution toward the end of the eighteenth century.

During the generation following Euler and the Bernoulli brothers, the French mathematician Pierre Simon de Laplace, born in 1749, extended Newtonian mechanics beyond the *Principia* and applied a more modern calculus he and others in Europe had developed and refined. Laplace has been called "the greatest scientist of the late eighteen and early nineteen centuries,"[64] and according to the historian Robert Fox: "The age of Laplace witnessed the definitive establishment of the discipline of mathematical physics with the techniques of mathematics being used to an unprecedented effect in the elaboration of theories that could then be subjected to the control of experiment."[65]

The generation immediately following Laplace certainly did experiment, but by their time much of the creative flux had swept across the Channel, returning the English to the fore. James Watt, a practical man, an engineer, had the foresight to create the first modern steam engine, and the mathematicians and physicists of the time applied the theoretical ideas of Newton, Leibniz, Laplace, and others to create the "the age of the machine."

The conflict between Newton and Leibniz exaggerated a national rivalry that has always existed between mainland Europe and Great Britain. Newton's aloofness damaged the generation of mathematicians who followed him, but having fallen behind their European contemporaries, English mathematicians were obliged to work hard to redress the balance, eventually embracing the best of both words already enjoyed by their counterparts in France and Germany. By combining the notation of Leibniz, the physics of Newton, and the mathematical ideas of both men, they helped take civilization from the plow to the communications satellite.

# TWO

## The Fanatic and the Tax Collector

*Foole:* The reason why the seuen Starres are no mo then
seuen, is a pretty reason.
*Lear:* Because they are not eight.
*Foole:* Yes indeed, thou woulds't make a good Foole.
—*William Shakespeare,* King Lear

*Paris, May 1794*

A CART CONTAINING A SMALL GROUP of disheveled men rolls slowly along the cobbled streets, squeezing its way through the mocking crowd toward the place de la Concorde in the center of Paris. In the cart stand a group of prisoners, including France's greatest scientist, the chemist Antoine Lavoisier, and beside him his father-in-law, Jacques Paulze. The French Revolution is at its most vicious, its most depraved, and Lavoisier is being taken to the guillotine.

A few minutes later the cart stops at the scaffold and the prisoners are led up a short ladder propped against the platform. One by one they are pushed roughly down toward the blade; their heads are then severed with one smooth slice of the razor's edge. Lavoisier is fourth in the line. In silence he kneels before the executioner and places his neck on the block, the guillotine whines, and his head rolls into a great tumbrel awash with the blood of those gone before him, their heads bobbing in the gore.

Naturally, Lavoisier had fought hard to avoid his fate. He had protested to the prosecuting council, Robespierre's Committee of Public Safety, claiming he had important scientific work to complete. But, to his face, the judge's only comment was "The revolution does not need scientists." The day after his execution, one of his scientific colleagues, Joseph Lagrange, commented that "it had taken only a moment to cut off that head and a hundred years may not give us another like it."

And, as Lavoisier's head somersaulted from the guillotine, some fifteen hundred miles to the southwest, his most bitter rival, the English chemist Joseph Priestley, sat in a cabin aboard a ship heading for New York mulling over the fate that awaited him and his family in the New World. Priestley had first clashed with Lavoisier almost twenty years earlier, and he still harbored great resentment toward the Frenchman whom he considered a plagiarist. But like Lavoisier, Priestley had become a hostage to political fortune.

Hounded out of England, fleeing persecution of a different kind from Lavoisier's, the English scientist was on his way to a new and uncertain future in America. Priestley had long been Lavoisier's greatest rival, and he still resolutely refused to accept the French chemist's radical scientific views, even though they had been accepted by almost all other scientists. Yet, in spite of his feelings about Lavoisier's science, Priestley was appalled by the news from Paris, not least because it could so easily have been his head in the tumbrel.

THE ROW BETWEEN Lavoisier and Priestley that spanned almost two decades toward the end of the eighteenth century stemmed from a combination of willful disregard on Lavoisier's part and intransigence on Priestley's. Joseph Priestley discovered oxygen but had no idea what he had found. Antoine Lavoisier, who discovered nothing but understood Priestley's findings, was almost single-handedly responsible for interpreting the *role* of oxygen in combustion and its presence in the atmosphere.

Yet the dispute went further than a matter of priority. Priestley seemed completely unable to accept the meaning of Lavoisier's work and the way it changed the direction of our understanding of chemistry. Lavoisier's findings were so revolutionary they swept away the accepted philosophies that had dominated chemistry before him and destroyed the very foundation of Priestley's worldview.

The two men met only once. They corresponded, but almost all records of this exchange have now been lost. They shared scientific friends, including the great Benjamin Franklin, but that is about all they had in common. Their native lands had a long tradition of rivalry. They had diametrically opposite religious faiths. One was immensely rich, the other relatively poor.

Lavoisier claimed he respected the individual roles he and Priestley had played, and Priestley was accepted as the first to discover oxygen. What was more important in this conflict was the *interpretation* of what both men knew, what Priestley revealed from experiment and what Lavoisier elucidated through theory.

ANTOINE LAVOISIER WAS born in Paris in 1743. His father, Jean Antoine, was a successful lawyer who had inherited a modest fortune from his uncle as well as his influential position as attorney to the Parliament of Paris. As a successful lawyer, Lavoisier senior was able to marry into a wealthy bourgeois family and took as his wife Émilie Punctis, daughter of Clément Punctis, secretary to the marquis de Château-Renault, the vice admiral of France.

The Lavoisier family were strict Catholics, conservative and well connected. Born a little over a year after his parents' marriage, Antoine had a comfortable and secure early life. The family lived in a smart house in a district called the Marais, a quiet, well-to-do area where minor nobility and successful business folk had built large, fashionable homes.

Antoine was three when his sister, Marie, was born, but two years later, in 1748, their mother died suddenly, leaving Jean Antoine a widower with two young children. Antoine's grandfather had died a few months earlier, so it was decided the family should leave their house in the Marais and move into the home of Jean Antoine's mother, Mme. Punctis, situated near Les Halles in the heart of Paris. Here Antoine stayed until he married twenty-three years later.

Attending the best school in Paris, the Collège des Quatre Nations (today the site of the Academy of Sciences and the Academy of Fine Arts), Lavoisier was remembered by his contemporaries and teachers as a bright student who from an early age showed promise in a variety of subjects but no particular bent toward science. He wrote well, was encouraged to enter literary competitions that were popular at the time, and even won second prize in a national writing competition in 1760.*

Lavoisier was about fifteen when he began to show an interest in science. His first love had been geology, and he had already embarked on a research tour of France with the eminent geology professor Jean-Étienne Guettard. But from an early age he was aware there would be almost no chance of earning a living solely from science. During the eighteenth century, there were only two types of scientists, those who had acquired a university position after shining academically—successors to men like Isaac Newton, Robert Hooke, and John Flamsteed—and those wealthy enough to pursue independent research almost as a glorified hobby.† During the previous century, Robert Boyle had provided the model for this second approach. A contemporary of Newton, Boyle, the fourteenth child of the first earl of Cork, never held a university post but was independently wealthy enough to finance the expensive experiments he conducted during the 1660s and 1670s.

Although Lavoisier soon became obsessed with scientific discovery and realized he had a great talent for it, he also had an inborn and greater love for money and security. Neither the son of an earl nor able seriously to contemplate the life of an impoverished scholar, Lavoisier realized he would

---

*Another young and intellectually precocious Frenchman, Maximilien-Isidore Robespierre, was also successful in these competitions. He and Lavoisier met only once, over three decades after each won literary prizes, when, after a brief ramshackle trial, Robespierre nonchalantly signed the chemist's death warrant.

†As we will see, Priestley was for a time an exception to both these types. For six years he was patronized by a wealthy peer in much the same way artists and thinkers were supported by wealthy statesmen and philanthropists two centuries earlier. By the late eighteenth century this was a strikingly anachronistic arrangement.

have to practice science as an adjunct to a more secure career. To this end he studied law so that he could follow his father into business and worked hard at the prestigious Mazarin College in Paris, but always with the private plan to establish himself in a well-paid position so he might pursue his intellectual dreams using private funds.

But then in 1766, as he was about to embark upon his legal career, his grandmother Mme. Punctis died, and Antoine's father immediately passed on his inheritance to his son. Antoine suddenly found himself a wealthy young man and almost (but not quite) rich enough to pursue science entirely at his leisure, truly in the style of Boyle.

By this time Lavoisier's family had become aware of his keen interest and obvious abilities as a scientist. His tutors had spoken up for him as a promising young researcher, and Antoine's father, who had doted on his son, had had no objection to his ambitions. Yet Antoine himself still had reservations. Part of him wanted to devote himself fully to science, but a stronger part envisioned a future with a wife and perhaps a family. Besides this, he had always known wealth; he had wanted for nothing and was naturally prudent. Would his inheritance sustain him in the style to which he had grown accustomed?

Lavoisier's heart was leading him to science, but his head was grounding him in the sensibilities of comfortable bourgeois existence. He resolved this dilemma, but the solution he found was also to lead directly to his untimely and tragic death at the guillotine. In 1768 he decided to join the Ferme Générale (or Tax Farm) as a glorified tax collector.

For almost a century and a half, France had operated one of the oddest and most unfair tax systems of any civilized nation. Created by Louis XIV's finance minister, Jean-Baptiste Colbert, a private "corporation" called the Tax Farm bought six-year leases from the government that allowed its members to collect taxes as they saw fit. The members of this organization, the "Farmers," paid an annual advance to the government. This was covered by the taxes they collected; the rest went to line their substantial pockets.

That this system was open to abuse was obvious, and it was indeed abused, heartily and fulsomely. The roles and rewards of the Farmers were stratified within the organization. Each gained profits according to his social status, and each could buy into the organization to a degree based upon what he could afford as an initial investment to pass on to the government. Unsurprisingly perhaps, one of the greatest beneficiaries of the system was the king himself, who received the entire income from one of the most lucrative positions within the organization and creamed off over one hundred thousand livres per annum.* Other members of the royal family, including the king's children and siblings, received in excess of fifty thousand livres each year, and some of the French nobility had become infamous for their excesses, extravagances paid for from their tax profits. One Comte Beaujon was rumored to lavish an astonishing two hundred thousand livres each year upon the services of prostitutes who dressed him in diapers, read him bedtime stories, and rocked him to sleep each night.

What was so insidious about the scheme was the way the rich not only managed to profit from taxation but also avoided having to pay anything like a fair proportion of their income to the state. The nobility and the members of the Tax Farm were exempt from two major taxes, the *taillable* and the *corvées royales*, used for the construction and maintenance of roads. Both of these were borne mainly by the peasants. The nobility and the Farmers were liable to some newer taxes, the *capitation* and the *vingtièmes* (equivalent to income tax), but very few of them paid these, opting instead to pay off the *intendant* or chief tax collector. Clearly this was a recipe for disaster, and within it the seeds of revolution were germinating fast.

Although he was perhaps less greedy than many, when it came to making money through very little effort, Lavoisier appears to have had few scruples, and with the death of his grandmother he found himself able to buy into the system, albeit at a low level. He borrowed two-thirds of the million and a half livres he required and paid the rest in cash. This investment offered a return of almost 150,000 livres per year, for which he had to do very little except maintain an office and make sure his tax collectors did their job efficiently. Even with interest payments on his loan (which was paid off within a few years) he was making a very healthy living. And of course because he was a member of the body that effectively controlled the taxation system, he knew all the ways there were to avoid paying tax himself.

---

*The livre was replaced by the franc as the national unit of French currency immediately after the Revolution. Its original value was contrived to match the British pound, but by the time Louis was securing for himself an annual one hundred thousand livres, such a figure was probably worth under ten thousand pounds. However, allowing for inflation, this would now be equivalent to perhaps ten million pounds (fifteen million dollars).

By the age of twenty-five Lavoisier was financially secure and ready to apply himself to science. Two months after acquiring his lucrative position as a Farmer, he put his name forward for election to the Royal Academy of Sciences.

The academy was an elitist and closed-off organization structured to consist of 252 members spread across six scientific disciplines: geometry, astronomy, chemistry, physics, anatomy, and botany. Within each of the subjects there were three grades of membership. The lowest level was made up of the assistants who aided those in the second level, the associate members. The highest level was the *pensionnaires*, made up of the most respected scientists in the country along with representatives from the highest echelons of the nobility (whether or not they were scientifically literate). Only the top two grades, the associates and the *pensionnaires*, could vote in elections. According to one historian: "To some extent the early Royal Society and the Académie des Sciences may be seen as typifying the English and French scientific traditions. The Royal Society grew out of individual initiative and received royal recognition only after the fact. From its inception, it drew heavily upon the landed gentry for its membership and treasury; as a result, the breadth of its interests wandered away from the narrowly scientific. The Académie des Sciences, by contrast, functioned more as a branch of the French civil service, with a high degree of regimentation and control exercised from above."[1]

For membership, preference went to those with high social status rather than academic credentials, but although he was extremely young for consideration, Lavoisier could offer both. When a position became vacant upon the death of the chemist Théodore Baron, Lavoisier's name, along with that of another chemist, Gabriel Jars, was offered for consideration. Jars was the more experienced scientist and some years Lavoisier's senior, but Lavoisier was seen as a promising young man, a rising star. Split by loyalties and oiled by bribes, the honored members of the academy voted Jars to an assistant's position and granted Lavoisier a provisional place with the assurance that when the next position became vacant he would acquire it automatically. By June that year he had been admitted as an assistant into the academy proper.

For the following five years, Lavoisier settled into a well-organized and comfortable existence. He dedicated one day each week to his scientific explorations and lived the life of a wealthy Parisian gentleman. Never a handsome man, he dressed well and had a charm pleasing to others of his class. He was usually reserved, occasionally frosty, and some believed he wore his intelligence and his education on his sleeve.

Having acquired wealth and a career path, he decided he should find a wife. At the end of 1771, he married Marie-Anne Paulze, the thirteen-year-old daughter of a wealthy farmer-general, Jacques Paulze, who was head of a collective of Tax Farmers and a parliamentary lawyer. Marie-Anne came with the not insubstantial dowry of eighty thousand livres, but by the time he met her (when he was twenty-eight), Lavoisier was already considerably richer than his future father-in-law. He left his grandmother's house, where his father remained, and he and Marie-Anne moved to an extremely smart house in rue Neuve des Bon-Enfants, a wedding gift from Lavoisier senior.

The Lavoisiers' was a happy marriage. Marie-Anne was multitalented and intelligent and assisted her husband with his experiments, wrote up his findings, and later illustrated his papers and books. She was also pretty, a good conversationalist who spoke several languages, and admired as a hostess.

When Benjamin Franklin lived in France as the American minister during the 1770s, he and the Lavoisiers became close friends, and the great American polymath worked on a series of chemical experiments with them in Paris. Mme. Lavoisier painted a portrait of their friend, and the letter Franklin wrote from Philadelphia in 1788 to thank the couple still survives. "I have a long time been disabled from writing to my dear friend by a severe Fit of the Gout, or I should sooner have returned my thanks for her very kind present of the Portrait, which she has herself done me the honour to make of me," he wrote. "It is allowed for those who have seen it to have great merit as a picture in every respect; but what particularly endears me, is the hand that drew it. Our English enemies when they were in possession of this city and of my house, made a prisoner of my portrait, and carried it off with them, leaving that of its companion, my wife, by itself, a kind of widow. You have replaced the husband; and the lady seems to smile, as well pleased."[2]

The Lavoisiers lived in rue Neuve des Bon-Enfants until 1775, when Lavoisier was offered a position on the Gunpowder Commission, an organization that had been created by a minister of the newly crowned Louis XVI with the purpose of reorganizing the production of gunpowder, a task that had recently fallen into disarray. As one of four commissioners responsible for rethinking the manufacturing processes of gunpowder and its delivery to the military, Lavoisier had the great advantage of being given a splendid residence at the Arsenal and obtaining rooms where he could build a private laboratory. And it was here he conducted his most important experiments and established the facts he then molded into a new chemical theory.

JOSEPH PRIESTLEY WAS in almost all ways the very opposite of Antoine Lavoisier. He was born in 1733 in a tiny Yorkshire village named Fieldhead. His family was poor; his father worked at the local mill, and his mother was a farmer's daughter. His early childhood was spent in a tiny, freezing stone cottage; life for the Priestleys was brutal and often short. After delivering six children in six years, Joseph's mother died in the bitterly cold winter of 1739; that year the River Thames froze over in London.

The Priestley household was shattered by the loss. Joseph's father could not look after all six children, so they were split up. Joseph was sent to live with his aunt, Mrs. Constance Keighley, who lived in West Riding in a village called Heckmondwike. This side of the family had for some time been obsessed with religion; Aunt Constance's husband had experienced a powerful religious conversion, which had led the couple to become Dissenting Christians.

Dissenters, or Nonconformists, were a group who, some seventy years earlier, had broken away from the Anglican Church over disagreements concerning the interpretation of the Book of Common Prayer. Originally comprising over two thousand clergymen who had been sacked by the established church because of their unwillingness to comply with tradition, the sect had grown and developed. By the 1740s they were seen as a significant radical opponent to orthodoxy, distrusted by most churchgoers and ostracized from many segments of society. In order to survive, members closed ranks and established their own schools and colleges, built their own churches, and formed self-sufficient trade units between communities. By the end of the eighteenth century they would include in their number powerful figures such as Erasmus Darwin (Charles's grandfather) and the Wedgwoods, one of the wealthiest families in Georgian Britain.

Life with Aunt Constance was a continuous round of religious commitment. In fact, there seemed precious little else in Joseph's world at the time. "There was hardly a day in the week in which there was not some meeting of one part of the congregation," he later remarked.[3]

The Keighleys, who had no children of their own, were considerably better off than Joseph's parents had been, and they sent him to better schools than he could have hoped for in his home village. He showed an early inclination toward languages, learned Greek and Latin and taught himself rudimentary Hebrew during his summer holidays. The first choice of career for such bright boys would usually have been the clergy, to propagate the Dissenters' doctrine, but Joseph was a sickly child, and, to further damage his chances of leading his own flock to heaven, at about age of ten he had developed a severe stutter. So the family decided young Joseph should be packed off to a family friend in Lisbon, where he might follow a career in business.

This, just as Newton's mother would have altered history if she had succeeded in forcing her son to stay home to look after the farm, in 1748, Priestley was almost lost to science. But then, shortly before the trip was finalized, Joseph's health improved dramatically (although his stutter did not), and he was sent to the highly regarded Dissenting academy at Daventry, where he was to begin his training as a minister.

At Daventry the young Priestley appears to have undergone his own religious conversion, and within a few months he had adopted the extreme doctrine of Arianism, the unorthodox Christian teachings Isaac Newton had followed all his adult life. In Newton's day Arianism was considered truly heretical and he could have been prosecuted, even executed, if his beliefs had been revealed. But by Priestley's lifetime, almost a century later, Nonconformists were usually perceived as something of an irritation and no longer lived in fear of the stake.

As Priestley grew older, he became more and more radical. When he first adopted Arianism, he was an academic, determined young cleric, but as the years passed his manner became more forceful and increasingly belligerent toward orthodoxy.

As with many rivalries in the history of science, the conflict between Lavoisier and Priestley was focused upon science but carried undercurrents of personal differences. Newton and Leibniz had held very different religious views, and this had added a gloss to their clashing attitudes toward the proselytizing of science, their different philosophies, and their nationalistic prejudices. The dispute between Lavoisier and Priestley likewise had strong nationalistic and political overtones, but it was also underpinned by great differences in their religious perspectives.

Lavoisier was not a strict or even a practicing Roman Catholic; in fact, he has been described as a "scientific humanist" who placed rationality and logic ahead of religious dogma. This stance would have been anathema to Priestley. For him, the only thing worse than a Catholic was an atheist, and among those who cared (like Priestley), Lavoisier could have been perceived as a man with a foot in both camps.

But their differences went still deeper; as Priestley was struggling to survive on a tiny allowance from his modestly placed aunt, a few hundred miles away in Paris, Lavoisier was still at school, living an existence filled with comfort and privilege. This difference in social standing became more exaggerated as Priestley struggled to maintain a modest living and Lavoisier became a wealthy tax collector.

In 1761 Priestley took a job as a language teacher at Warrington Academy near Liverpool but continued to pursue his interest in theology, becoming a lay preacher soon after. He spent some time each year in London, and it was probably during one of these stays that he was introduced to an intellectual circle centered on Benjamin Franklin, then in his mid-fifties and working to resolve political differences between his own government and that of the English.

By the time the two men met, Franklin was already world famous. He had been elected to the Royal Society in 1756, a few years after establishing a scientific reputation from a series of experiments into the phenomenon of electricity (including the famous flying of a kite in a thunderstorm). In London he kept rooms in Craven Street close to the Strand, where he entertained some of the most famous intellectuals of the day. Through Franklin, Priestley was introduced to this circle, and he was inspired by them to absorb the spirit of the age, to appreciate the power of experiment and the new methods of scientific investigation.

Influenced by his friends, during the early 1760s Priestley acquired a plethora of new interests and wrote a collection of short books on subjects ranging from electricity and magnetism to language and law. He even turned his hand to literary criticism. Although he made little money from his endeavors and his books were appreciated by only a tiny elite, he did gain valuable kudos from them. In 1762 he was admitted to the Royal Society. Elitist still, as it had been in Newton's day, it was certainly more democratic than its French counterpart, placing greater emphasis on the intellectual merits of its members than on their social standing.

That same year Priestley married. His wife, Mary Wilkinson, an iron-monger's daughter, came from a family as poor as his own. Unlike Lavoisier's wife, Marie-Anne, Mary Priestley had no interest in science, but she was a dependable partner, and, through the traumas and hardship that lay ahead of them, their marriage was a solid blend of mutual devotion and love gelled by a shared religious fervor and gritty Northern stoicism.

For Priestley, money was a perennial problem. As a bachelor he had eked out an existence on a salary of thirty pounds per year. Now with a growing family and a wife to support, the one hundred pounds per annum he earned from his teaching position supplemented with fifteen pounds a year from private pupils was insufficient, and in 1767 he accepted a position as a minister in Leeds, a move that gave him a better salary and a small but comfortable house. It was also a less demanding job that offered him free time to pursue his new scientific interests.

His earliest experiments were studies of electricity. He achieved little with them, but soon after this the family moved to Leeds, and Priestley stumbled upon a series of chemical curiosities that stimulated his interest and led him into exciting new areas of investigation. The Priestleys' home happened to stand next door to a brewery. The brewing process produces copious amounts of what was then called fixed air, and, as he passed the brewery each morning, Priestley saw this substance bubbling from the beer kegs and decided to investigate its properties. Drawing some of the gas into a sealed vessel, he conducted as many experiments as he could devise, including tests to see whether it dissolved in water and how it would affect a mouse placed in a gas-filled flask. He then wrote up his findings in a report he sent to the Royal Society in London.

And so began Priestley's lifelong obsession with gases. His first useful discovery came almost immediately, when he stumbled upon the idea of passing the carbon dioxide he had bottled from the brewery into plain water. By doing this he made the first ever soda water and started a craze for what G. K. Chesterton years later mockingly referred to as "windy water." Sadly for Priestley, although his invention created a fashion throughout Europe and America and survives today with every can of Coke sold, he did not earn a penny from his discovery. Instead, the idea was noticed and adapted successfully by others with greater business acumen.

But then, at this stage in his career, Priestley was probably not too disturbed by the fact that his work had been taken up by others who cut him out of the equation. And as we shall see, the root of his dispute with Lavoisier began with exceptional openness, illustrating that as a young researcher Priestley was not precious about his findings and held back little when discussing his discoveries. During the 1760s he appears to have cared little for priority. In those early days his attitude was dominated by the belief that science was for the benefit of all. In this sense he was the very embodiment of the Industrial Revolution; because he came from poverty (and remained poor most of his life) he saw science as a tool of the people, the precursor of what is now called technology. It was only when others took advantage of his magnanimous approach and failed, as he saw it, to give him any recognition, as even a contributor, that Priestley lost his calm detachment. Evidently, as a young scientist he cared more for chemical theory than the commercial rewards soda water might offer.


$\mathcal{I}$T IS A STAGGERING fact that before Lavoisier there really was no such science as chemistry. Just look at the world around us, contemplate the very fabric of our civilization, the materials we use in our homes, those we need to build our vehicles, to clothe ourselves, the foods we eat. The way we manipulate matter to create these things had its origins in Lavoisier's laboratory a little over two hundred years ago. Without the foundation he laid, we could have no plastics, no synthetics, no special materials for constructing lightweight machines, no smart fabrics, and because of this, no microelectronics, no space technology, no communications industry, no cars, no aircraft.

Some people wonder why it is that chemistry seems to have lagged far behind physics; after all, Newton had elucidated the fundamental laws that govern movement within the universe over a century before Lavoisier created a revolution in chemistry. Yet the reason for this is clear; physics deals with "primary principles," pure fundamentals, such things as mass, length, velocity; the chemist has to work with what might be thought of as "secondary principles," color, texture, odor, changes on a macroscopic scale such as chemical processes we may observe in the lab. These things are far more difficult to quantify, to put numbers to, or to build a logical framework around, so chemistry took longer to grow up; it took longer to adopt the vestments of formality and structure.

Indeed, the history of chemistry before the days of Lavoisier and Priestley is far from auspicious. The original chemists were the ancient Greeks. They used practical chemistry in their farming methods, in the construction of their buildings, and in the use of dyes.* The Greek theorists thought deeply about the fundamental structure of Nature. Aristotle formulated his concept of the four elements—fire, earth, air, and water—believing that all matter was composed of different proportions of these basics.

Aristotle's ideas held sway up to Newton's time, but, as we have seen, his was certainly not the only theory, even if it was the one that influenced most thinkers up to the seventeenth century. The ideas of Democritus were particularly visionary. For example, he claimed: "According to convention there is a sweet and a bitter, a hot and a cold, and according to convention there is color. In truth there are atoms and void." But neither Democritus nor any of the ancient Greek philosophers really understood what this statement meant—to do so required an atmosphere and a base knowledge like Lavoisier's. Lavoisier lived in an age during which more was known, the foundation of his thinking was broader. Besides this, although he had been conceptualizing along the correct lines, Democritus was sidelined by the dominant philosophy of Aristotle and ignored for two millennia.

Ancient chemistry languished among the Romans and found a second wind only with an associated science, what we now see as the predecessor of real chemistry, alchemy. Enwrapped as it was in mysticism and magic, alchemy was far removed from modern chemistry—a science intimately

---

*The Phoenicians of the fifteenth century B.C. used a dye called Tyrian purple produced from crushed sea snails, and the ancient Jews used another dye, Hyacinthine purple, a blue-purple dye otherwise known as *tekhelet*, and employed thirty-six hundred years ago for dyeing ritual prayer shawls.

linked as it now is with physics and, through physics, mathematics. The alchemists formed a "secret college," which transcended the millennia with roots in first century A.D. Alexandria. Gradually alchemical ideas found their way to Dark Ages Europe via the mystics, philosophers, and thinkers of the Near East. Alchemy was always a radical discipline, an art rather than a science.

Carl Jung was fascinated with alchemy and made an extensive study of the subject. He came to the conclusion that the alchemists were so obsessive in their search for the philosophers' stone they believed could transmute base metals into gold because this desire was actually a wish to transmute themselves, to free their own souls. "The alchemical stone," Jung wrote, "symbolises something that can never be lost or dissolved, something eternal that some alchemists compared to the mystical experience of God within one's own soul. It usually takes prolonged suffering to burn away all the superfluous psychic elements concealing the stone. But some profound inner experience of the Self does occur to most people at least once in a lifetime. From the psychological standpoint, a genuinely religious attitude consists of an effort to discover this unique experience and genuinely to keep in tune with it."[4]

The alchemists loved mystery and were driven to obfuscate and obscure their findings because this enhanced their image and encouraged others to believe alchemists were privy to the "secrets of the universe." To use a modern term, such behavior endowed them with "cult status." This is the very reverse of the modern scientific ethic. The job of the scientist and the thing that holds science together is the search for Truth and the need to reveal that Truth, to fit it into the greater scheme of things, not to deliberately obscure discoveries for one's own ends. The alchemists passed on their knowledge enshrouded in code and confused it with poetic imagery so that successive generations had to work hard to relearn what had already been discovered, almost as an initiation.

Yet, in spite of the fact that hundreds of years of effort and faith-led devotion set thousands of seekers upon the alchemical road to precisely nowhere, we do owe a debt to the alchemists. During its long history, alchemy provided us with two invaluable things. First of these are the results obtained by Sir Isaac Newton, who was as dedicated to alchemy as any mystic who ever lived and who reached his great discoveries in major part through his alchemical work. Second, alchemists invented a collection of useful devices and chemical apparatus, some of which are still used in modern laboratories; the retort, the distillation flask, the evaporation dish all came from the alchemist's dark cell. Along with these, alchemists devised valued chemical techniques—distillation, filtration, and the separation of liquids.

Men like Isaac Newton and Robert Boyle did much to drag alchemy out of the mire of mysticism and turn it from an art into a science. Boyle referred to other alchemists as "vulgar *stagyrists*."* Early scientists like Newton and Boyle employed parts of the ancient texts they obtained on the black market (alchemy was illegal and its practice punishable by death during the seventeenth and eighteenth centuries), but they were never fooled by dreams of limitless wealth or the production of an elixir for eternal youth. They took the seed of alchemy, applied logic and their considerable intellects, and built something altogether new. Newton gave us the *Principia* and the *Opticks*, and Boyle laid down some of the foundations of chemistry later adopted by Lavoisier.

Boyle built a laboratory in a small house next to University College, Oxford, and was assisted by Newton's later associate Robert Hooke. Fascinated by the properties of gases and the nature of air, he spent many hours each day researching and making notes from his observations. These he collected in his famous book the *Sceptical Chymist*, published in 1661 (the year Newton entered Cambridge as an undergraduate). Although many of its ideas were ill-conceived and confused, in stark contrast to the hundreds of thousands of books produced by generations of alchemists, Boyle's treatise may be seen as the first serious analytical chemistry text.

Boyle used simple air pumps built by Hooke to deduce how the volume of a gas and its pressure were interrelated. His work led to the formulation of the eponymous law of gases, taught in all secondary school chemistry courses. Along with another Oxford natural philosopher, his contemporary John Mayow, Boyle conducted clever experiments on air to

---

*A term pertaining to those who hankered after pure mysticism rather than natural philosophy.

show that respiration and combustion were impossible without it. Boyle had no idea that only a *part* of the air was responsible for these two processes and imagined air to be a single, simple substance. However, he rejected Aristotle's idea of the four elements, discarded almost all the pseudoscience of the alchemists, and set the study of chemistry on a firmer footing than anyone before him.

But despite these efforts, chemistry (or rather, *iatrochemistry*, as chemistry before Lavoisier is known) remained the backward poor relation of physics, progressing remarkably slowly. And to damage its reputation further, immediately after Boyle and Mayow, chemistry was led into a disastrous cul-de-sac, an intellectual blind alley that retarded the subject for almost a century. The reason for this—the phlogiston theory devised by George Stahl.

Stahl was born a year before the publication of Boyle's *Sceptical Chymist*. He was a bright student who became a successful medic and eventually ascended to the lofty post of physician to the king of Prussia. In his spare time he experimented with chemistry, conducting rudimentary researches in his rooms. From his observations of combustion and respiration, he created a startlingly new, but alas completely fallacious, chemical idea. This stated that when anything combusted, it gave off a substance he called phlogiston, which was absorbed into the atmosphere. Sulfur, Stahl declared, was almost pure phlogiston because when it burned it left little residue. Phosphorus was, he argued, in large part composed of phlogiston plus a powder, the residue of its combustion. Continuing with this line of reasoning, Stahl stated that metals were what he called calx plus phlogiston, because when metals were combusted they gave off invisible phlogiston and left behind a powdery substance, calx, sometimes dubbed "calcined metals."

From here he went on to claim that when animals respired they too produced this mysterious substance called phlogiston, which was absorbed by the air; and when a candle burned it also gave off phlogiston. He knew animals died after spending some time in a sealed container and claimed that this was because the air in the container had become saturated with phlogiston. The same idea could be used to explain why a candle or a fire stopped burning after a time in a similarly sealed vessel.

This theory is of course the very opposite of what really happens during combustion or respiration. We now know that the powder left after a metal has burned in air is what we call an oxide of the metal. This has been formed by the oxygen in the air (comprising one-fifth of the whole air) combining with the metal to form a compound made up of the metal and oxygen. But before Priestley no one knew of the existence of oxygen, and nobody had surmised that air was a mixture of gases, only one of which sustained combustion and respiration. And indeed, the misconceived approach of the phlogistonists should not be viewed too harshly. Unless one probes further into the question of how combustion and respiration occur, the first logical conclusion might be that something is "given off" rather than absorbed. After all, when a substance is combusted, flames rise from the material, and if this burning is conducted in a sealed vessel, charring may be seen, soot may appear, and perhaps water vapor on the inside of the vessel. When an animal is placed in a sealed container and it respires, moisture appears on the inside of the vessel.

So Stahl and his followers may have initially made an understandable mistake, but what is staggering is that the phlogiston theory lasted for three quarters of a century, from Stahl's first experiments until the idea was buried by Lavoisier. Furthermore, because until Lavoisier there seemed to be no alternative theory around, the idea survived a deluge of information and cogent argument. Because the real mechanism was not realized, experiments such as those Boyle had conducted were dismissed by the phlogistonists as misconceived—or, worse, the results were explained away as being due to careless technique.

As early as those experiments in Boyle's Oxford laboratory and before Stahl was even born, it had become clear that when a substance burned it gained weight. Boyle had shown this to be the case by conducting careful tests in which he weighed his combustibles before the experiment and the products after it. But the vast majority of scientists during the first three-quarters of the eighteenth century either ignored this obvious, demonstrable fact or they stuck with an ingenious, if meretricious solution to the anomaly. They declared that, because flames rise from burning material, phlogiston must have *negative mass*. So when it was evolved from a combusted material, the substance left behind would be *heavier* than the original material.

This was the state of chemistry when in 1774 Priestley discovered oxygen and Lavoisier began to theorize on the discovery. The subject then comprised a vast collection of facts, some confused, some supported by centuries of experiment. But crucially, none of what was known was held together by anything like a general theory. Embryonic chemistry, iatrochemistry, was a mishmash of unconnected islands of information. If one were to compare the level of understanding of the true nature of the chemical world in 1074 with that in 1774 one would find little difference, but a similar comparison between 1074 and 1784 would reveal a world of change. This revolution sprang from the work of a very small collection of men, including Henry Cavendish, Joseph Black, and Joseph Priestley in Britain, and Antoine Lavoisier in France. And in particular, the shift in understanding in those few years was accelerated by the chafing of Lavoisier's personality with Priestley's, two reactionary characters whose approach and intellectual natures lay at opposite extremes of the spectrum.

$\mathcal{I}$N 1772, WHEN LAVOISIER began to experiment seriously with gases, he wrote about what he was setting out to do, and to his great credit, he stuck with it. "However numerous may be the experiments of Messrs. Hales, Black, Magbridge, Jacquin, Cranz, Prisley and de Smith,"* he declared, "in this direction, they come far short of the number necessary to establish a complete doctrine. . . . I feel bound to look upon all that has been done before me merely as suggestive: I propose to repeat it all with new safeguards, in order to link our knowledge of the air that goes into combination with, or is liberated from, other substances, with other acquired knowledge, and to form a theory. The results of the other authors whom I have named, considered from this point of view, appeared to me like separate pieces of a great chain; these authors have joined only some links of the chain. An immense series of experiments remains to be made in order to lead to a continuous whole."[5]

This statement was to set the tone for Lavoisier's approach and reflected his meticulousness. He planned to dive into the complexities of a

---

*It is perhaps revealing that in his manuscripts Lavoisier almost never spells correctly the names of foreign chemists.

totally misunderstood subject, to wipe away the cobwebs before attempting a complete overhaul of the subject: In this sense he clearly did not lack ambition.

He started in October 1772 with an experiment that was as simple as it was potent. He took some phosphorus, which burns easily, and combusted it to produce what he called an "acid spirit of phosphorus" (what today we call phosphoric acid). His crucial finding was that the product of the combustion weighed more than the original phosphorus. This led him to the conclusion that the phosphorus had *absorbed* air to form the product.

To us this is obvious, but if we think for a moment of Lavoisier's chemical ancestry, it is easy to see that his conclusion was made only by taking a significant step away from convention and inherited wisdom. In fact, to almost all other chemists, this simple, logical conclusion would have been nothing less than scientific heresy. What Lavoisier was saying flew in the face of the entire phlogiston theory.

He was quite aware of how refractory his idea was. He also knew that before he could offer it to the world he would have to conduct many more tests and be able to show that his calculations fitted repeatable experiment. Aware that he would have to do, as he had said, "an immense series of experiments," but equally conscious that someone else might reach the same conclusions before he could publish, Lavoisier decided to stake his priority by completing two more tests, after which he would draft a memorandum to present, in confidence, to the secretary of the Academy of Sciences.

The first experiment involved repeating what he had already done, but this time with sulfur. He found that a similar thing occurred and another acidic substance (then known as vitriolic acid, what we call sulfuric acid) was formed. In the second experiment, he tried to create the reverse process, what we would call reduction. He heated lead in air and formed an oxide (what Lavoisier referred to as calx) and observed that, as before it increased in weight. He then reacted this oxide with charcoal, which changed it back to the original metal and at the same time liberated a large volume of gas (the oxygen it had absorbed when it was oxidized).

In his memorandum Lavoisier wrote:

About eight days ago I discovered that sulfur, in burning, far from losing weight, on the contrary gains it; that is to say that from a pound of sulfur one can obtain much more than a pound of vitriolic acid [sulfuric acid] making allowance for the humidity of the air; it is the same with phosphorus; this increase in weight arises from a prodigious quantity of air that is fixed during the combustion and combines with vapors. This discovery, which I have estimated by experiments that I regard as decisive, has led me to think that what is observed in the combustion of sulfur and phosphorus may well take place in the case of all substances and I am persuaded that the increase in weight of metallic calcs [what today we call metal oxides] is due to the same cause. Experiment has completely confirmed my conjectures. I have carried out the reduction of litharge [a calx or oxide of lead] in closed vessels, with the apparatus of Hales [a method invented by the scientist Stephen Hales], and I observed that, just as the calx changed into metal, a large quantity of air was liberated and that this air formed a volume one thousand times greater than the quantity of litharge employed. This discovery appearing to me one of the most interesting of those that have been made since the time of Stahl, I felt I ought to secure my right to it, by depositing this note in the hands of the Secretary of the Academy, to remain sealed until the time when I shall make my experiments known. Paris November 1, 1772, Lavoisier.[6]

Reading this at the start of the twenty-first century, we cannot help but wonder why it had not all been said before. Was it really true that others had not conducted this simple experiment, perhaps centuries earlier? The simple answer is that this experiment had indeed been conducted before Lavoisier. But what distinguished Lavoisier's effort from all the others was not the quality of the experiment (after all, he was only marginally more precise in his methods than those in previous centuries), it was the manner in which he formulated clear *explanations* for what he observed occurring. All Lavoisier's antecedents (and there had been many) had offered semi-mystical explanations for the process or else they had merely toed the phlogiston line.

And this was just the beginning. The following winter, in February 1773, Lavoisier was ready to write a second memorandum describing his work of the previous autumn, another set of experiments that took his ideas much further. This note to himself (which interestingly he misdated February 1772 and later changed) was written in his laboratory notebook and again kept private until he was ready to publish officially.

Two respected British chemists had described experiments that, to Lavoisier, seemed both contradictory and at odds with his own findings. During the 1750s a Scottish medical student, Joseph Black, had stumbled upon the discovery of carbon dioxide (a gas he called fixed air), and he later showed that this gas was produced during respiration and fermentation (as Priestley later found), and most important, it was evolved when charcoal was burned. In 1756 Black published these findings in a book titled *Experiments upon Magnesia Alba, Quicklime, and some other Alcaline Substances*.

However, Lavoisier was confused by these results because he also knew that some thirty years before Black's book appeared, an English vicar named Stephen Hales, who had dabbled in private researches, had described in a book called *Vegetable Staticks* how substances gave off "air." Hales had no idea what the nature of this material (or materials) could be, but from Lavoisier's reading of these experiments it was clear to him that Hales's "air" was different from Black's "fixed air."

Lavoisier's experiments showed him that combustion of different substances generated quite different gases. He had no inkling then what these gases were or how they differed, but it was clear that Black and Hales had been combusting totally different substances and obtaining quite distinct products. His predecessors had, Lavoisier wrote, "observed differences so great between the air liberated from substances and that which we breathe, that they deemed it to be another substance, to which they have given the name of 'fixed air.' . . . It is established that fixed air shows properties very different from those of common air. Indeed, it kills the animals that breathe it; whilst the other is necessary and essential to their preservation."[7] He then stated in detail the series of experiments he was planning, how they would be arranged, and how they would demonstrate his growing suspicion that Black's fixed air was entirely different from the gas involved in combustion.

Soon after writing this, in May 1773, Lavoisier allowed the memorandum of November 1, 1772, to be read before the academy, and he pressed on with experiment after experiment in which he combusted a range of materials, collected the gases, and measured the changes in mass of the materials involved. All of this he finally published in a book titled *Opuscules physiques et chymiques (Physical and Chemical Essays)*.

*Physical and Chemical Essays* was the first of Lavoisier's revolutionary contributions and was as far removed from conventional chemical wisdom as one would have expected from him. In the first part he surveyed the history of chemistry and on this occasion gave ample credit to the contributions of British chemists before him, even going so far as to say that France had until this time offered nothing to the development of this subject. In the second part, he described his revolutionary experiments and what they meant.

Energized by his findings, Lavoisier had returned to his experiments before anyone had had time to absorb this startling work and had begun writing descriptions of the next stage of his explorations. Ever since his earliest experiments, he had assumed that the air, as a single entity, was combining with the combusted substances, but he now began to suspect that the air might contain more than one component and that only part of the air was combining with the burning material.

He had been working quantitatively, carefully weighing the combustible material before and after the experiments, but now he began to concentrate upon the air itself. In April 1774 he found that between one-sixth and one-fifth of the air used in an experiment was consumed in the process of combustion, and he went on to find that 154 grams of phosphorus used up 89 grams of "air". What these figures meant was only to be revealed gradually during the following years, but the first set of results—the fact that only a small part of the air was used in combustion—supported his belief that it was made up of more than one gas and that only a part of the air was the combusting agent.

Lavoisier was building up to an announcement describing these most recent findings and had already booked a reading of his latest memoir before the academy on November 12, 1774, when word of his results was leaked. On the eve of his announcement, an Italian chemist, Cesare Beccaria, wrote from Turin informing Lavoisier that fifteen years earlier he had shown that there was diminution of the volume of air when tin was combusted and that his work had been documented clearly in the journal of the Turin Academy.

Lavoisier was shocked. He had taken great pains to verify what had already been written, but he had missed Beccaria's research. Clearly disturbed, he immediately told the academy journal (which was even then setting the type for his contribution) that they must accompany his report with an excerpt from Beccaria's letter along with a disclaimer that he, Lavoisier had been quite unaware of the Italian's work.

But he had overreacted. Upon close investigation of Beccaria's contribution, it was clear he had been stumbling around in the wilderness just as Black, Hales, and Priestley had done. Beccaria had had no idea what he was observing and was merely reporting a "curious" result of another "accidental" experiment.

But there was more misunderstanding to come. Soon after Lavoisier published his experiments in the *Observations* (the journal of the French Academy), a fellow French chemist, the elderly Pierre Bayen, then master apothecary to the French Army, had taken exception to Lavoisier's comment that no Frenchman had contributed to the understanding of gases and wrote an indignant letter to the editor of the journal pointing out that Jean Rey had described what he believed was Lavoisier's new discovery in his *Essays*, published in 1630: "It was a Frenchman who by the power of his genius first divined the cause of the gain in weight shown by certain metals when exposed to the action of fire and converted into calcs," Bayen fumed, "and that this cause is precisely the same as the truth of which has just been proved in the experiments described by M. Lavoisier at the last public assembly of the Academy of Sciences."[8]

Lavoisier had apparently missed Rey's work also, but, to be fair, this was not entirely surprising. Jean Rey had died an obscure physician and an even more obscure chemist some hundred and thirty years earlier. In the 1770s there were known to be in existence just seven copies of his book, *Essays*, and these could be found only in the libraries of a few specialists. It is almost certain Lavoisier had never even seen the book.

But, whether or not Lavoisier had known of Rey's work is quite irrelevant because Jean Rey had been yet another "lucky stumbler" who had had absolutely no idea what he was writing about. He had shown that a material gained weight when it combusted but thought fire itself was responsible for this adhesion. This was due to "heat driving away the lighter airs during combustion," he wrote, "allowing denser airs to adhere or stick to the material."[9]

This misunderstanding on the part of Bayen, who was, we must remember, a respected and senior member of the scientific establishment, illustrates just how revolutionary Lavoisier's propositions were. Men such as Bayen simply could not understand the difference between what

Lavoisier was saying and what the work of quasi-alchemists such as as Jean Rey had to offer. To the conservatives of the time Lavoisier was as mistaken as the ancients; as far as they were concerned, materials did not combine with the air but produced phlogiston, which was absorbed by the air.

It was into this confused choreograph, where the establishment could not conceive of an alternative to orthodox iatrochemistry (even if tradition clearly clashed with the provable facts), that our other protagonist, Joseph Priestley, entered, stage left.

A S LAVOISIER WAS beginning his experiments on gases and storing away his ideas in sealed memoranda, across the English Channel, Priestley had been far from idle. "I am going on with my experiments on air with uncommon success, for which I am not unthankful to the Giver of Knowledge," he wrote in 1771.[10] But helpful as the Giver of Knowledge may have been with Priestley's experiments, he was not quite so forthcoming with material comfort. The Reverend Joseph Priestley's growing family was proving to be an unmanageable burden on his salary as a Nonconformist minister.

After six years in Leeds, late in 1773 he accepted a position as librarian and "tutor" to William Petty, second earl of Shelburne. His Lordship, who had received only a modest education but was obsessed with news of the latest scientific discoveries, had offered Priestley the respectable sum of two hundred fifty pounds per year and the free use of a beautiful little house on his estate at Calne in Wiltshire. The position allowed Priestley plenty of spare time to preach, to write, and to continue with his scientific research. Shelburne even financed a laboratory and furnished it with the latest apparatus, actively encouraging his "Merlin" to investigate further the secrets of the chemical world.

Priestley later wrote of his patron's generosity: "He encouraged me in the persecution of my philosophical enquiries, and allowed me forty pounds per annum for expenses of that kind, and was pleased to see me make experiments to entertain his guests, and especially foreigners."[11]

By the time of his appointment at Shelburne, Priestley was becoming a respected figure within the scientific community. Indeed it may have been his enhanced reputation that had attracted the attention of William Petty. In 1772 Priestley had been awarded the Copley Medal, the highest honor offered by the Royal Society. This was primarily for his discovery of soda water, which at the time was thought to be a preventive for scurvy and was being used extensively by the Royal Navy.* Priestley had become interested in combustion during his time in Leeds and had since conducted ad hoc experiments with different materials, burning them to see how they changed when heated and noting any vapors evolved. Conventional in his thinking, Priestley had supported the idea of phlogiston. He did nothing quantitative in his experiments; he simply burned materials and reported what he observed.

He carried out his most important experiment on Monday, August 1, 1774, at Shelburne House. Using a giant lens to focus the sun's rays to generate sufficient heat, he combusted some mercury calx (mercury oxide) in a vessel, collected the evolved gas, and tested it. Using a candle, because he "happened to have one to hand at the moment,"[12] he found that the gas not only supported the burning but gave a greater brightness to the flame.

A little later, he wrote: "What surprised me more than I can express, was that a candle burned in the air with a remarkably vigorous flame . . . and a piece of red hot wood sparkled in it . . . and consumed very fast."[13] After placing a mouse in a vessel filled with the gas and noticing that it seemed invigorated, Priestley inhaled the gas himself and experienced a sense of exhilaration, prompting him to speculate that the gas might be used to help those with lung illness. But true to character, he cautioned: "It might not be so proper for us in the usual healthy state of the body; for . . . we might, as may be said, live out too fast, and the animal powers be too soon exhausted in this pure air."[14]

All this was possible, he assumed, because the gas was able to absorb unusually large quantities of phlogiston. It appeared clear to Priestley, the dedicated phlogistonist, that the gas was "pure air" or almost pure air, and he named it "dephlogistonated air," in other words, an air containing no phlogiston.

Priestley may have just been getting into his stride with these experiments, but his patron had important plans that would take him from his laboratory and his family. A few weeks after the experiment with mercury oxide, Priestley and Shelburne embarked upon a grand tour of the Conti-

---

*Soda water was actually no help for scurvy sufferers at all, but soon after Priestley had produced it, an entrepreneur persuaded the Admiralty of its preventive powers, and in desperation because of the ravages of the disease, it had clutched at this straw.

nent, taking in Germany, Holland, and France. In October the final leg of the tour brought them to Paris and a dinner engagement with Antoine Lavoisier. It was the only time Lavoisier and Priestley met, and for the latter it proved a pivotal occasion.

Marie-Anne Lavoisier was a gracious hostess, and after dinner conversation turned to science. Both Priestley and Lavoisier took their work very seriously; on this, if on no other level, they had much in common. Lavoisier said little about his findings, but Priestley seems to have been happy to talk. Later, he portrayed the scene in a letter to his wife in which he recounted a dinner "at which most of the philosophical people of the city were present." He went on to tell her how he had described his recent discoveries and "the company, and Mr. And Mrs. Lavoisier as much as any, expressed great surprise."

Perhaps he felt he needed to impress the company. After all, he was recognized as an important scientist at home, but that was all he had going for him. He was poor and Lavoisier was fabulously wealthy; Priestley had received the Copley Medal, but he was little more than a respected servant of the illustrious Lord Shelburne; and if it had not been for his intellectual reputation those gathered at Lavoisier's table would have viewed him as nothing but a peasant with ideas above his station. More than once that evening there appears to have been some confusion over what Priestley was actually describing because of his poor command of French and his broad Yorkshire accent. "I told them I had gotten it [oxygen] from *precipitate per se* [mercuric oxide], and also from red lead," he wrote home. "Speaking French imperfectly, and being little acquainted with the terms of chemistry, I said *plomb rouge* (red lead), which was not understood till Mr. Macquer said I must mean *minium*."[15]

From the tone of Priestley's report of that evening, it is easy to imagine that to allay any sense of inferiority he might have had, he showed off and launched into a vivid and detailed description of his latest experiments, along with the conclusions he had drawn. They were after all fresh in his mind because the excursion that had brought him to Paris had come just as he was making headway in his lab. But, for Lavoisier, Priestley's enthusiasm and openness must have come as a delightful surprise because the Englishman was offering a missing piece in the combustion puzzle.

Lavoisier had noted that different substances evolved different gases when they burned, but because he had been preoccupied with weighing and measuring them he had paid little attention to their chemical properties. And, although Priestley had calculated the relative volumes of the gases he had produced and compared them with the amount of material combusted, he had not been so interested in the quantitative aspect of his experiments.

Proud of the fact that he had "discovered" dephlogistonated air, Priestley had no idea of its true nature or even how he had produced it. He also had no conception of the true nature of the chemical processes involved in combustion or respiration. Yet, with his talk of *plomb rouge*, Priestley had unwittingly offered the key to Lavoisier.

A few days later Priestley was back in England, quite unaware of what he had set in motion. He continued his experiments where he had left off, testing the properties of the "pure air" he had isolated, and in March 1775 he published his findings in the Royal Society's *Philosophical Transactions*, the first public announcement of the production of dephlogistonated air. With this account Priestley gave a full description of the gas, thus claiming his priority for the *discovery* of oxygen, and then gave his explanation of how it was made and the role it played in his experiments.

At the time of my last publication I had not a large burning lens. . . . I have now procured one of twelve inches in diameter, and the use of it has more than answered my highest expectations. The manner in which I have used it, has been to throw the focus upon the several substances I wished to examine, . . . when confined by quicksilver, in vessels filled with that fluid, and standing with their mouths immersed in it. I presently found that different substances yield very different kinds of air by this treatment. . . . But the most remarkable of all the kinds of air that I have procured by this process is one that is five or six times better than common air, for the purpose of respiration, inflammation [supporting a flame], and, I believe, every other use of common atmospherical air [sic]. As I think I have sufficiently proved, that the fitness of air for respiration depends upon its capacity to receive the *phlogiston* exhaled from the lungs, this species may not improperly be called *dephlogistonated* air. The species of air I first produced from *mercurius calcinatis per se*, then from the red precipitate of mercury, and now from red lead. . . . A candle burned in this air with an amazing strength of flame, and a bit of red hot wood crackled and burned with a prodigious rapidity, exhibiting an appearance something like that of iron glowing with white heat, and throwing out sparks in all directions.[16]

Meanwhile, the morning after Priestley's dinner engagement chez Lavoisier, his host had returned to his laboratory with renewed fervor. First he repeated the experiment his English counterpart had described and reported his own set of properties of the new gas. He found it did not dissolve in water, it did not turn limewater cloudy (as carbon dioxide would), it did not react with alkalis, and most important, it could be reacted with metals to re-form oxides. Lavoisier published these findings in the *Observations* in May 1775, two months after Priestley's account.

But Lavoisier and Priestley were now traveling in completely opposite directions. Priestley had become the high priest of the phlogiston theory, and with his latest reports Lavoisier had publicly discarded the idea. He could see that when Priestley's dephlogistonated air reacted with metals it produced the same substance that evolved the gas when heated. In other words, using modern terminology, an oxide of a metal gives off oxygen when it is heated and the metal when combined with oxygen forms the oxide; all quite obvious to us, but a revelation to chemists of the eighteenth century.

metal + oxygen ⟶ oxide

oxide — (heat) ⟶ metal + oxygen

Although outwardly Priestley was not perturbed by Lavoisier's publications, privately he seethed. By this time Priestley was no longer the magnanimous young experimenter of his days in Leeds, and he had began to be influenced by the irritation of his powerful friends. Most notable of these was Shelburne, who for his own reasons claimed "the foreigner" Lavoisier had intruded into Priestley's domain and had literally stolen his ideas. Missing the point entirely, another peer, Lord Brougham, remarked: "Lavoisier never discovered oxygen until Priestley discovered it for him."[17] Whatever Priestley's mood in 1775, he certainly did clash with Lavoisier over the later *interpretation* of these experiments, and within a decade, to his friends at least, he was expressing his annoyance with the Frenchman for not giving him sufficient credit.

For almost ten years after their encounter, both men continued their researches and published their findings, but their explanations for what they were discovering diverged; Lavoisier forged a new chemistry and Priestley dug in his heels. Some, like Shelburne and Brougham, rallied to Priestley's side, ready to start a war of science between France and England, but a repeat of the damage caused by the feud between Newton and Leibniz was this time avoided because some of the more farsighted (and most able) British chemists quickly began to see that Lavoisier was following the correct path and that Priestley was charging up a blind alley.

Perhaps Lavoisier felt encouraged by this, because in his next paper to the French Academy he went much further. In an elegant set of experiments he reacted metals with air in sealed vessels and then allowed air into the vessels to replace what had been used. Because the amount of gas that needed replacing was less than the total volume of the container, it showed that only part of the air had been involved in the reaction with the metal; in other words, air is more than just a single substance.

To support this evidence he went on to examine the part of the air that was left after a reaction with metal. Branching into totally original territory, he then found that this gas could not support life, could not sustain a flame, and was quite unreactive. This led Lavoisier to conclude that air was composed of two distinct parts, a gas that supported life and combustion and a component that was a form of *asphyxiant*, or what he called a *mofette*.

We now know that air is actually a collection of many gases, but that, as Lavoisier's simple model showed, there are two major components: oxygen, which constitutes 21 percent of the whole, and nitrogen, which composes another 78 percent. The final 1 percent of the air is composed of several gases present in small amounts, including the noble gases and a little carbon dioxide and water vapor. It was not until many years after Lavoisier's experiments that these minor components were identified.

During the 1770s, the elder statesman of chemistry was the Scot Joseph Black. Unlike Priestley, Black was open to new ideas; he had long been disturbed by the inconsistencies and obvious failings of the phlogiston theory and quickly came to realize that Lavoisier's ideas were right. Black's approval provided the Frenchman with invaluable kudos and eventually led others to see the folly of the phlogistonists. But even then Priestley and his supporters were not easily silenced.

Although most of the scientific community was still confused by the new discoveries and many refused to support Lavoisier over Priestley, by 1780 the French chemist was gaining confidence and pushing the boundaries. During the delivery of a paper on his latest findings, an account describing how his "respirable air" was contained in all the acids then known, he coined for the first time a word produced by combining the Greek for "acid" with that for "beget"; that word was *oxygen*.

As the new decade dawned, a decade that would see revolution on both sides of the Atlantic and a change of equal importance in the world of science, Lavoisier's entire stance and attitude appears to have changed. As a young man he had questioned the validity of the tired and inconsistent theories that seemed to support chemistry, but he had at least shown respect for his older colleagues both in England and at home, even if their science, he realized, was outmoded. Now, as he approached forty, he decided the time was right for him to impress forcefully his carefully constructed views upon the scientific establishment; the phlogiston theory was in his sights, and he cared little if he tore chemistry apart to bring this dinosaur down.

"It is time," he wrote, "to lead chemistry back to a stricter way of thinking, to strip the facts, with which this science is daily enriched, of the additions of rationality and prejudice to distinguish what is fact and observation from what is system and hypothesis and, in short, to mark out, as it were, the limit that chemical knowledge has reached, so that those who come after us may set out from that point and confidently go forward to the advancement of science." Then, in a direct attack on the proclamations of Priestley, he added: "Chemists had made phlogiston a vague principle, which is not strictly defined and which consequently fits all the explanations demanded of it. Sometimes it has weight, sometimes it is fire combined with earth; sometimes it passes through the pores of vessels, sometimes these are impenetrable to it. It explains at once causticity and noncausticity, transparency and opacity, color and the absence of color. It is a veritable Proteus that changes its form every instant!"

And so, having lambasted and ridiculed the old, Lavoisier set out the new, declaring:

1. There is true combustion, evolution of flame and light, only insofar as the combustible body is surrounded by and in contact with oxygen;

combustion cannot take place in any other kind of air or in a vacuum, and burning bodies plunged into either of these are extinguished as if they had been plunged into water.

2. In every combustion there is an absorption of the air in which the combustion takes place; if this air is pure oxygen, it can be completely absorbed, if proper precautions are taken. In every combustion there is an increase in weight in the body that is burned, and this increase is exactly equal to the weight of the air that has been absorbed.

3. In every combustion there is an evolution of heat and light.

"Stahl's phlogiston is imaginary," he proclaimed, "and its existence in the metals, sulfur, phosphorus, and all combustible bodies is baseless supposition."[18]

This was written in 1783 and published in 1786. Three years later, at the end of the decade, as the French state teetered on the edge of revolution and the American War of Independence was about to alter the political map of the world, Lavoisier published his masterpiece, *The Elements of Chemistry*, in which he laid out his findings from two decades of research. With this he began to refigure the map of the chemical world.

And what of Priestley? As early as New Year's Eve 1775, a little over a year after visiting Lavoisier in Paris, he was beginning to feel that the younger chemist had not given him enough credit for his help. This much is clear from a letter to his friend and fellow amateur philosopher Thomas Henry, in which he comments: "He [Lavoisier] ought to have acknowledged that my giving him an account of the air I had got from *Mercurius Calcinatus*, and buying a quantity of M. Cadet while I was in Paris, led him to try what air it yielded, which he did presently after I left."[19] Gradually, Priestley's annoyance began to show; frustrated that he had been ignored as the progenitor of the Frenchman's experiments, he grew angry as Lavoisier began to dismantle chemical dogma and a theory that meant so much to Priestley.

Continuing with his own experiments, the English chemist never stopped formulating interpretations precisely opposite to those of his rival. In 1778 he was able to write with confidence to a colleague from the Royal Society, James Keir: "Among other things, which I do not mention to you because I hope you will soon see them in print, I think I have fully proved by the firing of inflammable in common air, that this and nitrous air contain, in equal bulks, equal quantities of phlogiston."[20]

And in letters to friends and colleagues, he revealed how his thinking, so vague and muddled compared with Lavoisier's, had been reached by an altogether different path from that of phlogiston's nemesis: "My work is so large," he wrote, "that I believe I must make two volumes of it. . . . As soon as ever a volume is finished, I shall send you the sheets, and shall hope for the favour of your remarks, for I am afraid of tripping on chemical ground. My work is between what is called *chemistry* and other branches of *Natural Philosophy*." And with thinly veiled pique, he added: "On this side [of the English Channel] I am pretty well received, but on the other are some that show a willingness to peck at me, and therefore it behoves me to be on my guard."[21]

His conclusions, drawn from experiments almost identical to those of Lavoisier, could hardly have been more different or more wrong: "I have also made some other experiments of considerable consequence since my return from London," he enthused to Josiah Wedgwood, "especially some that I think are decisively in favour of my theory of respiration, against a series of papers by Mr. Lavoisier in the last volume of Parisian Memoires."[22]

At the same time he was keeping tabs on his rival's efforts through a network of friends in Europe. To a colleague recently returned from Paris in 1783, Priestley wryly remarked: "I thank you for your intelligence from Paris. For my own part, I wish to see Mr. Lavoisier's *facts* unexceptionably ascertained by competent witnesses."[23]

As we have seen, Priestley was certainly not alone in supporting the phlogiston theory. He was a prominent member of a group called the Lunar Society, an informal version of the Royal Society whose members (almost exclusively Nonconformists) met monthly at the full moon to discuss the latest scientific ideas and to share their pet theories. The Lunar Society was based in the Midlands and included such figures as Josiah Wedgwood, Erasmus Darwin, James Watt, and Matthew Boulton. Among less well-educated locals they were seen as a group of rich eccentrics, and sometime during the 1780s the name of their group was adopted as a colloquialism so that anyone professing what might be considered peculiar beliefs or manners was henceforth known as a *lunatic*.

But the intellectual tide was beginning to turn against Priestley, and as it did he became increasingly entrenched within the Lunar Society, drawing upon his Nonconformist friends for support. Many of them were wavering over the new ideas crossing the Channel, but Priestley rallied them. In 1782 he wrote to Wedgwood: "Before my late experiments phlogiston was indeed almost given up by the Lunar Society, but now it seems to be re-established. Mr. Kirwan, in a letter I have received from him this day, says that he has given in a paper to the R. Society, to prove from my former experiments, that phlogiston must be the same thing as inflammable air, and also that dephlogistonated air and phlogiston make fixed air."[24]

Priestley's need for validation became exaggerated as Lavoisier's star was rising and his own scientific reputation began to crumble. By the early 1780s Priestley was clearly yesterday's man, and there can be no better illustration of this than a comparison of Lavoisier's pronouncements, readily intelligible and laudable to any chemist of the twenty-first century, and a typical comment of Priestley's from the same time. "Atmospheric air, or the thing that we breathe, consists of the nitrous acid and earth, with so much phlogiston as is necessary to its elasticity,"[25] he declared proudly as Lavoisier was formulating a modern chemical model.

Lavoisier's great treatise, *The Elements of Chemistry*, was published in France in 1789. It was translated into English a year later, by which time Lavoisier and Priestley had grown as far apart as they would ever be. In France and beyond, Lavoisier was now acclaimed as the most important chemist of the age. He was incredibly wealthy, an important figure on the political scene in Paris, and a leading member of the academy. An indication of his success may be seen in an account of a visit to Lavoisier's lab written by one of the many young acolytes who visited the master at the Arsenal:

> To the arsenal to wait on Monsieur Lavoisier, the celebrated chemist . . . I mentioned in the course of conversation his laboratory. . . . That apparatus the operation of which have been rendered so interesting to the philosophical world, I had pleasure in viewing. . . . I was glad to find this gentleman of considerable fortune. This ever gives one pleasure: the employments of a state can never be in better hands than of men who thus apply the superfluity of their wealth. From the use that is generally made of money, one would think it the assistance of all others of the least consequence in affecting any business truly useful to mankind, many of

the great discoveries that have enlarged the horizon of science having been in this respect the result of means seemingly inadequate to the end: the energetic exertions of ardent minds bursting from obscurity and breaking the bands inflicted by poverty perhaps by distress.[26]

But as Lavoisier was reaching his scientific peak, the political world in which he lived was about to change immeasurably, leaving his kind exposed and vulnerable. Although a dyed-in-the-wool conservative, during the early days of the Revolution, Lavoisier may have shared the views of the majority, who believed the liberal shift a good thing. As a logical thinker he would have known that the taxation system he had been party to and from which he had benefited greatly was a fragile entity and could not be sustained forever. But gradually factions within the revolutionary movement turned against one another, and before long those originally considered reformers were being perceived as "counterrevolutionaries," and many were dragged to the guillotine. By 1793 the Revolution had turned into the wanton slaughter of the Terror, a bloodletting spearheaded by the sociopathic Robespierre, who was himself later executed by other extremists who disagreed with his policies.

During the early days of the Revolution, Lavoisier had been publicly attacked by one of the leading revolutionaries, the journalist Jean-Paul Marat, who had been consumed with jealousy and hatred for Lavoisier ever since the chemist had blocked his admission to the academy. Marat had claimed: "I denounce this Coryphaeus* of the charlatans, Lavoisier, son of a land grabber, chemical apprentice . . . fermier général . . . administrator of the Discount Bank. . . . Would you believe that this little gentleman, who has an income of forty thousand livres a year, is now actually clamoring like a demon to be elected departmental administrator of Paris? Would to heaven he had been hung on a lamppost on the sixth of August."

And in another invective he spat: "Lavoisier is a putative father of all discoveries that are noised abroad. Having no idea of his own, he steals those of others, since he hardly ever knows how to evaluate them, he abandons them as lightly as he takes them up, and changes his systems like his shoes."[27]

---

*In Greek myth, the leader of a group of religious fanatics.

But soon things grew still worse for France. King Louis XVI was executed in January 1793, and the country slid rapidly into anarchy. By the summer of that fateful year, Marat too had been murdered, but his extremist friends were now at the height of their powers. In August the Royal Academy of Sciences was closed by order of a group that judged the fate of thousands of French citizens during the Terror—the absurdly named Committee of Public Safety.

A month later Lavoisier's home and laboratory were searched and his letters and papers studied by representatives of the committee. He protested against the intrusion and appeared before a succession of prosecuting groups to plead innocence of any crime against the French people. But it was to no avail. At the end of November, Lavoisier and his father-in-law, Jacques Paulze, were arrested and clapped in chains in the prison at Port-Libre. Lavoisier had spent his final four days of freedom wandering the empty rooms of the academy, contemplating his fate to the echoing tap of his solitary footfall.

Lavoisier's arrest shocked the French scientific community. But because the whole of Paris was gripped by fear and a single ill-advised word or badly timed remark could lead to execution, members of the disbanded academy remained utterly silent. Across the English Channel, Lavoisier's admirers could afford to be more vocal, but they were utterly powerless. Shortly before his arrest, the Royal Society had voted to award Lavoisier the Copley Medal for 1793 but then decided to cancel the announcement when they realized such news might give Lavoisier's enemies an opportunity to accuse him of being in league with English aristocrats, placing him in greater danger.

Priestley was in complete sympathy with the early revolutionaries in France, but he did turn against the extremists as they took to slaughtering tens of thousands of their countrymen during 1793 and 1794. However, when combined with his extreme religious views, his unorthodox political stance attracted enemies in England. Priestley's nonconformity had long been considered eccentric, but as the mood of the Revolution turned increasingly ugly and news of socially sanctioned murder across France spread the short distance to London, men like Priestley were viewed as potentially dangerous. Furthermore, Priestley had never been one to hide his feelings or to refrain from openly discussing his opinions, whatever the political climate.

Many of the aristocracy of France and England were related by marriage at least, and some lucky French nobility found safe havens in England during the Revolution. Because of these close ties between England and France, the 1790s was a period during which the stability of English society was itself threatened by the horrors of the Revolution across the Channel, and within this political climate men like Shelburne could not allow themselves to be associated with radicals such as Joseph Priestley. Six years after joining Lord Shelburne's service, Priestley was dismissed.

He was naïvely puzzled by this dismissal, but it was clearly linked to the fact that as he had grown older he had become even more of a religious radical, applying to his religious stance the same obsessiveness he had demonstrated by clinging to the dead horse of the phlogiston theory.

So Priestley was forced to return to preaching as a living, and he moved his family to Birmingham, where he could at least enjoy closer contact with his friends in the Lunar Society, upon whom he relied greatly for financial assistance. Now more than ever, he readily and unashamedly accepted money from Wedgwood and Darwin, money given, not lent. But although Priestley's friends remained loyal to the end, many within the English establishment grew to despise him almost as much as the *sans-culottes* loathed Lavoisier. One biographer has said of the popular attitude toward Priestley: "He succeeded in making himself an object of hatred and suspicion to the orthodox clergy . . . by the end of a decade of energetic religious controversy and political propaganda on behalf of Dissenters, Priestley had acquired an almost diabolic reputation among staunch upholders of church and state."[28]

As Lavoisier's fears of the Revolution grew, Priestley was also beginning to face persecution. Soon, because of his constant proselytizing of unpopular political and religious views expressed from the pulpit and in publications, he was dubbed "the most widely hated man in England." He even fell out with the Royal Society because they would not accept his nomination of a member who held radical religious views. In response, Priestley ended all communication with them and never again attended a meeting in London or offered work for publication in the *Philosophical Transactions*.

Things reached the point of crisis for Priestley and his family in 1791, some two years after he had taken a position in a Birmingham ministry. On the night of July 14, a group of Dissenters and political radicals attended a dinner to commemorate the second anniversary of the storming of the

Bastille. Contrary to popular myth, Priestley had not been invited to the dinner, but by now the most famous dissenting minister in the country and the most loathed, he was suspected of being the leader of a group intent upon revolution in England. A mob attacked the hotel, but the guests had left. Frustrated, they moved on to Priestley's home and laboratory, a few miles away. They set fire to his house, gutting it completely. In fear for his life and with his wife and children protected by friends in a safe house, Priestley fled to London.

The journey took him a week. He traveled alone, evading his enemies by stealth and living by his wits and the charity of friends. Soon after arriving in the capital, he was joined by his wife, and they set up home and a new ministry in Hackney, East London. For a while they felt safe, but their enemies were constantly snapping at their heels, and in April 1794, as Lavoisier approached his final days, the Priestleys packed their bags and, with money provided by rich friends, left for America. Joseph Priestley lived out his final years in peaceful isolation in Pennsylvania preaching, researching, and writing his memoirs.

$\mathcal{F}$ROM OUR PERSPECTIVE, more than two centuries after these events, it is still possible to cut away the hyperbole and the prejudice and assess how these two men played their roles in the advancement of a major scientific principle. What is clear from the events described here is that Antoine Lavoisier discovered nothing. He was a fine experimental chemist, but he was not first to isolate a single gas and many of his experiments were based upon the work of others. In stark contrast, Joseph Priestley was all experiment. He was the first to isolate oxygen even though he remained far from understanding what he had produced or indeed what he had done. Together, the work of the two men created a gestalt because they lived at the same time, they were interested in the same evidence, and Lavoisier needed Priestley's observational data.

It is sometimes said that in the world of chemistry Lavoisier's work may be compared with Newton's in physics, but this is going too far. Certainly Lavoisier radically overhauled chemistry and laid a firm theoretical foundation for generations to follow, but there is a major difference between Newton's and Lavoisier's roles. Newton accomplished alone all the steps—

revelation, experimentation, elucidation of a theory, and derivation of proof—to the law of gravitation and the laws of motion. The discovery of oxygen and its function was the work of two men. Lavoisier, who defined, measured, and formulated, and Priestley, who stumbled upon revelation and made lucky guesses.*

And Priestley was indeed a lucky guesser, self-taught and proud of it, a man who had gleaned almost everything he knew about chemistry from a lecture course at Warrington Academy, where he had taught during the 1760s, and from conversations with friends. "I was led," he later admitted, "to devise an apparatus and processes of my own, adapted to my particular views. Whereas, if I had been previously accustomed to the usual chemical processes I should not have so easily thought of any other."[29] Elsewhere he wrote: "For my part I will frankly acknowledge, that, at the commencement of the experiments recited in this section, I was so far from having formed any hypothesis that led to the discoveries I made in pursuing them, that they would have appeared very improbable to me had I been told of them; and when the decisive facts did at length obtrude themselves upon my notice, it was very slowly, and with great hesitation, that I yielded to the evidence of my senses."[30]

Lavoisier stumbled too—all chemists of the time did—but he has always been granted the greater honor as a chemist because he saw the right direction in which to travel and formulated an approach toward his target rather than merely feeling his way in total darkness. This does make Lavoisier the superior scientist. Priestley certainly understood what being a scientist really involved and wrote: "Every experiment, in which there is any design, is made to ascertain some hypothesis. For an hypothesis is nothing more than a preconceived idea of an event."[31] And later he commented: "Hypotheses, while they are considered merely as such, lead persons to try a variety of experiments, in order to ascertain them. In these experiments, new facts generally arise. These new facts serve to correct the hypotheses

---

*To be strictly accurate, it is now thought that a Swedish chemist named Karl Wilhelm Scheele actually discovered oxygen some years before Priestley, but he had very little idea what he had found and his suggestions were ignored. Priestley also had little idea what he had found but as a member of the Royal Society, he could successfully proselytize his claims.

which give rise to them. The theory thus corrected, serves to discover more new facts; which, as before, bring the theory still nearer to the truth."[32]

Yet oddly, Priestley did not apply this logical approach to his own efforts. He was sloppy and imprecise and often vague about his methods. Furthermore, he reasoned in a conservative, linear fashion and did not offer anything like the radical stance he gave to his religious and political views. Although Priestley applied to chemistry the fanaticism he displayed toward religion, he did not seem capable of a single original thought or a radical explanation for anything he did scientifically. The historian Sir Oliver Lodge went as far as to say: "In theory he had no instinct for guessing right, such as the great men of science have had—an intuitive feeling for the right end of the stick; he may almost be said to have held a predilection for the wrong end."[33]

Ironically, as far as chemistry is concerned, Lavoisier was the unconventional one. The son of a lawyer and materially wealthy all his life, happy in his marriage and secure in his multifaceted career, he applied his thinking laterally to chemistry and deserves great credit for pushing forward the limits of his subject. With his *Elements of Chemistry* he laid the cornerstone of modern chemistry. In his lifetime he instigated a methodical naming system for chemicals, which remains fundamental today, and, as he promised he would do, he swept away the old vagaries and established modern, precise thinking and, most important, a methodology for this science.

Lavoisier was spurred on by the rivalry with Priestley and the other chemists who held doggedly to the outmoded notion of phlogiston. In the *Elements of Chemistry*, he made a barely veiled attack on phlogistonists, writing:

In the study and practice of the sciences it is quite different; the false judgments we form affect neither our existence nor our welfare; and we are not forced by any physical necessity to correct them. Imagination, on the contrary, which is ever wandering beyond the bounds of truth, joined to self-love and that self-confidence we are so apt to indulge, prompts us to draw conclusions that are not immediately derived from facts; so that we become in some measure interested in deceiving ourselves. Hence it is by no means to be wondered, that, in the science of physics in general [meaning all of science], men have often made suppositions, instead of forming conclusions. These suppositions, handed down from one age to another, acquire additional weight from the authorities by which they are supported, till at last they are received, even by men of genius, as fundamental truths.[34]

Lavoisier assumed that the Englishman placed faith in phlogiston beyond anything reason or experimental evidence could ever show him. This greatly irritated Lavoisier, the rationalist, so that although he may have realized nothing he could do would change his rival's mind, he could not stop pursuing greater experimental precision and further refinement of his chemical theory in order to prove himself right. Like that of all great scientists, Lavoisier's primary drive was the search for fundamental truth, but he was also motivated by a need to show that science could conquer the blind faith and blinkered vision of men like Priestley. If it had not been for this, Lavoisier would not have worked so hard or achieved so much before the executioner's blade ended his life.

Priestley remained deaf to progress, which is an unforgivable sin for any scientist. Even if we ignore his slapdash practices, which were, we must remember, not so different from those of any other chemist of the time (including some of Lavoisier's efforts), and if we forgive him for not seeing that the phlogiston theory was a fairy story, his stubborn refusal to accept the glaringly obvious facts cannot be ignored. Quite simply, Priestley's ego and his obsessive character overruled his scientific judgment, and to a degree this must tarnish his reputation. Lavoisier possessed an ego every bit as powerful, and he certainly did not give his contemporaries sufficient credit for their contributions. To an extent, he was dismissive of Priestley and sure the Englishman had tripped over his discovery without really knowing what he had revealed; but at least Lavoisier never allowed dogma to override reason.

And so, thanks to Antoine Lavoisier, a man inspired to reeducate his rivals, chemistry became a rational science. He formulated a language for the subject and opened the way to creating a structured, logical framework that described how chemicals behaved and interacted. "Lavoisier's novelty was to reform the language of chemists," one science historian has observed. "He purged chemistry of its alchemical names and redefined central concepts."[35]

Half a century after Lavoisier, John Dalton was able to suggest a version of atomism that was gradually refined and developed during the nineteenth century to become a precise tool with which scientists began to understand how compounds were formed and reacted, and how atoms were arranged.

The Russian chemist Dmitri Mendeleyev charted the positions of the elements into what became known as the periodic table, and researchers of the late nineteenth century expanded the canon of chemical knowledge enormously because they had a logical foundation upon which to build their ideas.

By the beginning of the twentieth century, chemistry was progressing as rapidly as physics, and links between physics and chemistry were coming to be understood, so that the former could successfully underpin the latter, providing a symbiosis that led to still more rapid advance. Chemical knowledge and biological insights also began to fuse, giving the world organic chemistry and biochemistry. Such combinations were to create some of the most profound developments in both pure, theoretical science and technology.

WITH DELICIOUS IRONY, circumstance almost led both our heroes to the guillotine. Toward the end of the French Revolution, Joseph Priestley and the political thinker Thomas Paine were made honorary French citizens for their support of the Revolution, and each was offered a place on the Revolutionary Council in Paris. Paine accepted and moved to Paris, Priestley declined. But a few months later, when the political climate shifted a degree, Paine found himself on the wrong side and was imprisoned by reactionary forces and sentenced to death. The night before he was due for the guillotine, chalk crosses were inscribed upon the doors of the prisoners' cells in his block to indicate who should be led to the cart the following morning. But somehow Paine's cross was accidentally rubbed away during the night, and when the soldiers came to lead the condemned away, he was left unmolested. Shortly after, political power shifted again and Paine was freed to play his role in history. Priestley might not have been so lucky, and Parisian crowds may have seen the heads of two chemists rather than one roll into the bloody tumbrel that cruel month of May 1793.

# THREE

## Of Monkeys and Men

Our ancestors are very good kind of folks; but they are
the last people I should choose to have a visiting acquaintance with.
—*Richard Brinsley Sheridan*, The Rivals

*Oxford, June* 1860
*I*T WAS A WARM SUMMER EVENING in Oxford, and many of the
most important figures in the British scientific community were gathered
for the annual meeting of the BAAS, the British Association for the
Advancement of Science. But late on Saturday a talk at the newly built
Oxford University Museum turned what may have been just another
roundup of scientific achievement for the year into one of the most famous
public battles between science and the Church, and for many it marked the

beginning of the modern relationship between the two institutions, a facing off that would define attitudes for generations.

The meeting itself promised little. A second-rate social scientist named John Draper delivered the first speech. His subject was the possible application to modern society of Charles Darwin's theory of evolution described in *On the Origin of Species*, which had been published a little over six months earlier, at the end of 1859. But what turned this gathering into a seminal event was the presence of the bishop of Oxford, Samuel Wilberforce, who intended to deliver a speech of his own, a response to Draper's pro-Darwin paper.

Wilberforce, or Soapy Sam as he was known by some, was an ultraconservative, a Tory Anglican bishop who had taken a stance against many of the incendiary new ideas of contemporary scientists and in particular the latest ideas of men like Charles Darwin, who offered a scientific explanation for the origin of living things completely at variance with religious orthodoxy. Thirteen years earlier, at the 1847 meeting of the BAAS, Wilberforce had publicly humiliated the best-selling science writer Robert Chambers, who had dared to offer alternatives to the most sacred cows of the Victorian religious establishment.

And it seems the intellectuals gathered in Oxford that weekend in 1860 were spoiling for a fight. Hundreds were turned away at the door of the packed museum, and almost one thousand students, dons, and scientists who have traveled from London and from distant outposts of the empire were crammed together under the vaulted roof.

Wilberforce cared little for Draper or his ideas, really his presence at the meeting was aimed solely at attacking Darwin and his new theory, for the clergy had spoken of little else during the past half year. And there can be no doubt that the bishop had been coached for the event by one of Darwin's bitterest enemies, one of the protagonists in this tale, Richard Owen, with whom Wilberforce had shared digs the previous evening.

And, of course, the huge audience knew why Wilberforce was there; for many he was the reason they were there too. The room was packed with Darwin's many opponents and his eager supporters. Draper was a poor speaker and his arguments weak. With a talk lasting almost two hours he had bored the audience so much that when he finally sat down and Wilberforce rose to his feet, there was a palpable relief, a hush of expectancy.

Charles Darwin himself was not at the meeting but sixty miles away in a clinic in Richmond, resting after the first wave of excitement his book had solicited. He never spoke publicly and had lived the life of a virtual recluse for more than two decades. Richard Owen, Hunterian professor at the Royal College of Surgeons, the voice of the scientific establishment, sat in the meeting but did not utter a word throughout. Instead, the battle between the two most important biologists of the age, and a row that would rumble down through the centuries, took on its public face via spokesmen. Wilberforce acted as the voice of Richard Owen, and for Darwin there were present two of his most loyal and able supporters, the director of Kew Gardens, John Hooker, and a man who soon gained the nickname Darwin's Bulldog, Thomas Henry Huxley.

Wilberforce spoke uninterrupted for half an hour. He was an eloquent and persuasive speaker, imbued with enthusiasm, passion, and self-confidence. He loathed Darwin's ideas and saw evolution as a serious threat to the very fabric of Church and State, a poison to be neutralized. As he concluded his speech, his skillful delivery and confidence overwhelmed him, and he made a huge mistake. Turning to Thomas Huxley, with whom he had clashed on many previous occasions, he asked with a thin smile whether it was on his grandfather's or grandmother's side that Huxley had descended from an ape.

The remark was met with a voluble blend of laughter from supporters and outrage from Wilberforce's enemies. Considering the sanctimonious Victorian view of the family, a suggestion that a rival's grandmother was an ape was a rather dangerous comment to make. According to a witness, Huxley turned white with rage; another reported that a woman in the audience, one Lady Brewster, fainted. Huxley himself later remarked that he had not been angry but thought the bishop had spoken with "inimitable spirit, emptiness, and unfairness," and that when he made the remark about his lineage, Huxley, the arch-agnostic, merely whispered sarcastically to a friend beside him: "The Lord hath delivered him into mine hands."

Huxley stood, witnesses later recounted, barely managing to keep his temper, but then, staring straight at his enemy, he said evenly: "It would not have occurred to me to bring forward such a topic as that for discussion myself, but I am quite ready to meet the Right Rev. prelate even on that ground. If then the question is put to me, would I rather have a miserable ape as a grandfather or a man highly endowed by nature and possessed of great means and influence and yet who employs these faculties and that influence for the mere purpose of introducing ridicule into grave scientific discussion I unhesitatingly affirm my preference for the ape."[1]

Some claim that Wilberforce's comment, close as it was to indecency, was actually meant as a joke to lighten the atmosphere after Draper's incredibly tedious talk, but the Darwinians certainly did not take it that way, and Soapy Sam had no reputation for levity. Whatever his intention, the crowd responded with partisan enthusiasm. Wilberforce's men were stunned into angry silence while the Darwinians cheered and shouted with delight.

Wilberforce, or Sam Oxen as Huxley and his colleagues dubbed him when they were tired of Soapy, was a slick public speaker, but Darwin's men were out in force that evening. After Huxley, had said his piece and returned to his seat, John Hooker, a future president of the Royal Society, picked up the baton. He declared to the gathering that Wilberforce could not have even read Darwin's masterpiece (despite the fact that the bishop had only recently been paid the enormous fee of sixty pounds to review it for the conservative *Quarterly Review*) and that he was completely ignorant of the issues under discussion.

"My blood boiled, I felt myself dastardly.... I swore to myself I would smite that Amalekite Sam hip and thigh if my heart jumped out of my mouth," Hooker later wrote Darwin.[2]

By the end of the evening both camps thought they had won the day, and the audience spilled out into Oxford and filled the pubs with endless talk of what they had witnessed. Both Huxley and Hooker wrote to Darwin immediately telling of their battles, and Wilberforce returned to his rooms believing he had secured another round in his ongoing war with science.

If it had not been for the enormity of the subject under discussion that June evening, this vocal duel would have passed into history and been forgotten, but at issue was nothing less than the meaning of human existence. Charles Darwin's theory of evolution via natural selection, explained so eloquently in his best-selling book, was the most radical and most important scientific development since Copernicus proposed his alternative to Ptolemy in *De revolutionibus orbium coelestium*. And some would argue that because of the fact that Darwin's ideology so intimately linked with the human condition, with the very essence of what it means to be a human being, the theory of evolution is more important than any cosmological revelation. Wilberforce knew this, and Owen, Darwin's former friend turned greatest scientific rival, knew this too. Both were quite aware of what was at stake. Huxley, Hooker, and Darwin's other disciples had known both the value and the incendiary nature of their master's theory and reveled in it.

The inflammatory 1860 meeting of the BAAS may have been the most public evocation of the dispute between the orthodox and the revolutionary thinkers of the age, but it was really only the first lancing of the boil. For at least a decade, arguments over the idea of evolution and its significance for humanity had raged between factions of the scientific community and the Church. And the row between Wilberforce and Huxley that night was the latest stage in that war of words. It is a conflict that rages still today, remaining perhaps the most emotive issue in science because it is directly about you and me and the rest of humankind.

DARWIN HAD JUST TURNED FIFTY when *On the Origin of Species* was published. He knew that its publication would cause a stir, but he believed this inevitable reaction would be restricted to the rarefied world of the scientific establishment. He had been working on the book for some twenty years, refining his ideas and honing his theory into perfected form. He could have published long before but had held back for two reasons. First, he was greatly concerned the work would be badly received by his peers, and second, he was sure his published theory would upset his deeply religious wife, Emma, whom he adored.

Some have argued that by waiting Darwin did in fact write a much better book and that his theory of evolution via natural selection had been allowed to mature and many of the details to be clarified. It has also been suggested that when it did appear, society was better prepared for the paradigm shift it provided. What is certainly true is that by the time the book was published Darwin had managed to secure around him a cadre of devotees who defended their hero unstintingly and acted as his mouthpiece. This was just as well, because for many years. Darwin had lived in isolation in rural Kent, surrounded by his books, his specimens, and his growing family. He suffered irregular but nevertheless debilitating bouts of illness left undiagnosed. Shunning all forms of publicity, he was constitutionally unable to face public speaking. As he commented to Hooker after the BAAS debate: "It is something unintelligible to me how any one can argue in public like orators do. . . . I am glad I was not in Oxford, for I should have been overwhelmed. . . . I honour your pluck, I would as soon have died as tried to answer the Bishop in such an assembly."[3]

But Darwin had not always been a recluse. Indeed, in his youth he had been a great traveler and explorer, and it was because of his free-ranging spirit and sense of adventure he had discovered firsthand the facts that would later support his theory of evolution.

Darwin came from a family of innovators and intellectuals. His grandfather, the liberal and forward-thinking Erasmus Darwin, had been a leading light of the Industrial Revolution and a friend to both Josiah Wedgwood and Joseph Priestley. The Wedgwoods had remained close to the Darwins, their large country estate lay nearby, and the children from both families played together and later intermarried.

Charles was born in 1809 into a wealthy family who owned a vast house, The Mount, surrounded by several hundred acres of Midland England near Shrewsbury. For such privileged people, the beginning of the nineteenth century was an exciting time to be English. The Industrial Revolution was still young, the social repercussions it brought still nascent. Darwin's childhood saw Britain victorious over Napoleon, vast social change at home, and the beginnings of the empire spreading like spilled ink across the globe. Industry was about to provide the support for economic growth, and science was the tinder for this spreading flame.

But Darwin's childhood was not as rosy as it might have been. Along with every one of our lead characters discussed so far, Charles Darwin lost his mother when he was young; Susannah Darwin (née Wedgwood) died from cancer when he was eight. This tragedy left his father, Robert, a prominent doctor, permanently scarred, turning him into a lifelong depressive who never remarried. The loss of Darwin's mother cast a shadow over The Mount and suffused Charles's early life with sadness.

He attended Shrewsbury School and was then encouraged to study medicine at Edinburgh University. But within a few months of arriving there Charles had discovered a love of biology and geology and had begun to neglect his medical studies. Much to his father's chagrin, by the end of his first year it had become clear the boy was not going to follow in his footsteps. Although he was incensed by what he regarded as his son's willful disregard for his wishes, Robert Darwin was a good enough father to realize he could not force his son into a career he would despise. When he agreed to Charles's leaving his medical studies, a place was found for his son at Cambridge.

Here young Darwin found his niche and formed a close and lifelong friendship with a brilliant professor of botany named John Henslow, who had acquired the chair of botany at the extraordinarily young age of twenty-six. Henslow had noticed Darwin's abilities soon after he arrived in Cambridge and had opened his eyes to the magic of botany before going on to guide Charles through his degree course. A few weeks after Darwin left Cambridge, one of his former tutors talked admiringly of him to a colleague at the Admiralty. Soon word of his talents as a biologist had reached a young naval captain, Robert Fitzroy, who was about to begin a circumnavigation aboard his new commission, the *Beagle*, as part of the South American Survey, a government-funded project to chart the continent of South America and its environs.

Fitzroy wrote to Darwin offering him a position as ship's naturalist, but to Charles's dismay, upon hearing of the scheme his father rejected the idea out of hand. According to Darwin, his father's objections were as follows:

> 1. Disreputable to my character as a Clergyman thereafter.*
>
> 2. A wild scheme.
>
> 3. That they must have offered to many others before me, the place of Naturalist.
>
> 4. And from its not being accepted there must be some serious objection to the vessel or expedition.
>
> 5. That I should never settle down to a steady life hereafter.
>
> 6. That my accommodation would be most uncomfortable.
>
> 7. That I was again changing my profession.
>
> 8. That it would be a useless undertaking.[4]

---

*Presumably Darwin's father assumed his son would enter the clergy, as he thought he would be good for little else. It was also traditional for such families as the Darwins to place their sons in the Church if they had not taken up careers in the military or as doctors or lawyers.

Fortunately for Charles and for posterity, he had long since learned how to manipulate his father and asked his former tutor John Henslow for support. This came in the form of a letter outlining to Robert Darwin the potential merits of the offer. To gain further validation Charles had talked to his uncle Josiah Wedgwood II, his father's closest friend, and had also convinced him to write a letter.

Charles's efforts were rewarded. Within a few weeks he received his father's agreement, if not his blessing. On December 27, 1831, HMS *Beagle* set sail from Devonport. Darwin's father; brother, Erasmus; and their four sisters, Marianne, Caroline, Susan, and Catherine, were utterly convinced they would never see Charles again.

The fragments of Darwin's intellectual life finally coalesced on the voyage of the *Beagle*, and he found a form and shape to his career as a scientist. During the five-year mission he learned an enormous amount, not just about the species of plants and animals he investigated but about methods of working, a mental discipline derived from the need for strict, methodical practices aboard a cramped ship that traveled through often inhospitable regions. In a letter to his sister Caroline, Darwin said of the *Beagle*: "The vessel is a very small one; three-masted, and carrying ten guns: but everyone says it is the best sort for our work, and of its class it is an excellent vessel; new but well-tried, and half again the usual strength."[5]

The *Beagle* took two months to reach the coast of South America, where it was scheduled to stay, traveling up and down for the next two and a half years, from All Saints Bay, Bahia (now Salvador), over to the Falkland Islands and beyond the southernmost tip of the continent to Tierra del Fuego. Darwin was entranced by the luxuriant variety of life in the jungle bordering the coast of this vast continent and immediately set about studying the seemingly endless diversity of species, many of which had never been seen by humans before.

His drawing ability was poor, so he relied upon written descriptions and began keeping notebooks, a collection that by the end of the voyage grew to over seventeen hundred pages. These provided the background for his personal account in the journal, *The Voyage of the Beagle* (published in 1839), and provided some of the evidence to support his later theories.

A few months after reaching the coast of South America, during a solo venture inland, Darwin found his first fossil remains in a bay called Punta

Alta. The find consisted of a thighbone and some teeth from a huge mammalian creature. He removed them from the cliff face in which they were buried and carried them back to the ship. To his surprise and annoyance, Captain Fitzroy saw in them absolutely no scientific value.

But later, others did appreciate them. As Darwin collected samples, they were carefully packed and sent home to England. Before he left, Professor Henslow had agreed to act as official receiver for Darwin's finds, but from his position some six thousand miles from home, Darwin had absolutely no idea in what condition the samples would reach London or if anyone was even taking notice of them. At this time he had no inkling just how important his work would be, for both practical and theoretical science. He certainly had no original thoughts about evolution or the part he might play in its development and was completely unaware of how highly Henslow and his colleagues regarded the samples he was dispatching home. He was also faithfully sending off sections of his journal as he wrote them. These were being passed around the family and cooed over by his relatives, but he had no idea that his observations and descriptions of his experiences were also being passed on to his academic friends and received with great interest by the scientific community.

During the early 1830s Darwin was in his prime, healthy, fit, and energetic. Each new experience brought a thousand and one questions; some he could answer there and then by experiment, others had to wait until he could consolidate his experiences and find time to devise a cohesive scheme to account for what he had witnessed. During the voyage he gave little thought to the longer-term importance of what he was doing and assumed that he would simply return to an academic life after it was all over.

As the *Beagle* rounded Cape Horn, the crew received long-awaited news from home. They had left a Britain facing the worst political crisis in a generation. The recently elected Whig government was trying to force through the Reform Bill—a radical plan to alter the social framework of the country, to change the electoral system, and to create a more powerful middle class. But after its third reading the bill had been rejected by the House of Lords. The country had erupted in protest; there were mass demonstrations, seventy thousand people marched in London, and a Tory duke and a bishop were attacked in the street. Things became so bad the newly crowned king, William IV, was forced to suspend Parliament. And as Darwin traveled through alien waters on the far side of the world, back home a variety of bills were finally passed to try to stabilize the country.

It is clear from his journals that by this stage of the voyage, some two years after leaving England, Darwin was feeling homesick and isolated. But among the depressing news from home, he also heard how well his work had been received in England. According to Henslow, his samples were the talk of scientific society and his correspondence detailing his discoveries had been received with enthusiasm. Furthermore, Henslow had arranged for Darwin's fossils, in particular the fossil bones Fitzroy had thought useless baggage, to be displayed before the Cambridge meeting of the British Association for the Advancement of Science earlier that year; they had been identified as the remains of a long-extinct animal called a *Megatherium*. This, Henslow claimed, had already established Darwin's name within the scientific community, a useful foundation for his return.

About this time Darwin wrote in a letter home: "I am convinced it is a most ridiculous thing to go around the world when by staying quietly, the world will go round you."[6] But he was not entirely right in believing this, for his most important finds were yet to come. After a further eighteen months traveling along the west coast of South America, in September 1834 the *Beagle* arrived at Chatham, one of a small group of rocky windswept islands, the Galápagos.

To many, Darwin's voyage aboard the *Beagle* is remembered solely for his visit to the Galápagos Islands and the discoveries he made there. The remainder of his five-year journey is generally viewed as irrelevant. This is quite untrue, but it demonstrates just how important Darwin's discoveries on the islands became and how significant they were to his development of the theory of evolution.

Darwin thought the Galápagos ugly and inhospitable, which they are, but they are also a veritable gold mine for both the naturalist and the geologist. The climate makes them a haven for reptiles of all varieties, and the islands are home to dozens of species of lizards, and turtles. The islands are also volcanic, which, according to Darwin, gave much of the terrain the look of an industrial wasteland. When he arrived it was baking hot—temperatures sometimes reached the nineties in the shade—there was little fresh water, and the only human inhabitants were two hundred exiles deported from Ecuador living on Charles Island. Most important to Darwin, the islands were populated by many species of finches and tortoises. Although Darwin did not at the time realize the significance of the various species of both these creatures, differing as they did from one island to another, when he began to analyze his samples and to piece together his findings, their importance became paramount and acted as the experimental backbone of his theory.

The *Beagle* stayed in the Galápagos for five weeks before heading west toward Tahiti and on to New Zealand and Australia. From there in the spring of 1836, the crew set sail upon the six-month journey home, reaching Falmouth in October 1836, a little under five years after the ship had left those shores.

Darwin returned to England with 1,383 pages of geology notes, 368 pages of zoology notes, a catalog of 1,529 species in spirits, and 3,907 labeled skins, bones, and miscellaneous specimens, as well as a live baby tortoise from the Galápagos Islands. His diary amounted to 770 pages, and parts of his journals sent ahead of him had already been published in Britain. He returned to England a little older but a great deal wiser, and because of the material he had been sending home, he had already won the respect of his peers and elders. And, although others would never have imagined it, he knew quite well that his real labors were about to begin. For the remaining forty-six years of his life, he would not leave England but would process the observations he had made and coagulate them into a definable whole, held together by a central, all-pervading concept, the theory of evolution.

DARWIN'S EARLIEST TENTATIVE STEPS toward this work were delayed. First he moved to London and found a bachelor flat in Great Marlborough Street close to where Euston Station stands today. The London-based publisher Colburn had been contracted to complete his memoir of the trip, *The Voyage of the Beagle*, which became a best-seller soon after its publication in 1839. That year he married his childhood sweetheart Emma Wedgwood, and the couple moved to a little terraced house in Upper Gower Street, Bloomsbury, then the epicenter of intellectual London and close to the newly established "Godless" University College.

Almost from the month he returned to England, Darwin suffered a succession of illnesses, symptoms of which included bouts of intense tiredness, vomiting, headaches, and stomach cramps. It was suspected initially that he had contracted a disease from his voyage, but his symptoms seemed totally unconnected; they would come and go without discernible pattern, often forcing him to bed for weeks at a time. There have been many theories to explain this, including speculation that he did indeed contract a tropical infection. Others have suggested he was hyperallergic or that he had some form of immunodeficiency. The best current thinking on the subject is that he suffered from a succession of stress-related allergies. Never able to find a cure or even relief from his collection of odd symptoms, as he grew older Darwin simply learned to live with illness and to work around it.[7]

Much of the foundation for what became the theory of evolution was formulated in the house in Bloomsbury. For three years Darwin lived comfortably, collating his collection of samples and drawing the first sketches of a theory to explain some of the oddities he was beginning to see emerging from his notes. He had been elected a fellow of the Royal Geological Society and then the Royal Society soon after his return, and through his new contacts he met and became friends with Richard Owen, who then took many of his samples for analysis at the Royal College of Surgeons, where he had recently been appointed Hunterian professor. But what Owen was to make of the specimens was to bear almost no relationship to Darwin's interpretation. Before the voyage the two men shared a relatively orthodox position on biology. But after the *Beagle* voyage Darwin's and Owen's thoughts began to diverge rapidly: Soon they would be separated by a huge intellectual gulf.

Darwin's mind was turned toward evolution by an array of experiences and observations combined with the ideas of others working in what, on the surface, may seem unrelated fields of study. The first and perhaps most important experience Darwin applied to his development of the theory of evolution via natural selection came from his stay on the Galápagos Islands. In his autobiography, he describes how he had been so impressed "by the manner in which closely allied animals replace one another in proceeding southwards over the [South American] Continent; and . . . by the South American character of most of the productions of the Galapagos archipelago, and more especially by the manner in which they differ slightly on each island of the group."[8]

He is describing here the fact that although many of the species found on the islands had come from a common source—almost certainly the mainland of South America—they had adapted differently on different islands according to the conditions. The most obvious example is the finch, a bird found in large numbers on the Galápagos Islands. Darwin noticed that the beaks of the finches had developed with pronounced differences on widely scattered islands within the archipelago. Where the birds had to find food by burrowing into crevices, the beaks were long and thin, whereas on other islands, where seeds that formed their staple diet were tougher to crack but easier to find, the birds had heavy, powerful beaks. Yet clearly both types of finches possessed a common ancestor. In his autobiography Darwin said of this finding: "Such factors as these could be explained on the supposition that species gradually become modified"—they had developed over many generations.[9]

This was a fascinating observation, but he needed to gather as much information as he could to verify his findings. To help him, he turned to a group he knew had the best available data on selection and the effects of selective mating. Writing an eight-page booklet listing forty-four specific questions, he sent it to dozens of breeders as well as to everyone he knew within the scientific community who had some knowledge of practical biology. A typical query was "If the cross offspring of any two races of birds or animal, be interbred, will the progeny keep as constant, as that of any established breed; or will it tend to return in appearance to either parent?"[10]

What he learned from the replies supported his own observations, but there were other influences that in the end proved just as important to the progress of his thoughts. Key to the development of Darwin's explanation for evolution was his interpretation of the work of an economist, Thomas Malthus, whose *Essay on the Principle of Population* was published in 1798. Darwin had begun what he called his "secret notebooks" in July 1837, while living in Great Marlborough Street, and their contents grew rapidly during the following few years. The ideas of Malthus begin to appear in the notebooks soon after Darwin had read *Essay*, and he interspersed extracts with his own embryonic theories.

Malthus had been intrigued by the fact that human population appeared to be increasing geometrically; that is, it doubled in size during a specific period of time, doubled again during the next period of the same length, and so on. Malthus calculated that during the first part of the nineteenth century this doubling period was about twenty-five years, or once per generation. But he reasoned that this could not always have been the case or the planet would have been overrun by humanity long before. For the same reason, no other species on Earth could have followed a geometric growth pattern. Reasoning that there must then be a counterforce to that of reproduction and growth, some factor or factors that reacted against the onward rush of life, he concluded that species reached a state of equilibrium in which they would sustain themselves. No one species could overwhelm another, but instead each operated in harmony with the others.

Of course, modern humans have destroyed this pattern. With the development of civilization and later technology (still young in the days of Darwin and Malthus), we have managed to defy this law of Nature and to increase our numbers geometrically rather than maintain our population in a state of equilibrium. Darwin, however, was not so much interested in this phenomenon as he was in the idea of the counterforces that kept populations of species in check. Malthus had written: "The natural tendency to increase is everywhere so great that it will generally be easy to account for the height at which the population is found in any country. The more difficult, as well as the more interesting part of the inquiry is to trace the immediate causes which stop its further progress.... What becomes of the mighty power ... what are the kinds of restraint, and the forms of premature death, which keep the population down to the means of subsistence."[11]

And on the very afternoon Darwin read this, he noted in his secret journal, which he now boldly titled *Notebook on Transmutation of Species*. "On an average every species must have same number killed year with year by hawks, by cold etc—even one species of hawk decreasing in number must affect instantaneously all the rest. The final cause of all this wedging must be to sort out proper structure ... there is a force like a hundred thousand wedges trying to force every kind of adapted structure into the gaps in the economy of nature, or rather forming gaps by thrusting out weaker ones."[12]

This was the first great breakthrough for Darwin and led him far along the path in which he could fit together the elements of what he had observed on the *Beagle* voyage. It was clear to him that members of a species competed for supremacy, and here was the driving force behind the idea of evolution.

It is important to realize that Darwin did not *create* the idea of evolution. The concept that species may change over long time periods had been proposed as early as the fifth century B.C. However, no one until Darwin had the slightest idea *how* this could happen. It is one thing to suggest a phenomenon might occur or an effect might be possible and quite another to show *how* such things happen. This is the difference between a "good idea" and a scientific explanation or a genuine theory. Many thinkers, from the ancient Greek Empedocles to Darwin's contemporaries, thought in terms of life on Earth changing because of some form of competition between species. It was up to Darwin to apply the theories of Malthus, to blend them with his own observational experience, and then to develop the idea of evolution via natural selection.

In essence, a modern definition of Darwin's theory of evolution would be: Over many generations species evolve or develop in part because only the best-adapted members survive to reproductive age and pass on their genes to the next generation. The way a species develops depends upon two factors. First, when two individuals blend their genetic material, there is a shuffling of genes. This is genetic variation and comes about through both the complex biochemistry of reproduction—the random chances of certain genes from one partner combining with those of the other—and random mutation. This accounts for the vast diversity of individuals within species and why (if we consider natural breeding as opposed to the results of cloning) no two individuals are ever identical.

The second factor, which plays an equally important role in evolution, is the effect of selection, the fact that, in general, only the best-adapted members of a species will survive to reproductive age. This mechanism is controlled by Malthusian factors, and from it the popular expression "the survival of the fittest" was derived. Darwin wrote: "Natural Selection, as we shall hereafter see is a power incessantly ready for action and is immeasurably superior to man's feeble efforts, as the works of Nature are to those of Art. . . . I use the term Struggle for Existence in a large and metaphorical sense, including dependence of one being on another, and including (which is more important) not only the life of the individual, but success in leaving progeny." He went on to declare: "A struggle inevitably follows from the high rate at which all organic beings tend to increase. Every being, which during its natural lifetime produces several eggs or seeds, must suffer destruction during some period of its life, and during some season or occasional year, otherwise, on the principle of geometrical increase, its numbers would quickly become so inordinately great that no country could support the product. . . . it is a doctrine of Malthus."[13]

Within Darwin's own experience evolution could be observed in the finches of the Galápagos Islands. The original community of finches had come from the mainland but, as different members of the community settled on different islands with slightly varying environments, over many generations they adapted to the conditions they found there. On some islands where a long beak was advantageous, those that *by chance* (or because of random genetic fluctuations) had slightly longer beaks than their contemporaries fared better and were more likely to survive to maturity and then to mate. Those of their descendants who also by chance had inherited the genes that encoded longer beaks also did better, and they too had an enhanced chance of passing on their genes in competition with their fellow finches.

This is natural selection at work, or what Darwin referred to as "preservation of favourable variations and the rejection of injurious variations," leading him to add: "As all the inhabitants of each country are struggling together with nicely balanced forces, extremely slight modifications of the structure or habits of one inhabitant would often give it an advantage."[14]

From this line of reasoning, Darwin realized two crucial results. First, it explained why species found niches in the biosphere, fine-tuning their relationship with the environment. For example, how some species of hummingbirds have special, extralong beaks that allow them to extract nectar from only one flower, how giraffes reach their preferred leaves, or how pandas have evolved a digestive system that requires only bamboo and water. The other, grander consequence of Darwinian selection is that species that are today quite separate have derived from a common origin and that in fact all life on Earth, including human beings, has arrived where it is by following quite different evolutionary paths, but paths that may be traced back to a common source.

This did not come to Darwin all at once. What he had learned from the voyage aboard the *Beagle* took time to filter through his mind. If we look at Darwin's "secret notebooks" from around 1837–38, it is clear that the essence of the theory did indeed come quickly—possibly during a short period of frantic intellectual activity shortly before his marriage. But from these basic tenets he had to let his thoughts themselves enter an evolutionary process before he could begin to construct a sound theory. As well as this, he had to acquire a body of evidence to support his ideas and be able to define clearly the mechanism by which natural selection operated—a mechanism to describe the mechanism. All of this could come only with time and hard work.

It is clear from Darwin's personal letters and private diaries that he was becoming increasingly anxious about the social and personal ramifications of what he was doing. It can be no coincidence that his illness started around the time he began contemplating his theory and that his outward physical symptoms were a direct result of the tension his realizations were causing him. Darwin was fully aware from the moment he set out on his intellectual journey that not only would his theory be considered revolutionary by the establishment and refuted outright by the orthodox Church but it would cause his future wife great distress. This latter possibility disturbed him most of all, and throughout their lives together Darwin tried to protect Emma from the upset his views might cause her.

But there was another reason for keeping quiet about what he was thinking, at least for a while. Even though he had produced impressive work on the *Beagle*, Darwin's reputation in the scientific world was based solely upon his efforts as a geologist. He had been trained in that subject, and the bulk of his official findings on the voyage were geological. During the 1830s and 1840s, Darwin was certainly not perceived as an important biologist. If he had attempted to publish a half-baked theory based upon raw observational data and an ill-defined mechanism, he would almost certainly have been laughed out of the scientific community just as he had taken the first steps into its outer circle.

So Darwin turned to other things, but these were also studies related directly to his "secret" ideas. He spent seven years researching barnacles and finding links between different species. He bred orchids and made a detailed study of how orchid types could be cross-fertilized to produce original and beautiful hybrids. He spent hours each day studying the development of seeds and how environmental conditions altered their growth. He published papers on his findings in learned journals that specialized in each of his enthusiasms. At the same time he wrote tracts on geology based upon his vast collections of notes made during the voyage of the *Beagle*.

And his efforts eventually paid dividends. During the 1840s Darwin became known as the one of the world's leading geologists, and through his researches on barnacles, orchids, and seed development during the 1850s, he developed a reputation as an eminent biologist. In this way he was preparing the ground for the eventual exposure of his grand theory.

In 1842 Charles and Emma Darwin moved to a large country home, Down House in Kent, where they started a family. Within ten years they had six children, and as Darwin's family grew, the world changed just as rapidly. The Georgian era had ended in 1837, with the ascension to the throne of Queen Victoria, and the British Empire had entered its period of greatest expansion and power. The Industrial Revolution had finally begun to forge enormous social change, and the earliest skeletal rail system was created to link distant parts of the country and oil the wheels of British industry more efficiently than could have been imagined even a generation earlier. James Watt, a friend of Erasmus Darwin and cofounder of the Lunar Society, had done much to develop steam power at the end of the previous century. George Stephenson had built and equipped the Stockton & Darlington Railway during the early 1820s, and within twenty-five years there were passenger and freight services crisscrossing the country; one of the most important, the Liverpool & Manchester, opened in 1830.

Darwin came from a liberal dissenting family; he was never a political animal, neither was he a bulldog patriot, but the mood of the time, the onward surge to greater dominion, and the all-embracing fervor for industry, machines, production, the accruing of great wealth and international power, lay as an undercurrent to all life in Britain during the mid-nineteenth century; it affected everyone. This impulse reached a peak with the Great Exhibition of 1851, which Darwin and his family, along with millions of others from the four corners of the empire, attended. The exhibition demonstrated all that was great about Britain and placed a heavy emphasis upon mechanization, the modern concept of the stratification of labor and mass production. Such things almost certainly crept into Darwin's subconscious mind. Industry we may now perceive as a manmade copy of Nature's paradigm. Nature is a vast production line, a quality-controlled conveyor belt, a global factory in which models are developed, constructed, modified, perfected, and churned out by the million. Darwin's model for natural selection explained how all creatures found their niche, how they evolved, and how all species interrelated to produce the megalithic biosphere of planet Earth. Industrialized man was following the ancient pattern for all to see.

Science was also changing rapidly during the two decades Darwin pondered and evolved his ideas. The scientific world of the 1850s was very different from that of the 1830s, and this was particularly apparent in the relationship between orthodox Christianity and science. Slowly, some intellectuals were realizing that an acceptance of religious orthodoxy was capable of inhibiting scientific innovation, that the traditional genuflecting to the Church stifled and smothered. In many ways this tale of conflict between Charles Darwin and Richard Owen is a perfect model for the larger ongoing conflict between science and orthodox Christianity. When Darwin finally published his great work, almost a quarter of a century after his first thoughts on evolution, it was met with consternation, rebellion, and in some quarters hatred and dismissal, but this was nothing compared with what might have been had he been more hasty.

Furthermore, some of the ground had been prepared for Darwin by way of a popular science book, *Vestiges of the Natural History of Creation*, which was published anonymously in 1844.* The author possessed only a sketchy scientific background, and his book, reflecting his personal interests, ranged from genuine science to cranky ideas lifted from mysticism, mythology, and a hotchpotch of comparative religion, muddled geological theory, and pseudoscientific ideas. Using this collection of concepts and disciplines, the author attempted to explain the origin and development of life on Earth. Not surprisingly, *Vestiges* was savaged by the scientifically literate, although such dismissal did nothing to dent sales; the public loved it, and the book was a huge best-seller in its time. Darwin said of it: "The prose is perfect but the geology strikes me as bad and his zoology far worse"[15]

Although much of the book was far-fetched, it at least contained a kernel of truth, and Chambers attempted to convey a little of the state of science during the 1840s. Most important for Darwin, it was read by the influential, wealthy, educated middle classes. And although for serious scientists the book became an object of mockery, *Vestiges* did much to shift the

---

*The author was soon exposed and turned out to be a middle-aged gentleman, Robert Chambers, founder of the famous publishing house that still bears his name.

views of the educated nonscientist, an effect that filtered down into the collective conscious of the literate working classes.

But it was this very popularity and the cosmopolitan nature of the book's readership that frightened many within the establishment. In a delicious prequel to the Lady Chatterley trial over a century later, one overzealous reviewer of 1844 said of *Vestiges* that he could see only "ruin and confusion in such a creed . . . it will undermine the whole moral and social fabric, bringing discord and deadly mischief in its train."[16]

Darwin could hardly have imagined it, but when the *Origin* was published, it too touched all sections of society and caused a stir within the working classes. This excitement was spurred on by the press. In these pieces Darwin's words (particularly his few guarded comments about human beings) were often misquoted and exaggerated.

By the early 1850s it was clear even to the prevaricating Darwin that he would have to publish his ideas before too long; Victorian society had never been so well primed. He was encouraged in this by his friend the great geologist Charles Lyell, who had written a definitive book on his subject, *Principles of Geology*, published in three volumes between 1830 and 1833. Lyell seems to have been the first of Darwin's circle to raise the specter of priority for evolution via natural selection. Writing in May 1856, he told Darwin: "I wish you would publish some small fragment of your data . . . and so out with the theory and let it take date and be cited and understood."[17]

But even the rather otherworldly Charles Darwin must have been aware that others would be snapping at his heels and that he ought to put his stamp on a theory he had first formulated during the late 1830s. In fact, someone he knew quite well was about to blow the lid on the entire theory.

ALFRED RUSSEL WALLACE WAS BORN into poverty. His father had been an unsuccessful solicitor who had died young, leaving his family in debt. The eighth of nine children, Alfred had been forced to leave school at age thirteen, and while holding down a succession of poorly paid jobs as a laborer and joiner, he had studied the latest scientific publications and taught himself the natural sciences, focusing on biology and geology. He had read Chambers's *Vestiges* with unalloyed enthusiasm, consumed Darwin's *Journal*, pawed over Lyell's *Principles of Geology*, and absorbed every word of Malthus's *Essay*. When the depression of 1843 struck, Wallace lost his job and decided to travel to the New World in the hope of making a fortune as a freelance naturalist.

Although this sounds like an odd career move, it actually promised great riches. There was a large and growing market for exotic specimens, and Wallace had both the knowledge (admittedly all gleaned from textbooks) and the enthusiasm to succeed. He also had a friend who was an experienced, if amateur naturalist, and the pair, Wallace aged only twenty-five and his older colleague, Henry Bates, set sail for America in 1848.

And they were successful. The pair traveled through the wild, uncharted wastes of the American continent, an experience that furnished Wallace with firsthand knowledge of nature similar to what Darwin had acquired a dozen years earlier, and he collected an impressive and very valuable set of specimens. But after four years he had had enough and decided to head back to Europe to cash in his treasure. Parting amicably with Bates, he set sail for Southampton.

Then disaster struck. The ship was gutted by fire, everything was lost, and it took ten days for the crew and passengers to be rescued by a passing ship. But even this did not stop Wallace. He simply picked himself up, dusted himself down, and lived on the poverty line in England for three years while he wrote a book about his adventures called *Narrative of Travels on the Amazon and Rio Negro*. Then, somehow, he found the funding for another voyage.

This time Wallace journeyed east to Borneo, where he was to spend the next eight years studying the indigenous fauna and flora. As Wallace sailed away from England in 1854, Darwin was reading his account of travels in the Amazon and becoming increasingly impressed with Wallace's intelligence and detailed observations, most of which tallied with his own views. Wallace returned to England at the height of the furor over evolution with no fewer than 125,660 specimens, all of which survived the passage.

The two men began a correspondence, and during the following four years they exchanged ideas and opinions on the subject of evolution.* Darwin, the senior scientist and a man whom Wallace greatly honored, offered

---

*Wallace was living in Ternate, one of the Spice Islands (now Indonesia) when he contracted malaria. As he was recovering he began to think about how the drive for survival could influence the evolution of species.

encouragement and guidance, but, it seemed never to occur to him that Wallace was quite independently reaching conclusions similar to his own about how evolution could work.

Perhaps Darwin did have an inkling of Wallace's thoughts, because by the time the exchange of letters was in full flow, Darwin had begun to make serious plans to publish his own work. He had written a sketch of the idea of natural selection some fifteen years earlier, later noting in his autobiography: "In June 1842 I first allowed myself the satisfaction of writing a very brief abstract of my theory in pencil in 35 pages."[18] He had also talked to Hooker and another confidant, George Waterhouse. Two years later, he had expanded this document to a manuscript of some fifty thousand words and had a fair copy made by a local schoolmaster. Darwin obviously knew he was on to something important because he wrote a letter to Emma in which he stipulated that if he should die, she must arrange for his manuscript to be published.

Because of this there was absolutely no question of priority concerning the theory of evolution via natural selection, and Darwin, if he cared at all, knew this. However, on June 18, 1858, he received a letter from Wallace that changed everything. The letter was like any of the dozens that had passed between them during the preceding four years except that enclosed with it was a paper titled "On the Tendency of Varieties to Depart Indefinitely from the Original Type." In the letter Wallace asked Darwin for his opinion of the paper and a view on its worthiness for publication.

Darwin was staggered by what Wallace had written. The same day he wrote to Lyell: "I never saw a more striking coincidence; if Wallace had my MS sketch written out in 1842, he could not have made a better short abstract! Even his terms now stand as heads of my chapters. Please return me the MS, which he does not say he wishes me to publish, but I shall of course, at once write and offer to send it to any journal."[19]

But really Darwin should not have been so shocked. All the signs had been there, he had simply been blind to them. Wallace had even published a paper some two years earlier in the *Annals and Magazine of Natural History* in which he had surveyed the evidence for evolution based upon his own observations and experiences, but at that stage he had offered no theory to explain what he had noticed. Furthermore, Lyell and others had warned

Darwin that Wallace was a man to be watched. "Your words have come true with a vengeance," an angst-ridden Darwin confessed to Lyell.[20]

Fortunately for Darwin, he had listened at least in part to Lyell's advice and had developed the version of the theory he had written in 1844. By the time Wallace's paper arrived, Darwin had all but completed the manuscript for what became *On the Origin of Species* and was ready to submit the book to a suitable publisher and to have an abstract read before an appropriate audience.

But Darwin was no Newton. He immediately decided that Wallace deserved as much recognition as possible and that the two of them should publish jointly and simultaneously. He arranged through Lyell for both Wallace's paper and his own abstract to be read before the Linnean Society and then published under both names in the society's *Journal*.

Wallace's work actually made no difference to the reception of Darwin's accomplishments. Like so many great discoveries, the work of the two men was independent, and in this event, because of the characters involved, at no time was there any suggestion of plagiarism or any form of ungentlemanly conduct over either party failing to credit the contribution of the other. It was accepted immediately that Darwin had priority, and he remained by far the more important figure in evolutionary biology. Wallace was certainly a great evolutionary biologist and should be honored for his effort. Later in life he and Darwin diverged enormously in their interpretations of how evolution worked. Darwin held the line followed by most scientists today—that there is no plan or pattern to natural selection and that there is no God or controlling influence to guide our genes, to weed out the weak or to promote the strong. Wallace possessed a deep spiritual vein and later drew considerable opprobrium upon himself because he tried to create a blend of religion and evolutionary biology by conjuring up a role for a "guiding Creator." Even so, Darwin helped support Wallace materially, and although they disagreed strongly on an intellectual level, the far richer, far more successful and respected Darwin never deserted his colleague. When later in life Wallace faced bankruptcy, Darwin used his influence with Queen Victoria to secure for him a royal pension.

Unlike almost any other discovery of similar magnitude, the theory of evolution was never subject to any form of priority dispute. But this is not the essence of the story surrounding Darwin and his theory, for he had plenty of other battles to fight, some every bit as bloody as those instigated by rivals spurred by greed and jealousy in the rooms of the Royal Society, the French Royal Academy, or other learned institutions around the world. The greatest battle Darwin faced over his discovery stemmed solely from the fears, anxieties, and irrational fury it precipitated.

DURING THE 1850S, Richard Owen was considered the most eminent biologist in England after Darwin himself. Born in Lancaster in 1804, he was five years Darwin's senior. He had pursued a highly successful academic career, and, unlike Darwin, he had stuck with a medical course at Edinburgh. But before he could begin a career as a surgeon, he was offered a much-coveted position as curator of an important set of specimens called the Hunterian Collection at the Royal College of Surgeons in London. At the same time he established a medical practice and split his days between the two jobs until he was made Hunterian professor at the college in 1836, a few months before Darwin arrived home aboard the *Beagle*.

At the time Darwin was introduced to Owen through their mutual friend Charles Lyell, Owen was attracting attention as one of the most promising young biologists in London. Two years earlier, at the age of thirty, he had been made a fellow of the Royal Society and was already becoming recognized as a prolific and versatile anatomist. From his position at the Royal College of Surgeons, he could devote himself to research and had at his disposal hundreds of thousands of specimens gathered from around the world, including a substantial collection of fossil remains and the preserved bodies of exotic creatures brought back to England by adventurers who had provided the template for the careers of men like Wallace and Bates.

Owen was tall and exceptionally thin, his face birdlike; he had huge eyes and a long, sharp nose, accentuating high, protruding cheekbones. He was, in sum, a rather stiff, wooden character. His writing illustrates his eye for detail and his precision, but it is also as dry as old leather. An example comes from one of his many books, *Anatomy of Vertebrates*, in which Owen takes a swipe at Darwin and subtly questions his priority for the theory of evolution using a sentence with the proportions of one of his dinosaurs. "Of course, to every competent judge, the difference between a theory founded on the application of the principle of the contest for existence to the preservation of extinction of certain species and that of a theory of the origin of all species partially based upon the same principle, must have been obvious; nor was any pretension advanced, in the letter rectifying the date of the 'idea,' to the ample and instructive degree in which it had been worked out, and doubtless as an original thought, by the accomplished author of the *Origin of Species*."[21]

In old age Darwin wrote of him: "I often saw Owen, whilst living in London, and admired him greatly, but was never able to understand his character and never became intimate with him. After the publication of the *Origin of Species* he became my bitter enemy, not owing to any quarrel between us, but as far as I could judge out of jealousy at its success. Poor Dear Falconer [the paleontologist Hugh Falconer], who was a charming man, had a very bad opinion of him, being convinced that he was not only ambitious, very envious, and arrogant, but untruthful and dishonest. His power of hatred was certainly unsurpassed. When in former days I used to defend Owen, Falconer often said, 'You will find him out some day,' and so it has passed."[22]

But Owen's personality was perfectly suited to the task of sorting through ancient bones, piecing together often tiny scraps of physical evidence with which to derive an overview, and the intricate processes required to construct a taxonomy, a job most other scientists would have found intolerably dull. Owen was extremely conservative, an Anglican who perceived the growing atheist movement in Britain as quite literally the work of the Devil and who must have known of and loathed the religious inclinations of the Darwin family and their famous friends. Unlike Darwin, Wallace, Hooker, or Huxley, Owen never undertook a voyage of discovery overseas but spent almost his entire working life closeted in his rooms in Lincoln's Inn Fields to the west of the City. It was a working life not so very different from Darwin's after he moved to Kent, but unlike Darwin's Owen's lifestyle was enlivened considerably by social and professional commitments in the capital, which made him very much more the establishment scientist of the day than Darwin could ever have been.

Owen had very clear views about the way species interrelated and how the hierarchy of life was structured. For many years he had been a Creationist and had followed the teachings of the French anatomist Georges Cuvier.

Cuvier, the grand old man of French biology, had died in 1832, soon after Darwin had left aboard the *Beagle*. An antievolutionist, Cuvier had held the view that all species were created by God and remained unchanged unless they became extinct as the result of catastrophes such as the biblical flood. He expressed these ideas in two books *Leçons d'anatomie comparée* (1805) and *Recherches sur les ossements fossiles* (1812). Although these works are now viewed as discourses upon a long-redundant theory, both were treated almost as holy texts by the young Owen.

But as he began his own research, Owen slowly moved away from such fundamentalist views, and before Darwin had begun even to consider the idea of evolution and the transmutation of one species into another, Owen had formulated his own model, based upon a blend of what he had observed and what he derived from his strict Anglicanism.

But by its very nature this line of reasoning went only so far. What held Owen back from any radical theory was his abhorrence of the idea that any species could transmute into any other. He once wrote: "The man who is willing to believe that his great-great-great etc. grandfather was a Baboon, and his great-great etc grandmother a Chimpanzee, will not be converted by whatever manifestation of the *mens divinior* may shine in a refutation of such an opinion."[23] To Owen, the very notion of transmutation (a doctrine that would, by inevitable consequence, involve humans) was unchristian and attempted to undermine his belief in human uniqueness.

But Darwin was not the only one to see things differently. As early as the 1830s, there had been a strong and growing underswell of anticlerical feeling in England. By the time Chambers's *Vestiges* was published in 1844, the Atheists had become a recognized group who were beginning to form dangerous liaisons with political radicals, most especially the Anarchists.

Back in the 1820s, during Darwin's schooldays, Thomas Arnold (who was later made the headmaster of Rugby school) warned that medical students were turning away from the Church in droves, becoming "materialist atheists of the greatest personal profligacy."[24] A generation after this statement, political and religious extremists were to be seen demonstrating together on the streets of London, and Owen, no armchair defender of the faith, became a volunteer in the urban gentlemen's regiment, the Honourable Artillery Company, who supported the police during riots.

For Owen, religious beliefs clearly had a greater priority than intellectual reasoning, a handicap the dissenting Darwin certainly did not suffer. And so, to accommodate both the observational evidence he had gathered and his religious sensibilities, Owen was forced to find an alternative to transmutation of species. And he found it in the ideology of a German biologist named Johannes Müller, who postulated that there was no external force altering the development of a species but that, instead, each organism possessed a nebulous *organizing energy* guiding growth and development to a predetermined plan.

Today we can see that, at its core, what Owen was grappling for may be related, at least in part, to the science of genetics, although neither he nor Müller would have considered it that way. Owen visualized this organizing energy as concentrated in the "embryonic germ" and dissipated into the tissues or cells of living things. He believed this energy gave life to organism and because he believed that the *life force* was different for each type of creature and that each species was self-contained, he refused to accept the idea that one species could change into another. For Owen, the life force or organizing energy could never be expanded to allow an organism to transmute into one that was more complex.

Today we know there is a template for *individual* organisms that is indeed built into the embryonic germ (the nucleus of each cell). This individual code is found in the paired chromosomes carrying the instructions for the growth and development of a particular human. However, this is not Owen's organizing energy but biological material through which information is passed from one generation to the next. We may think of it as the carrier of a computer program that instructs the detailed construction of an individual organism.

Darwin's theory of evolution considers the development of life from the other end of the telescope. To Darwin, the individual organism is utterly meaningless, and Darwinian evolution shows that all living things are solely at the mercy of two factors, the random shuffling and mutations of genes and the forces of natural selection. In Darwin's universe there is no guiding hand inside or outside the individual because *there is no plan*, no objective other than the drive for survival. Nature is mindless, ungoverned, a free spirit, and because of this, Darwin tells us, life is cruel, violent, and utterly meaningless. This then is a model for all of Nature. Thomas Hobbes touched upon this truth some two centuries earlier when he said of the human condition (a tiny element of Nature's grandeur) that it is "solitary, poore, nasty, brutish, and short."[25]

If Darwin had given Owen even a clue to his thinking during the 1840s and early 1850s, their friendship would have dissolved far sooner. But almost until the day the *Origin* was published, he kept his radical views to a tiny cadre of close associates. However, while Darwin kept quiet, studiously avoiding conflict, the Hunterian professor was already making an enemy of a young man who would one day act as Darwin's mouthpiece and pour boiling oil over Owen's theories.

$\mathcal{T}$HE CIRCUMSTANCES OF Thomas Huxley's birth and childhood could not have been more different from Darwin's. His father, George Huxley, was a penniless schoolteacher who drank; his mother, Rachel, already forty when Thomas, their seventh child, was born, was a Cockney who carried a restless spirituality that her son never forgot and, in an odd way, adopted. Destined for a career as a cleric or a teacher perhaps, Huxley was saved by his great intelligence and academic brilliance. He taught himself Greek and studied advanced texts on subjects as diverse as geology and mathematics. In 1841, at the age of sixteen, he won a scholarship to Charing Cross Hospital Medical School and lived as an impoverished student close to Euston Station in North London, perhaps a mile from the house Charles Darwin shared with Emma before they left London in 1842. Parodying the invective of the establishment and the Church, Darwin was later to call Huxley "my good and kind agent for the propagation of the Gospel i.e. the Devil's gospel."[26] But one can only wonder what Huxley and Darwin would have made of each other if they had crossed paths on a street in Bloomsbury.

Perhaps they did, but it was to be many years before the career paths of the two men would intersect. Huxley excelled at medical school. He won a string of prizes and saw his first paper published when he was just twenty years old. However, his scholarship expired before he could graduate, and he volunteered as a surgical assistant aboard a Royal Navy frigate, the *Rattlesnake*. So, like Alfred Wallace and John Hooker, Huxley followed in Darwin's footsteps and traveled the world, making scientific notes on all he saw, and he too sent home papers and specimens. Returning in 1850 and again mirroring Darwin's career some fourteen years earlier, he found that in absentia he had become an important scientific figure. Feted by the scientific community, at the extraordinarily young age of twenty-six he was elected a fellow of the Royal Society and soon after secured a research position financed by the Royal Navy. Ironically, he never did complete his medical degree, and throughout his life the only academic credentials he could offer were a growing collection of honorary doctorates awarded by some of the esteemed institutions of the world, including degrees from the universities of Oxford, Cambridge, Berlin, Edinburgh, and Dublin.

A few years after returning to England, Huxley gave up his research post and was appointed a lecturer at the tiny Royal School of Mines in London. He soon became a leading figure there and transformed the almost totally unknown school into one of the most respected in the country. Huxley married in 1855 and supported a family of seven children on a professor's income supplemented with sporadic earnings from writing. He and his adored and adoring wife, Henrietta, established a clan that would in later generations include many scholars and literary figures, most famous of whom was Thomas's grandson the novelist Aldous Huxley.

Domestically secure, Huxley struck out at the many scientific targets that had for long hovered in his sights, and during the 1850s he became the leading Young Turk who, with a group of like-minded academics, spearheaded scientifically argued attacks upon the religious establishment. He took the dissenting line to an extreme, horrified orthodox churchmen and the laity with a form of "scientific humanism" refined to the point of evangelism. Huxley put the word *agnosticism* into our vocabulary, and more than anyone else from the Victorian era he may be seen as the motivating force behind twentieth-century skepticism, proclaiming that we can say nothing of the world beyond the material one in which we know we exist. "I took thought," he wrote, "and invented what I conceived to be the appropriate title of 'agnostic.'"[27]

Huxley did not want to destroy the Church, but he loathed bigotry and believed that science had to coexist with Christian teachings while at the same time bearing no allegiance to it; science must be completely unshackled. "It is clear to me," he commented to his friend the cleric Charles Kingsley, "that if that great and powerful instrument for good or evil, the Church of England, is to be saved from being shivered into fragments by the advancing tide of science—an event I should be very sorry to witness, but which will infallibly occur if men like Samuel [Wilberforce] of Oxford are to have guidance of her destinies; it must be by the efforts of men who, like yourself, see your way to the combination of the practice of the Church with the spirit of science."[28]

Naturally, Huxley was respected and admired by all who took an anti-establishment view. The American historian and philosopher John Fiske, who was to popularize what became known as social Darwinism—the dubious application of Darwin's work to the evolution of society—said of him: "I am quite wild over Huxley. He is as handsome as an Apollo. . . . I never saw such magnificent eyes in my life. His eyes are black, and his face expresses an eager burning intensity. . . . He seems earnest,—immensely in earnest,—and thoroughly frank and cordial and modest. And, by Jove, what a pleasure it is to meet such a clean-cut mind! It is like Saladin's sword which cut through the cushion."[29]

But there were many who were actually scared of Thomas Huxley, the Devil's Disciple, horrified by his lashing tongue and his vitriolic pen. The novelist and commentator on the theory of evolution Samuel Butler said of him: "Men like Huxley . . . are my natural enemies."[30] Some grew to view him as the Devil incarnate rather than his mere "disciple." One bishop's wife was overheard to say of him in bewilderment: "But I hear that he is a devoted husband and an affectionate father."[31]

And of course Huxley and Darwin were made for each other. They met in 1853 at a gathering of the Royal Geological Society in London and immediately began to exchange ideas, to send each other specimens and advance copies of their latest papers. And gradually Darwin began to realize that in Huxley he had found not only a kindred spirit but someone to whom he could offer his most extreme ideas in order to gauge a reaction.

Darwin probably first told Huxley of his mechanism for natural selection around 1855, but initially Huxley was not convinced. Naturally, he had no objection on moral or religious grounds, and by this time he had cultivated a loathing for Owen and his associates (a loathing wholeheartedly reciprocated). Because of these Huxley would have been quite open to any theory that further diminished the Hunterian professor, but he maintained reservations about some scientific details of the theory of evolution.

Darwin became aware of this early in 1856, and from his citadel in Kent he surveyed the scene before deciding the best way to win Huxley over. By this time Darwin's relationship with Owen had cooled considerably, and it had become clear the Hunterian professor had heard rumors about him, rumors he did not like. In the typical manner of Victorian gentlemen, nothing was said, at least not at this juncture, but the two men were beginning to circle each other warily. Soon Darwin could admit to his friends that he hated Owen's "mysticism and murky, Platonic thinking."[32]

And Darwin was aware of how his young friends, Huxley, John Hooker, and the physicist John Tyndall, were creating a stir in London. They had begun to bait their superiors, honing their attacks on the Oxbridge Mafia, the entrenched elder statesmen of science. Huxley fought harder than the others combined, but he was also hotheaded, and his passion often ran away with him. Publicly satirizing Owen's fascination with all types of vertebrates and invertebrates, Huxley had dubbed him "a queer fish." Then, lampooning his preoccupation with classification, he went on to declare that the Royal College of Surgeons' chief anatomist was "not referable to any known Archetype of the human mind."[33]

The motivation for Huxley and the others was clear. They wanted to create a new class of scientist, to turn science into a "real profession." Until then, science had been practiced by poorly paid professors or moneyed gentlemen (ironically typified by Darwin himself). Overwhelmed as he was by a need to validate his theories and unconcerned about money, Darwin naturally saw Huxley's campaign as a side issue, but it was also one he could utilize to his benefit. Huxley and Hooker, he believed, were the key, because in their efforts to gain recognition they were anxious to bring down the antiquated ideas of men like Owen. As well as being rebellious and high-minded, they were determined and they had guts. Darwin knew that if he could win them over to evolution, then half the battle would be won. By 1855 Hooker was already convinced, but if his recent proclamations on the subject were anything to go by, Huxley was still confused, and this confusion was, if anything, leading him away from where Darwin wanted him.

Concluding that he needed to quickly nudge his pawns into position, in April 1856 Darwin convened a "conference" at Down. Ostensibly gathering friends and their families, he really intended to sound out the feelings of his closest colleagues and discover on what grounds and over which issues they might object to his theories if his work was to be published.

Huxley, it transpired, had misunderstood some of the central tenets of current thinking on evolution, but he did have reasoned objections to some of Darwin's theories. He pointed to the incomplete nature of the fossil record and the fact that many ancient fossils showed little change compared with their modern relatives. Darwin said little at the time but noted down the objections and worked out answers: the fossil record was more complete than Huxley realized; its mosaic character would be filled in given time, Darwin argued; and in any case, an incomplete record did not automatically augur against it.* Later he wrote: "I look at the geological record as a history of the world imperfectly kept, and written in a changing dialect; of this history we possess the last volume alone."[34] As for the apparent insubstantial change between certain ancient and modern relatives, Darwin explained this by saying that these species were evolving only slowly because they had long ago found their niches.

Of course, Huxley did come to accept fully Darwinian natural selection soon after the weekend at Down. First he was driven to evolution by his growing hatred of Owen, finding joy in taking the view diametrically opposite his enemy's. But this in itself would have been a poor foundation for supporting Darwin and was only the start of the shift in his thinking. Gradually, Darwin made Huxley understand where his thinking on evolution had been misguided; he answered all his objections and eventually nurtured in him a fundamentalist zeal for the new theory. By the time the *Origin* was published, Huxley had become the most ardent supporter of the theory after Darwin himself.

Darwin's masterpiece appeared in the shops at the very end of 1859, and almost immediately the war between Owen and his supporters and Darwin and his erupted into public view. The first print run of 1,250 copies of the *Origin* sold out on the day of publication, and Darwin was astonished to learn that travelers passing through Waterloo Station were snapping it up. As a result, the publisher, John Murray, ordered an immediate reprint of

---

*Today we know that the gaps in the fossil record and the sudden appearance of new species at certain points in the distant past is a complex matter but may be explained by dramatic changes in the earth's environment. One example is the appearance of many new species about 570 million years ago, a change called the Cambrian Explosion. This is thought to have been precipitated by a sudden surge in oxygen levels in the earth's atmosphere.

3,000. All of this delighted Darwin's friends but greatly irritated Owen. And then, when a glowing anonymous review of the *Origin* appeared in no less a vehicle than *The Times* newspaper (a review Owen and other interested scientists of the day knew had come from Huxley), Owen's irritation turned to unalloyed fury.

Huxley had managed to persuade a scientifically illiterate staff writer to let him compose the piece, and he reeled off the most brilliant, witty, and expansive review anyone could have wished for. Writing for what Huxley flippantly dubbed "the educated mob who derive their ideas from the *Times*," he conjured up images of everyday experience to hammer home the central theme of natural selection. Those who did not make Nature's grade fell by the wayside "like the crew of a foundered ship, and none but good swimmers have a chance of reaching the land," he wrote.[35] Darwin was enthralled, writing to Huxley: "The old Fogies will think the world will come to an end. . . . I should have said that there was only one man in England who could have written this Essay and that *you* were the man."[36]

Owen too was perfectly clear who the "only one man in England" was. In April 1860 his own review appeared in the *Edinburgh Review*, and he was merciless. He slammed Darwin's attack on Creationists expressed in the *Origin*, asking how life could originate if not for a Creator. Parading his own ugly theory, to which he gave the equally ugly name the "continuous operation of the ordained becoming of living things," Owen declared that Darwin's evidence was insubstantial and that he was dealing in hypotheses, not verifiable ideas. "We have no objection to this result of 'natural selection' in the abstract," he thundered, adopting the royal "we," "but we desire to have reason for our faith. What we do object to is, that science should be compromised through the assumption of its true character by mere hypotheses, the logical consequences of which are of such deep importance."[37]

With his write-up in the *Quarterly Review*, Samuel Wilberforce waded in but showed astonishing ignorance, pointing out that the animal remains found in Egyptian tombs did not differ at all from modern specimens. "There has been no beginning of transmutation in the species of our most familiar domesticated animals," he noted.

With this comment, Wilberforce ignored or failed to understand two of the central tenets of evolution via natural selection. First, evolution requires many more generations than those between pets of the ancient

Egyptians and animals of today. Second, evolution operates far more slowly if organisms have already found their niches. Throwing in theological idioms and dogmatic statements by the dozen, he added: "Man's derived supremacy over the earth; man's power of articulate speech; man's gift of reason; man's freewill and responsibility; man's fall and man's redemption; the incarnation of the Eternal Son; the indwelling of the Eternal Spirit,—all are equally and utterly irreconcilable with the degrading notion of the brute origin of him who was created in the image of God."[38]

Wilberforce was genuine in his beliefs, but perhaps we should gauge these comments against another voice; one of Wilberforce's many critics, the scientist and writer William Irvine, has aptly pointed out that "his chief qualifications for propounding on a scientific subject derived, like nearly everything else that was solid in his career from the undergraduate remoteness of a First in mathematics."[39]

Indeed such comments as Wilberforce's drew only laughter and contempt from Darwin and his supporters, but Owen's criticisms hurt so much Darwin lost a night's sleep over his review, and he called his erstwhile friend's comments "spiteful . . . extremely malignant, clever, and . . . damaging."[40] And Owen's review, the two men rarely spoke to each other.

Writing to Lyell, Darwin proclaimed: "It is painful to be hated in the intense degree with which Owen hates me."[41] And to Henslow, he lamented: "Owen is indeed very spiteful . . . The Londoners say he is mad with envy because my book has been talked about," he added in typical Darwin fashion, "what a strange man to be envious of a naturalist like myself, immeasurably his inferior."[42] Two years later he was still fuming, writing to Hooker: "In simple truth I am become quite demonical about Owen—worse than Huxley. . . . I shall never forget his cordial shake of the hand, when he was writing as spitefully as he possibly could against me."[43] Later he called Owen "one of my chief enemies, the sole one who has annoyed me."[44] And to Huxley he announced: "I believe I hate him more than you do."[45]

Owen's attacks cut deep because Darwin knew the man was an exceptionally good scientist who had contributed much with his anatomical studies stretching back a quarter of a century. But, on a quite different level, the greatest misunderstandings were reflected in the comments of men like Wilberforce, people who were utterly ignorant of science and misinterpreted both deliberately and unintentionally almost everything Darwin wrote. Even today a majority of people believe Darwin claimed that humans had descended from apes. Yet Darwin studiously avoided mentioning humans in his scheme. In the original edition of the *Origin*, he wrote: "Light will be thrown on the origin of man and his history."[46]

The idea that humans and other primates had a common ancestry was first declared openly and publicly not by Darwin but by Thomas Huxley. He was aware that by drawing humans into the equation he would open up the real debate, the gossamer veil would be ripped away. Huxley was spoiling for a fight with orthodoxy, and the more Darwin was misunderstood, the harder Huxley fought.

In his time Huxley became immensely famous in all sections of society. He was of course known and reviled by many in the religious establishment and disliked by the upper class, whom he attacked and ridiculed, but he was also seen by the interested layman as "one of us," a "man of the people." And he was the perfect representative to convey Darwinian evolution to the public. He had fingers in many pies and a battery of influential contacts, from rebellious lords and newspaper editors to presidents of learned societies. Having become, as he put it, "sick of the dilettante middle class" as early as 1855 Huxley had established what became his famous Workingmen's Lectures at the Royal College of Mines, which offered basic science directly to the people. Writing to his wife, Huxley quipped: "My working men stick by me wonderfully, the house fuller than ever. By next Friday evening they will all be convinced that they are monkeys."[47] A dozen years later, after he had become a household name and Darwin's ideas were still very much in contention, Huxley could fill St. Martin's Hall in Central London and see two thousand people turned away at the door.

Owen could not compete and was thoroughly disgusted by what he considered Huxley's crude popularizing. His natural home was among the ruling class, the highest echelons of the Church and the Conservatives in all strands of society he had spent a career befriending. These people were extremely powerful in their own way, but Canute-like they stood in vain on the seashore. The world of science was evolving more rapidly than any species ever seen on Earth, and Owen found himself batting for the wrong side.

Although we live in a very different age still and the conflict between science and religion has largely been fought out, it is easy to see how, during the Victorian era, both Church and State were threatened by Darwin's ideas and Huxley's thrusting rhetoric. Today we commonly dispute scientific progress on ethical grounds, but such debates stem almost exclusively from the *application* of science, practical innovation, and technology, not from fundamental theoretical concepts. For scientists at least, the idea that we should question theory by referral to any form of religious edict is absurdly anachronistic. But in Victorian times, as in all ages since humans had begun asking questions, religion did inhibit, and for some nonscientists it does to this day.

"When I try to imagine how puzzling, how counter-intuitive, how strange and threatening this idea seemed to some people," the biologist Daniel Dennett has recently said of Darwinian evolution, "I conjure up a little fantasy. Suppose some scientists came out tomorrow and said, 'Well, here's what we've discovered: there's been evolution among numbers. You know number seven wasn't always a prime number, it started out as number four.' I think, 'This is crazy!' As far as I know, it is crazy, but that reaction of frank, indignant disbelief was part of the negative reaction to Darwin's ideas. So he had to overcome that and he also had to overcome the fear that people had that he was demoting all that they held dear."[48]

Richard Owen personified that fear. He was devout, conservative, and hugely respected. He had achieved great things by clawing his way through life and by applying his superb brain and would not give up his beliefs easily. He had personal and social reasons for reacting against Darwin and Huxley. He could never face the idea that he was an insignificant organism lost amid the vastness of Nature, and he salved the same fears and nightmares expressed by his rich and materially successful friends. They did not want to hear what Darwin and Huxley had to say, and Owen covered their ears. But the great radicals of the nineteenth century, Darwin and Huxley, did not care one iota for the place of humanity in the greater scheme of things. If the concept of God meant anything to them at all, like many great intellectuals before them, they would have thought in pantheistic terms, perceiving God as an aspect of Nature itself. For them, science was more powerful than any religion.

And, many still close their ears to Darwinian evolution and prefer to ignore it or else claim it to be false. Darwinian evolution goes against the grain; many feel it is unpleasant, demeaning, and degrading, but this makes it not wrong, just hard to swallow.

To scientists, Darwinian evolution is a fundamental truth, a given; there is no question concerning its validity, it is a fact of life. And the revealing of this basic tenet of existence was a staggering feat if for no other reason than that Darwin knew nothing of the underlying process that allowed his mechanism to operate—the science of genetics. It was twentieth-century science that produced the theoretical framework for how natural selection occurs. By analogy, it was as though Darwin had explained how a television works, how the cathode ray focuses a stream of particles that cause fluorescence on a screen, without knowing the structure of the atom.

DARWIN AND HUXLEY WERE never reconciled with Owen or Wilberforce. As Huxley's career continued to flourish and he grew to be regarded as a great administrator as well as an innovative researcher, Owen's stagnated and he began to be viewed as a man left behind by the onward march of progressive science. For thirty years after the publication of the *Origin*, Huxley and Owen clashed publicly over the consequences of the theory of evolution, matters that began to dominate biology. And until he died in 1882, Darwin cheered from the ringside as Huxley struck his blows.

In 1861 Huxley was elected to the council of the Royal Zoological Society and Owen resigned. A year later Huxley attempted to block Owen's acceptance into the council of the Royal Society but failed. In a letter to the council he stormed: "No body of gentlemen should admit a member guilty of wilful and deliberate falsehood."[49] In his own best-selling book, *Man's Place in Nature*, published in 1863, Huxley fired broadsides at Owen, and when Darwin published the triumphant follow-up to the *Origin*, *The Descent of Man*, 1871, Huxley said of it: "It pounds the enemy into a jelly, although none but anatomists would know it."[50] The same year Owen attempted to oust the Darwinian president of the Linnean Society, George Bentham. The Darwinians closed ranks and outmaneuvered their enemy, and after this battle had been won, Darwin confided to Hooker. "What a demon on earth Owen is."[51]

But the Darwinians did not win all the prizes. Of course today, for biologists, Darwin is second only to God, and for many he might rank still higher. Huxley is remembered as the great popularizer of Darwinian evolution and a fierce radical who transformed Victorian science, but Richard Owen is almost forgotten. Indeed, if students of history or science hear of him at all it is either for his founding of the Natural History Museum in London or in a footnote detailing how the theory of evolution appeared in the world. But, if posterity has no place for Owen's ideas about evolution, he had his revenge on Darwin and Huxley in the only way he really could— by helping to deny them official honors.

Huxley we can understand never being offered a knighthood. For all his genius and energy, he had more enemies than friends in high places. But Darwin? Charles Darwin was from an upper-class family and had single-handedly created one of the great theories of science, yet the Liberal prime minister Gladstone was not inclined to place him on the Honors List. Most people considered Darwin's ideas extreme, views that challenged popular sensibilities, but only one man could have been spiteful enough and powerful enough within the social framework of the time to influence the decision makers, and he did, exacting his personal revenge. How ironic that Isaac Newton, Arian occultist as well as epoch-making scientist, should have been knighted by an orthodox monarch when almost two centuries later Darwin was ignored. Not surprisingly perhaps, two years after Darwin's death, Richard Owen, friend of both Gladstone and Queen Victoria, became Sir Richard Owen.

And what of our other protagonist in this saga, Bishop Samuel Wilberforce? Never tiring in his efforts to fight the good fight against Darwin's "blasphemy," Wilberforce took every opportunity he could to attack Huxley and, through him, the creator of the theory of evolution via natural selection. But Soapy Sam bowed out of the battle early. In August 1873, on a riding trip with the leader of the House of Lords, Earl Granville, he began bragging about his horsemanship and showing off his skills, only to be thrown by his mount. He landed badly and smashed his skull on a rock, dying instantly. When Huxley heard the news, he apparently turned to his friend John Tyndall and retorted: "For once, reality and his brain came into contact and the outcome was fatal."[52]

Apart from his fear of reproach and rejection, Darwin delayed publication of his work because he had seen no rivals. However, this did not in any way slow scientific progress. Instead it allowed Darwin to refine his theory close to perfection. Although Wallace proposed an almost identical theory, his ideas were only given form almost two decades after Darwin had quietly laid the foundations. By the time Wallace offered his ideas to Darwin, everything had been prepared for presentation.

Richard Owen's hatred of Darwin's ideas and the bigotry expressed by Wilberforce and others did not at all influence Darwin's creation of his theory of evolution, but they did have an enormous impact upon the way others interpreted it and the way the public received and digested the news.

We may now see that the clash between the Darwinians and the opponents of natural selection was the first time pure science had been thrust into the public arena. As I have mentioned, the newspapers of the day wrote extensively about Darwin's ideas, about the characters and personalities, and about the clashes and the often furious exchanges between the two camps. This hype extended into all areas of the popular press. Cartoonists lampooned Darwin, often drawing him as part ape, part man, and satirists poked fun at both sides of the intellectual schism. Soon after the Oxford debate that had done much to publicize Darwin's theories, *Punch* carried the following verse:

> *Then Huxley and Owen*
> *With rivalry glowing,*
> *With pen and ink rush to the scratch;*
> *'Tis Brain versus Brain,*
> *Till one of them slain;*
> *By Jove! It will be a good match.*[53]

Such a reaction came in part from a need to neuter Darwin's radicalism, to eradicate the pain acceptance of his theory must inevitably bring. The historian Ernst Mayr makes this clear when he comments: "No Darwinian idea was less acceptable to the Victorians than the derivation of man from a primitive ancestor.... The primate origin of man ... immediately raised questions about the origin of mind and consciousness that are controversial to this day."[54] And so the satirists and the cartoonists were called upon to do their work. In response, Huxley, Hooker, and the others did everything they could to proselytize the truth of Darwinism, all of which succeeded in heightening the profile of the theory of evolution and attracted many important scientific thinkers who continue to extend Darwin's ideas.

During the past century and a half, Darwinian evolution has become established as the single most influential theory in the history of science and is itself evolving at a phenomenal rate. Darwinian ideas have found their way into a growing array of scientific disciplines from agriculture, psychology, and medicine to cosmology.[55]

Darwin's theory has never stopped being controversial. During the last quarter of the nineteenth century, principles of evolution via natural selection began to be applied to a bewildering array of sociological ideas and used, without any form of valid scientific foundation, to help excuse racism and anti-Semitism. Thus emerged the pseudoscience of social Darwinism. Proto-Nazis tried to give their ideologies a veneer of scientific respectability by calling upon what they claimed Darwin had said, applying the laws of the natural world to the modus operandi of human regimes. Ironically, Darwin and Huxley were vehemently antiracist, and after Darwin's death Huxley fought hard to prevent his friend's ideas being hijacked by political radicals. In his final years Darwin gently refused to allow Karl Marx to dedicate *Das Kapital* to him.

Many people who have studied what Darwin wrote still feel uncomfortable with his theory, and this is quite understandable. But a significant number of people, some of whom have taken the trouble to understand the science and others who have not, simply refuse to accept the concept of evolution via natural selection. Fortunately, only a few of these individuals have any real power in our society, but those who do have struggled to repress Darwinian thought. In some parts of the United States there are continuing efforts to ban the teaching of Darwinian evolution, and in others activists have pushed through laws forcing teachers to give equal importance to the Creationist argument, to present what has been called creation science as an "alternative" to evolution via natural selection.

Darwinian evolution began as a theory to explain a fundamental aspect of the way life behaves. Because humans are part of the natural order, and more specifically because of human self-awareness, Darwin's ideas are predictably controversial. Huxley and Darwin knew the reason for this, so of course did Owen and Wilberforce. All four men were entangled in an emotional web just as they were pinned by the cold rapier of science. Evolution is about you and me, and it offers an agonizing reality. Emotion as well as logic drove the men who fought over Darwin's revelation.

Today the rationalist sees Darwinian ideology as transcendent over our emotional frailty. We no longer have grounds to question Darwinism, except, crucially, to judge its scientific value by the evidence provided by life itself.

# FOUR

## The Battle of the Currents

*Why, Sir, you may soon be able to tax it.*
*—Michael Faraday replying to then*
*chancellor of the exchequer, William Gladstone, when asked*
*what possible use could be made of electricity*

*State of New York, August* 1890

THE ELECTRIC CHAIR stood, dull oak in the bleak gray basement of New York State Prison at Auburn. The prisoner, an ax murderer named William Kremmler, and a total of twenty-six officials, doctors, and witnesses entered the room just before dawn, and a small team of medics prepared the chair before Kremmler settled himself onto its wooden slats. A leather mask was placed over the prisoner's head and straps tightened around his ankles and wrists. The brine-soaked electrodes were applied, one inside the mask pressed onto a shaved area of Kremmler's scalp, the other protruding from the back of the chair and jammed against his spine. The straps were tightened and the wiring checked.

Eighteen months earlier, in March 1889, Kremmler had shattered his girl-friend's skull with twenty-six blows of his ax. Confessing to the crime almost immediately, he had been tried and sentenced to death. Like many drunken, unemployed drifters trapped in a love triangle, Kremmler had slid into vio-lence and committed a grisly murder for which he was given the ultimate sen-tence. But what distinguished him from the thousands before and after him and the reason that he sat in this roughly hewn chair was the way the state was to dispatch him; he was to be the first man executed in the electric chair.

The generator to provide the fifteen hundred volts that would pass through Kremmler's body stood across the courtyard of the prison, and leads ran from it over the roof, down the side of the building, and under the floor of the basement execution room. Here they were fed into the chair. After all was made ready, Kremmler said good-bye to the prison chaplain, and with complete resignation he sat motionless in the chair waiting for his most modern death.

A few moments later, the prison warden, Charles Durston, threw the switch that opened the chair to current. Then, as the first surge of electric-ity pulsed through Kremmler's body, everyone in the room heard a loud crack, and the prisoner jolted in his seat, his entire body rigid, as though all his muscles had spasmed simultaneously. If it had not been for the restrain-ing straps, the prisoner would have been thrown clean across the room. Two of the medics invited to observe and report stepped forward gingerly and leaned in toward the figure in the chair to get a closer look. They could hear Kremmler grinding his teeth and see where the nail of one of his index fingers had cut into his hand.

After the current had passed through Kremmler for seventeen seconds, the two doctors agreed the prisoner must be dead and signaled to the war-den to shut off the power. At first no one dared speak. Then one of the medics poked Kremmler's skin and they watched as the red patch turned white. Now certain the man was dead, one of the doctors, a strong advocate of execution using the electric chair, A. P. Southwick, exhaled deeply and took a step back. Breaking the silence, he declared: "There is the culmination of ten years of work and study. We live today in a higher civilization from this day."[1]

But his colleague, a Dr. Louis Balch, was not so pleased. As Southwick was proudly claiming success, Balch had been watching Kremmler's stiff hand where his nail had drawn blood. The prisoner's blood was still flowing, which could mean only one thing. Kremmler's heart was still beating; the man was still alive.

Suddenly, Kremmler shuddered, and foam poured from his mouth and down his chin. Recoiling, horrified, the doctors heard the prisoner sigh and saw his chest move a little; Kremmler was trying to draw breath.

First to regain his composure was, Dr. Edward Spitzka, a well-known forensic specialist of the day. "Start the current again," he shouted to the warden. "For God's Sake kill him and be done with it."[2] Contacts were realigned and the dynamos across the courtyard were activated once more. In the control room, the switch was thrown.

No one had expected this, and now no one knew quite how long to keep the current pumping through Kremmler. To add to the confusion and the horror, the dynamos were overburdened, and before long their pulleys began to slip; the power surged and then dropped, causing Kremmler to convulse and spasm violently in the chair. The screeching of the pulleys could be heard from across the courtyard, adding to the dull electric hum in the basement and the relentless, agonized grinding of the prisoner's teeth.

One of the witnesses gagged, and a reporter fainted. The man in the chair began to fume, producing a cloying acrid smell that mixed with the sickly sweet scent of burning human flesh. After two minutes smoke became visible from Kremmler's head and his hair caught alight. Black smoke plumed from his ears, and then a flash of blue light appeared at the base of his spine.

Many of the witnesses seemed rooted to the spot, so transfixed by the terrible spectacle they could no longer function logically. The current had now passed through Kremmler for a full two and a half minutes, and the screeching of the pulleys had risen to such an intensity it seemed the entire system would short-circuit. Finally snapping out of his shock, the warden ran back to the control room and yanked the switch, breaking contact once more. Kremmler's charred and smoking body slumped limp against the leather straps. The prisoner was dead, his body stiff from the cooking effect of the current, baked from the inside, as a witness told it, "like overdone beef."[3]

As he left the building, a remarkably calm Dr. Southwick remained enthusiastic, his support for this new form of execution apparently vindicated. "I tell you this is a grand thing, and is destined to become the system of legal death throughout the world," he announced to the waiting newsmen. "This is the grandest success of the age. After the execution I turned to Warden Durston, congratulated him, and said that I was one of the happiest men in the state of New York."[4]

But Kremmler's death in the electric chair was not simply about finding a new method of execution many hoped would be more humane than hanging. Those who had advocated its development and refinement and worked hard for many years to bring about this obscene spectacle had ulterior motives that had nothing to do with the welfare of the condemned and everything to do with money. The execution of William Kremmler was the climax in a vicious dispute between one of the most famous and accomplished inventors of the day, Thomas Edison, and George Westinghouse, a farsighted businessman who championed the scientific brilliance of a man who was in almost all ways the very opposite of Edison, the Croatian genius Nikola Tesla.

This disastrous first use of the electric chair was one of the last rounds in a conflict that had raged for over three years through two of the major research laboratories in the United States, across the pages of the national press, and into the very homes of the richest and most powerful people in the land. This rivalry had been dubbed by newspaper editors "The Battle of the Currents," and the winner would provide the means by which we today power our lives. Their successful method of delivering electricity into our homes, offices, and factories helped mold the future and along the way earned, for a lucky few, unimaginable wealth.

BY THE TIME the conflict between Thomas Edison and Nikola Tesla began, electrical theory had become relatively sophisticated, but application of this theory was extremely limited. However, after the resolution of this priority dispute, electricity became a fundamental aspect of everyday life because out of conflict had come a means to utilize electricity's incredible power.

The ancients observed the effects of static electricity, seeing tiny sparks crackle from wool or fur when it was rubbed with certain materials like amber. The English natural philosopher William Gilbert wrote about this in his book *De magnete*, published in 1600, and coined the word *electric* from the Greek *elektron*. But displays of static electricity and magnetism were perceived as little more than party tricks and largely ignored by men such as Newton or Boyle.

A century and a half after Gilbert, during the middle of the eighteenth century, Benjamin Franklin was one of the first to make significant progress in his experiments with electricity. He gained fame from his observation that a metal-tipped kite conducted electricity in a lightning storm (and was lucky not to have been electrocuted in the process).

It had been supposed for some time before Franklin's researches that there were in fact two kinds of charges: that associated with amber, which had been given the name *resinous*, and the charge present on fur or wool, called *vitreous*. Franklin too subscribed to this idea but renamed the two forms negative and positive electricity. And, although he did not understand many of the fundamental principles of electrical theory, he knew that when materials are both given negative charges, they repel, as do two materials with positive charges, and he went on to demonstrate that if negatively charged and positively charged materials were brought close together, they attracted each other.

Even though these ideas circulated among the learned of the time and were the subject of frequent discussion and experiment, little progress was made in understanding how these electrical qualities arose until late in the eighteenth century, when two Italian experimenters Luigi Galvani and Alessandro Volta, made significant discoveries based upon close observation of electrical phenomena.*

Galvani inserted strips of two different metals into the leg of a dead frog and watched the muscle twitch, a simple experiment that when repeated in the rooms of learned societies across Europe caused a sensation and even fil-

---

*Galvani's name has been immortalized in the use of the word *galvanize*, and, of course, Volta's contribution to the field was commemorated with the electrical unit the *volt*.

tered into the imagination of the educated lay public, inspiring Mary Shelley to write her famous novel *Frankenstein*.* Volta took Galvani's experiment one stage further and found that an electric current was produced when different metals were allowed to stand in solutions such as brine. Long before the voltmeter was invented, Volta decided whether or not a particular combination of metals and solutions produced a current by placing his tongue against the metal strips and judging the relative power of the subsequent shocks.

Volta's simple device was the first battery, or voltaic pile, and it works because of a few simple, natural processes. Some metals keep possession of their electrons more jealously than others. If strips of two different metals are placed in a suitable solution (a conducting medium such as brine), electrons will leave one of the metal strips and be drawn to the other. When the electrons arrive at the second metal strip, they have to go somewhere, so if a wire made from copper or some other conducting material is connected between the strips, electrons will flow back through this wire to the first metal strip. This system may be visualized as an "electron pump," and it allows electrons to pass through a solution and then on through a wire, creating a circuit. If a device is placed in the circuit, it will receive the electrons (or electric current) passing through the wire. This is a simple thing to do: just cut the wire and link the two ends to a lightbulb. The bulb will glow because the electrons flowing through the filament meet resistance from the material in the filament, the filament heats up, this provides energy for the atoms of the gas in the lightbulb, and they produce electromagnetic radiation or light.

Batteries like Volta's represented a great leap forward in the uses of electricity, and over the following century they were gradually improved to become practical sources of energy. But an explanation for how batteries actually worked came only after a sound chemical theory was established during the second half of the nineteenth century. The use of electricity at this time is an example of how a phenomenon may be observed, and in a limited way harnessed, without a full understanding of how it works.

However, as useful as they remain, batteries have quite limited applica-

---

*This experiment showed that electricity could be conducted through the fibers found in animal tissue.

tion. They are of course used in cars (where a device called a secondary cell provides the energy to start the engine) and in small electrical devices. But, to supply the consistently high currents needed to light homes and run large machines, a more powerful source of electricity is needed.

This was known about during the 1870s, when both Edison and Tesla were dreaming of a method of delivering high currents to domestic appliances, to light up the world and turn the wheels of industry. But the theory they used to produce such an apparent miracle had been elucidated some forty years earlier by the English scientist Michael Faraday.

Today considered the father of the modern theory of electricity, Faraday was an individualist who practiced a doctrine called Sandemanianism, which encourages its followers to shun public attention, personal wealth, or institutionalized honors. He came from a poor background and was almost entirely self-taught. Singled out and championed by one of the most important scientists of the English Enlightenment, Humphry Davy, Faraday became the great man's assistant and eventually outstripped him as an influential scientist. Fascinated with the ways science could be applied to improve everyday life, Faraday became obsessed with the interaction between electricity and magnetism, two phenomena that had been conceptually linked for centuries but about which little was understood.

The most important thing Faraday discovered was a property now known as electromagnetic induction, which he investigated in a series of experiments conducted in 1831 at the Royal Institution.* A few years earlier, as Faraday was beginning his experiments, a Frenchman, André-Marie Ampère (after whom the unit of current, the *ampere* or *amp*, was named), showed that an electrical current flowing through a wire produced a magnetic field around the wire. In effect this wire was behaving like a bar magnet—when current flowed through it, it moved the needle of a compass. In

---

*Michael Faraday was the first to publish his work on this property and has therefore become recognized as the discoverer of electromagnetic induction. But an American scientist, Joseph Henry, was actually the first to conduct these experiments (a year before Faraday, in 1830). Because Henry did not publish, Faraday has always been given priority. Although Henry (who coincidentally was professor of mathematics at Auburn Academy in New York State, where the electric chair was first used) challenged this, his contribution has been largely forgotten.

his experiments Faraday discovered that precisely the opposite phenomenon was also true. He moved a magnet toward a wire and found that a current was produced in the wire.

According to the nineteenth-century physicist Sir J. J. Thomson, "Faraday said of electromagnetic induction that it was a babe, and no one could say what it might do when it grew to manhood." But it is possible that even someone as visionary and imaginative as Faraday could not have realized just how prescient this statement would turn out to be. The idea that a magnet can produce a current and a current can produce a magnetic field lies at the heart of almost all practical electronics. This observation and the theory that explains it is the very essence of what both Edison and Tesla applied in practice and what they fought over. Two particular devices sprang from the work of Faraday and Ampère that were then employed and developed by Edison and Tesla. The first is the dynamo, which produces an electric current by spinning a coil of wire in a magnetic field; the other is the electric motor, which translates electrical energy into mechanical energy by making an armature wrapped with metal wire spin in a magnetic field. A dynamo and a motor are basically the same device, but one creates electricity from the phenomenon of electromagnetic induction, the other creates motion from the same natural process.

The rivalry between Thomas Edison and Nikola Tesla centered on both these devices, the dynamo (or generator) and the electric motor. Edison worked almost exclusively with dynamos and generators that used a form of electricity called direct current (DC), and Tesla was an advocate of AC, or alternating current. Direct current is a form of electricity that flows continuously, whereas AC constantly changes direction, backward and forward.*

It had been known since Faraday's time that a DC generator and a DC dynamo were relatively easy to make. The construction of an AC generator was more difficult because the electricity it produced changed direction rapidly, requiring the use of a commutator to harness the current.† Most

---

*The AC we use today changes direction about one hundred times per second.

†A commutator is an arrangement of slip rings that allows the current to flow continuously because different parts of the slip rings carry the AC produced by the generator in its different phases.

important, until Tesla's groundbreaking work, no one was able to build an AC motor because the AC kept changing the direction of the rotating arm or driveshaft of the motor.

It would seem obvious then that DC would provide the best way to deliver electricity to any appliance. Indeed, the first practical dynamos used DC, but this form of delivery has associated problems. The most important of these is matching safety with power.

In order to provide power for lighting and appliances, a large current is needed, and this must be combined with the correct voltage. If these are out of balance, the system does not work well. To understand how electricity works, it is always good to use the analogy of water. Think of electricity traveling along a wire as water flowing through a pipe. To clean a car with a hose pipe, you must have enough water (analogous to the current, the *amount* of electricity) and a powerful enough jet or head of water (in electrical terminology this is analogous to the voltage).

With electricity, the situation is complicated further. First, the amount of current that can be delivered depends upon the thickness of the wire used—the heavier the cable the greater the current it can carry—but electricity "leaks" from wires, and the thicker the wire, the worse the leakage. To use DC to provide electricity to a town, Edison needed to start with a huge current that would gradually seep away and leave a trickle to power the town's appliances. This means he required massive copper cables, and even then there was a limit to how far he could transport electricity. Edison's methods could not deliver enough electricity to power a single home more than half a mile from a generator, and if today DC was used throughout the world, all the copper ever mined would not provide a fraction of that needed to make the delivery cables. This means Edison's entire delivery scheme was based upon the idea that each building (or, at best, each group of buildings) would require its own power station so that the distance the electricity traveled was dramatically reduced; a relatively small current, a safe voltage, and narrower wires could then be used.

Tesla's rival system, AC, did not suffer from the same restrictions. The voltage of an alternating current can be lowered using transformers. This means it is possible to produce a current at a very high voltage that can be carried long distances (the leakage of current at high voltages is much reduced). The voltage can be stepped down between the power station and the domestic site and then stepped down again close to the appliances. This is the method used today. Electricity leaves a power station at up to 400,000 volts; it is stepped down to 11,000 at a substation and then down again to 240 volts for use in the home (110 volts in the United States). This means the voltage people might be exposed to is at a safer level, yet the current is sufficient for our needs. Furthermore, AC can be delivered to an appliance far from a large power station.

To us, the distinction between AC and DC is clear and the superiority of an AC delivery system obvious, but to those working during the 1880s and fighting to ensure the success of their preferred methods, the relative merits and failings were clouded by personal and financial issues and an almost blind need to prove themselves right. The two men at the center of this battle were very different characters from very different backgrounds, and each approached science in a polar opposite way. Only one of them could be right, only one of them could herald the Age of Electricity, and, as in all scientific disputes, for Edison and Tesla there would be no second prizes.

THOMAS EDISON WAS A BIG, bullish man who did not tolerate fools gladly. Opinionated, aggressively patriotic, and obsessed with money and success, he could claim little in the way of a formal education yet rose from midwestern obscurity to become one of the most famous people in American history. In 1913 one newspaper described him as "this big, smiling, white-haired, blue-eyed sixty-six-year-old boy."[5]

His parents, Nancy and Samuel Edison, were hardworking, highly principled, and politically minded people. They had lived in what is today called Ontario in Canada but emigrated to the United States in 1839 after the conservative supporters of British influence in the province had driven out the liberal leaders whom the Edisons supported. Arriving in the United States, they set up home in a little town in Ohio called Milan, where their son, Thomas Alva, was born in February 1847. For a while the family prospered there. Sam Edison owned a mill and later a grain-shipping business and was at his most successful about the time Thomas was born. They lived in large house, and Edison had fond memories of the town, writing in later life: "The town in its pristine youth was a great lumber center, and hummed to the industry of numerous sawmills. An incredible quantity of lumber was made there yearly until the forests nearby vanished and the industry with them. The wealth of the community, invested largely in business and in allied transportation companies was accumulated rapidly, and as freely spent during those days of prosperity in St. Clair County, bringing with it a high degree of domestic comfort."[6]

But by the time Edison was ten the family's fortunes had reversed, and Sam Edison's company went out of business. The reasons for this failure are unclear; the business could have simply been another casualty of exhausted natural resources and may have been further damaged by the Panic of 1857, when many small businesses in the Midwest collapsed.

Even more unclear is the way Edison junior was educated in Milan. Some reports suggest he was thought educationally subnormal by his teachers. Others claim this is simply misinterpretation and that a comment in which Edison was referred to as "addled" had been made off the cuff by a disgruntled headmaster after Edison's school bill had gone unpaid.[7] Thomas Edison himself was proud of his unconventional education and once declared: "School? I've never been to school a day in my life! Do you think I would have amounted to anything if I had?"[8]

This was of course an exaggeration and typical of the man's bombast. But it is true that Edison did not attend school often and was educated principally at home by his mother, a former teacher. From an early age Edison demonstrated an unshakable contempt for academic formality. He never trusted those he knew to be exceptionally academic and was strongly of the opinion that "doing" was infinitely superior to "thinking." "Doing the thing itself is what counts," he once told a reporter.[9]

Nevertheless, Edison's mother introduced him to a diverse world of knowledge, and he appreciated this his entire life. In a speech to an audience of schoolchildren in 1912, he declared: "My mother taught me how to read good books quickly and correctly, and as this opened up a great world of literature, I have always been very thankful for this early training."[10]

As a young boy he showed a leaning toward science and engineering and devoured the science books he was given, claiming that he learned physics from Richard Parker's *School Compendium of Natural and Experimental Philosophy*: the book used by science teachers at the local Union School. Such an interest later prompted his father to say of him: "Thomas Alva never had any boyhood days; his early amusements were steam engines and mechanical forces."[11]

His wholehearted commitment to an autodidactic course dominated especially by science may also have led Edison to foster strong religious views that were extraordinarily insightful for a midwestern schoolboy of the late nineteenth century. About God and orthodox religion he was reported to say: "Whatever truth Christianity did contain had existed in books of nature long before the coming of Christ. Therefore the morals of Christianity are good because of their harmony with nature."[12] And he later commented in a letter to a friend: "The old order of things are rapidly passing. The schoolhouse, the newspaper, and the advancement of scientific investigation, will in not many years make these beliefs [religious orthodoxy] seem ridiculous."[13]

Having rejected both religion and an orthodox education, Edison became remarkably self-reliant and placed his faith in two things: innovation and money. At the age of fifteen and with just a few dollars he started his first business, a vegetable stall in the nearest large town to Milan, Port Huron.

The Edison of this time was described by a friend as having "a lively disposition always looking on the bright side of things, full of a most sanguine speculation as to any project he takes in his head."[14] A surviving photograph of him at the age of fourteen shows a smiling, confident, ruddy-faced boy who looks like he had to be forced to sit still for the picture to be taken before he sped off to strike a new deal or forge another important business relationship.

During the next few years this energy and tireless enthusiasm led Edison from one small business venture to another. He made a little money, but he did not find his vocation until, at the age of sixteen, he discovered the existence of a new technology, telegraphy, a form of electronic communication that was exciting and seemed to offer vast commercial potential.

Many of the elements for the creation of a telegraphic system had been in place in Faraday's time, but the first telegraph was not produced until 1858, when a line was laid between Euston Square and Camden Town in London. But, although the innovation took time to appear, it expanded at a phenomenal pace. Just a few years later, by the time of the American Civil War, telegraph lines were being laid on the floor of the Atlantic Ocean, and across the continent of America a rapidly growing network was speeding up communication and fueling many new commercial enterprises.

Telegraph operators were an adventurous band of young men. Spurred on by the fluid economics of the time, a period during which an entire state could be booming while a neighbor faced depression, telegraph operators traveled to where work could be found. Telegraphy was at the cutting edge of technology, akin perhaps to the Internet of today. Operators were poorly paid, and tramping telegraphers, as the freelancers were known, commanded little respect, but Edison was probably drawn to the work because of the independence and opportunities it could offer him as well as the excitement of working with such new, beguiling technology.

In 1864 Edison settled in Indianapolis and was a senior operator for Western Union. By now he had become fascinated with the entire science of telegraphy and had reached the conclusion that there was much about the system he could improve. He had no form of backing or even a laboratory, but in his spare time he began inventing and testing new devices and systems, all funded by his earnings from Western Union.

His earliest patents, the first of a record 1,069 during his lifetime (and a further 24 awarded posthumously), were for improvements to the relays and other internal workings of conventional telegraph equipment. In 1866, before his twentieth birthday, he invented an automatic vote recorder. This consisted of a paper disk that rotated on a spindle. The dots and dashes of a message in Morse code were embossed on the disk, and these lifted a lever up and down on a transmitter. In this way an electrical impulse was sent to a receiver, which then reproduced the indentations on another disk.

This was a significant improvement upon the old method of communicating the count of votes, and it quickly attracted the interest of Edison's superiors at Western Union. With their encouragement he went on to adapt this device to produce a paper tape ticker, which found extensive use in the communication of stock prices around the country. Patented in 1869, this invention earned Edison thirty thousand dollars.

And with this success came further sponsorship. His first supporter was the president of Western Union, William Orton. Under Orton's guidance Edison was quickly relieved of all his routine operator duties and offered the facilities of the research and development division of Western Union. Here he was encouraged to apply his talent to any improvements to the telegraph system that might occur to him. Within two years Edison had become the most prolific and respected developmental engineer in the company.

Obsessed with his twin gods—the thrill of invention and the desire to make money—Edison seems to have cared for little else. He dressed shabbily, his hair was a constant mess, and he ate and slept without discernible pattern. Showing little interest in anything unconnected with his ambitions, he was driven to succeed, to make money, which he then used to help realize his inventions. Contemporaries remember him as a man who constantly needed to tinker, improve, and reevaluate every aspect of each machine he saw or touched.

His obsessiveness is clear in a newspaper account that appeared when Edison was at the peak of his fame in which the journalist described how the inventor met his first wife, Mary Stilwell:

> One day when standing behind the chair of one of his female employees, Miss Mary Stilwell, that young lady turned round and exclaimed, "Mr. Edison, I can always tell when you are behind or near me." "How do you account for that?" mechanically asked Mr. Edison, still absorbed in his work. "I don't know, I am sure," she answered, "but I seem to feel when you are near me." "Miss Stilwell," said Edison, turning round now in his turn and looking the lady in the face, "I've been thinking considerably of you of late, and if you are willing to have me I'd like to marry you." "You astonish me, . . ." exclaimed Miss Stilwell. "I know you never thought I would be your wooer," interrupted Mr. Edison, "but think over my proposal, Miss Stilwell, and talk it over with your mother." Then he added in the same off-hand, business-like way, as though he might be experimenting upon a new method of courtship, "Let me know as early as possible, as if you consent to marry me, and your mother is willing we can be married by next Tuesday."[15]

Perhaps also less interested in romance than in success, Mary's mother did agree, and the couple were married within weeks, on Christmas Day 1871.

Edison had access to the best laboratories Western Union could provide, but his ideas were soon extending far beyond the confines of the technology of the telegraph. He later used the experience he had gained at

Western Union as a powerful springboard for some of his most famous inventions but realized soon after leaving his job as a telegrapher that the only way he could truly fulfill his dreams was to establish his own laboratory.

The problem for Edison, and a difficulty he faced for almost all his career, was lack of money. No matter how much he earned from his inventions, he could always spend more on developing new ideas. He was a very modern inventor in the sense that he realized the need to secure financing from the business community to fund his researches, but he almost always found that he was still underfunded and had to spend time and energy in attempts to convince businesspeople of the value of his ideas. He frequently needed to supplement the cost of his experiments and the development of his designs from his own savings, and because of this he always suffered the stress of risking his own solvency to deliver what was required of him.

During his time at Western Union Edison had earned an enviable reputation as a designer and engineer, and he could ask his own price for many of his inventions as long as they served the company. In 1876, Alexander Graham Bell exhibited his now famous design for a telephone at the Philadelphia Centennial Exposition. Western Union executives immediately saw the huge potential of the device and wanted to infiltrate the market. They turned to Edison, who devoted himself to finding ways to improve Bell's design and patent some original aspect of the telephone so that Western Union could legally access the market.

His contribution was the carbon granule microphone used in the mouthpiece of the telephone. It represented a vast improvement on Bell's original method and provided Western Union with the financial rewards and Edison the credit for helping in the early development of one of the most important inventions of recent times.

Although Edison's work on the carbon granule microphone was received with delight by Western Union, it quite naturally infuriated Bell and his backers. But there was nothing they could do about it. Edison and Western Union had acted within the law because the device differed enough from anything Bell had produced to escape patent infringement. Encouraged by this, by the end of 1876 Edison had moved on to much bigger and better things when he secured financial backing to leave Western Union and establish his own fully equipped laboratory.

Menlo Park, close to Newark, New Jersey, was then a semirural district that appealed to Edison's desire for space, clean air, and accessibility to the big cities. The laboratory itself was an ordinary-looking, white, two-story affair close to the local railway station. One of the many journalists who made a pilgrimage there described it as "looking a little like a country school-house pulled out three times its length."[16] Yet its exterior appearance was a perfect chimera, for inside it was certainly no country schoolhouse. In 1878 a writer for one of the science magazines of the time, *Popular Science Monthly*, enthused to his readers: "On the ground floor, as you enter, is a little front-office, from which a small library is partitioned off. Next is a large square room with glass cases filled with models of his inventions. In the rear of this is the machine shop, completely equipped, and run with a ten horse-power engine. The upper storey occupies the length and breadth of the building, 100 × 25 feet, is lighted by windows on every side, and is occupied as a laboratory. The walls are covered with shelves full of bottles containing all sorts of chemicals. Scattered through the rooms are tables covered with electrical instruments . . . microscopes, spectroscopes, etc. In the center of the room is a rack full of galvanic batteries."[17]

It was here, in his own laboratory, said in its day to have been the best-equipped in America, that Edison fulfilled his potential. In the space of just seven years, from 1876 to 1883, he was to create some of his most famous and influential inventions.

First came what is perhaps Edison's best-known creation, the phonograph, the machine later called the gramophone and today the stereo or sound system. Edison first visualized the phonograph when he happened to hear a crude musical note produced by the spinning disk in one of his Morse code transmitters. This made him think it should be possible to store musical notes on a disk and play them back with a "receiver" contained within a single machine.

Edison's phonograph made an immediate media impact and quickly captured the public's imagination. This was the first time music could be "trapped" or stored on a record (originally a paper disk) that could be played back at leisure. Until then the only form of music had been live performance. Edison's earliest version of the phonograph evolved rapidly into a system using tinfoil, then wax, then cylinders, and finally, by the 1920s, Bakelite disks.

Fueled by public interest, millions of these disks and the machines they were played on were sold during Edison's lifetime, and the device made him a household name. But Edison's restlessness and energy meant he never stopped thinking about new machines and constant improvements to his inventions. Within a year of producing the first phonograph, his interests had shifted to the problem of supplying public lighting and lighting for the home in a more efficient and safer way than using gas.

Edison had had firsthand experience of a device that might make domestic lighting a reality. During a visit to the laboratory of the inventors William Wallace and Moses Farmer in July 1877, he had seen their latest creation, a lighting device they called an arc-lamp. The two men had also designed and built an eight-horsepower dynamo capable of lighting eight 500-candle lamps. But Wallace and Farmer's bulbs burned for only a short time before the filaments needed replacing, and more important, they required so much current to light a single house that using a DC system huge cables would have been needed to supply the electricity. Having seen these primitive lightbulbs in action, Edison was quite convinced electricity could replace gas as a domestic and general lighting system, and he was confident he could improve Wallace and Farmer's system radically.

Edison once said: "Everybody steals in commerce and industry. I've stolen a lot myself. But I know how to steal."[18] In producing his first lightbulb Edison clearly stole the idea from others, especially Wallace and Farmer, but he possessed a masterly talent for taking embryonic ideas and refining them into inventions that successfully met a need.

His greatest obstacle was that to bring about this revolution an entire set of new technologies would be needed, technologies that up to then had barely been contemplated. But at the end of the 1870s, if anyone could muster the forces needed to replace gas with electricity, it was Edison. By employing a method that has now become commonplace, he set about acquiring financial support from big business in order to fund the necessary research to develop his scheme. With the help of several important bankers, most prominently the magnate J. Pierpont Morgan, late in 1878 Edison was able to establish the Edison Electric Light Company.

Edison had based his pitch to his backers upon the idea that it would take him no longer than a year to perfect a general-purpose lighting system along with the generators to power it. He claimed this could then be used in houses as well as public places, thus offering a potentially huge return on a relatively small initial investment. However, even the worldly-wise Edison may not have realized just what those returns would be or that within three years of his proposal to produce the first practical lighting system it would be returning revenue of $150 million per annum.

But it seems Edison's motivation for creating a system of lighting better than that provided by gas was only in part fueled by a desire to make money. An impulse at least as powerful was the need to prove himself right and others wrong. It was almost as though he found excitement and fulfillment in generating conflicts with seemingly more powerful forces. According to one historian, Edison's efforts gave him "an opportunity to define himself through conflict."[19] His first battle had been a skirmish with Bell. Next, he pitched himself against the powerful suppliers of the gas lighting equipment and the massive corporations who provided gas. Edison thrived on the very idea of taking on these colossi and beating them. When a journalist on the *New York Tribune* commented: "If you can replace gas lights, you can easily make a fortune," Edison replied, "I don't care so much about making my fortune as I do getting ahead of the other fellows."[20] One of Edison's many biographers has remarked: "To stand as a leader among the world's foremost inventors, to make again and again a great impact on society and industry—even to 'change the world' if possible—meant everything."[21]

Whatever his motives, Edison worked harder on this new project than anything he had attempted before, and it was a mammoth undertaking. The comment for which Edison is today most widely remembered is his claim that "genius is 1 percent inspiration and 99 percent perspiration." This was genuinely a philosophy that guided him through all his efforts and never more so than in his development of the first practical lightbulb.

First he had to devise a better filament, one that would burn consistently for long periods. Second, he had to construct practical, cheap glass bulbs that would hold pressurized gases. Third, he needed an appropriate gas or blend of gases to use inside the bulb and a way to contain these at a precise pressure.

Edison worked slavishly, but his methods were far from subtle. Acting as a supervisor, or manager, he established teams of workers at Menlo Park to investigate every aspect of creating a practical lightbulb, and each group reported its findings back to him.

One team worked on the dynamo system that would, he hoped, carry electricity to the appliances. Another concentrated on finding precisely the correct blend of gases for the bulb, while Edison himself studied the literature dealing with metal alloys and their various properties, working his way through huge piles of scientific journals late into the night to find the perfect filament for his bulb. He knew the carbon arc was one of the weaker aspects of Wallace and Farmer's system, so at first he focused on resolving this drawback. Initially he used a platinum wire filament, but this produced a light for little more than an hour and he was forced to search for a more efficient material and a precisely balanced atmosphere inside the bulb in which the metal could burn and produce a bright glow.

By December 1879, fifteen months after embarking upon the effort, Edison had what he believed were solutions to all the problems. He and his teams had tested no fewer than six thousand substances for the filament, including boron, iridium, and molybdenum. Finally, he settled on a carbonized bamboo fiber that produced a suitable iridescence for up to a thousand hours. He had personally designed the bulb to be used and employed a famous German glassblower named Ludwig Boehm to produce a small first batch. Next, he and his co-workers devised ingenious systems to evacuate the air from these bulbs and to replace it with a combination of gases that would produce the best glow with the newly tested filament.

During this process Edison also stumbled upon what later became known as the Edison effect. This describes how electricity flows from a heated filament (such as the one he had in his bulbs) to an electrode, but not in the reverse direction. This discovery led to the first thermionic valve, the predecessor of the transistor and a key component in the production of almost all electrical devices up to the early 1960s. The entire research effort to produce Edison's lightbulb cost forty-two thousand dollars, but this was returned to the financial backers almost immediately, and within just a few years it was multiplied thousands of times.*

Edison's first demonstration of a long-lasting lightbulb elevated his fame still further. He was soon dubbed "the Wizard of Menlo Park," and the popular press made him into a tangible hero of the machine age, a veritable "Hercules of Science." As news of Edison's invention spread, thousands of people traveled to his laboratory to witness the spectacle of electric light and to meet the man who had made it possible. These admirers were behaving almost as pilgrims visiting a religious shrine. A journalist of the day wrote a sequel to H. G. Wells's *War of the Worlds* titled *Edison's Conquest of Mars*,[22] and the usually rational *New York Herald* ran a piece in which the writer proclaimed: "Invisible agencies are at his [Edison's] beck and call. He dwells in a cave and around it are skulls and skeletons, and strange phials filled with mystic fluids whereof he gives the inquirer to drink. He has a furnace and a cauldron and above him as he sits swings a quaint old silver lamp that lights up his long white hair and beard, the deep lined inscrutable face of the wizard, but shines strongest on the pages of the huge volume written in cabalistic characters. . . . The furnace glows and small eerie spirits dance among the flames."[23]

However, as impressive as it undoubtedly was, producing the first practical electric lightbulb was just one step in the process of creating domestic lighting. Edison was a wise enough businessman to know that to capture the imagination of the public with such revolutionary technology, he would have to supply a complete package, a delivery system, wiring, even the bulb holders and adapters. Furthermore, all this had to be coupled with a workable system to deliver to a building electricity that powered the lights without relying upon the use of a huge current. His idea was to provide every building with its own DC generator, each one autonomous. But this was the weak link in his scheme; it was clumsy, expensive, and inefficient. He knew this, but had no better solution.

To find an alternative to this crucial aspect of the entire system required an intellect capable of taking intuitive leaps and finding lateral connections,

---

*Indeed, it is almost impossible to calculate how much money has been generated by Edison's invention or what will continue to be earned from it.

something Edison had no aptitude for. But that type of mind was possessed by Nikola Tesla.

$O$NLY FOUR PORTRAIT PHOTOGRAPHS of Nikola Tesla survive, and these individually and collectively illustrate the arc of his life and career. The first shows a young man recently graduated from the University of Prague, clean-shaven and lean. He wears a floppy bow tie and presents an expression filled with promise and optimism.

The second, taken soon after Tesla arrived in the United States in 1884, presents the image of a confident, fashionable man who now parts his hair in the middle and sports a rather rakish mustache. He smiles faintly at the camera, with the visage of a man who believes he is about to conquer the world.

The third comes from a period some fifteen years later, when Tesla had undergone a series of deflating failures and glorious successes. Here is a man who has fulfilled many of his scientific ambitions and stunned the world with his genius. By now he has earned and spent a great deal of money and is only just emerging from the bitter rivalry with our other protagonist, Thomas Edison.

The final, rather sad portrait shows a sixty-year-old Tesla, still almost three decades from his death in 1943, but bitter and bowed, a man whose odd personality and untamable genius have conspired to ruin his ambitions and dreams.

Tesla was born in 1856 in the tiny village of Smiljan in Croatia. His father had been a military man who later turned to the priesthood. Nikola was the fourth of five children; he had three sisters and a brother, Dane, who was seven years his senior. His early childhood appears to have been happy, peaceful, almost idyllic.

In his autobiography *My Inventions*, Tesla claims his first experience of electricity was when he was three years old and he drew off a static charge from the family cat. This he claimed captivated him and inspired his later boyhood thoughts about what electricity could be and how it might be used. But by far the most important event in his childhood and one that shaped many aspects of his personality was the loss of his brother when Nikola was five.

Dane was his parents' favorite and seen as the "genius" of the family. In him they had invested great hopes, and they doted on him. When Dane died from a fractured skull after being thrown by a horse, Nikola resolved to fill the void his brother had left. He was the only surviving son; he knew he had always played second fiddle to Dane, but he believed that he could win the love of his parents, that he could somehow replace his dead brother.

Writing of the impact of his brother's death, Tesla recalled: "I witnessed the tragic scene and although fifty-six years have elapsed since, my visual impression of it has lost none of its force. The recollection of his attainments made every effort of mine seem dull in comparison. Anything I did that was creditable merely caused my parents to feel their loss more keenly. So, I grew up with little confidence in myself."[24]

Elsewhere in his memoirs Tesla recounted how his father in particular could appreciate nothing he did and seemed to care little for his surviving son:

> I had made up my mind to give my parents a surprise, and during the whole first year [at school] I regularly started my work at three o'clock in the morning and continued until eleven at night, no Sundays or holidays excepted. As most of my fellow students took things easily, naturally enough I eclipsed all records. In the course of that year I passed through nine exams and the professors thought I deserved more than the highest qualifications. Armed with their flattering certificates, I went home for a short rest, expecting a triumph, and was mortified when my father made light of these hard-won honors. That almost killed my ambition; but later, after he had died, I was pained to find a package of letters which the professors had written him to the effect that unless he took me away from the Institution I would be killed through overwork.[25]

Tesla's indefatigable drive led him to great academic success; he attended the Real Gymnasium in Karlstadt, Croatia, then graduated with honors from Graz Polytechnic before winning a scholarship to the University of Prague. But such success would have been of little use if he had not also been a naturally gifted scientist with a voracious and sometimes wild imagination. A glimpse of this may be seen in his own description of an important episode in his childhood.

> During my boyhood, I had suffered from a peculiar affliction due to the appearance of images, which were often accompanied by strong flashes of light. When a word was spoken, the image of the object designated would present itself so vividly to my vision that I could not tell whether it was real or not.... Even though I reached out and passed a hand through it, the image would remain fixed in space.
>
> In trying to free myself from these tormenting appearances, I tried to concentrate my thoughts on some peaceful, quieting scene I had witnessed. This would give me momentary relief; but when I had done it two or three times the remedy would begin to lose its force. Then I began to take mental excursions beyond the small world of my actual knowledge. Day and night, in imagination, I went on journeys—saw new places, cities, countries and all the time I tried hard to make these imaginary things very sharp and clear in my mind. I imagined myself living in countries I had never seen, and I made imaginary friends, who were very dear to me and really seemed alive.
>
> This I did constantly until I was seventeen, when my thoughts turned seriously to invention. Then, to my delight, I found I could visualize with the greatest facility. I needed no models, drawings, or experiments. I could picture them all in my mind.[26]

Whether or not this and the story of his fascination with electricity at the age of three are apocryphal, Tesla did acquire a very early interest in the possibilities of electricity and invention, and as an adult he demonstrated a remarkable ability to visualize problems. In parallel with his academic advancements, his imagination led him into strange and inspired territory.

At the Real Gymnasium he had the good fortune of being taught by an inspired and inspiring master, a Professor Poeschl, who quickly appreciated the boy's ability. One day Poeschl brought to a lesson a new dynamo sent from Paris, but as he was attempting to demonstrate the machine, the brushes started sparking dangerously so the experiment was ruined. Tesla, who was something of a favorite with Poeschl, proposed that a dynamo must be built that had no brushes, but the professor pointed out that without the brushes the dynamo could not function. Undeterred, Tesla then claimed it must be possible to design a dynamo that operated using AC. According to Tesla, the professor merely responded with the comment "Mr.

Tesla may accomplish great things, but he certainly never will do this. It would be equivalent to converting a steadily pulling force, like that of gravity, into a rotary effort. It is a perpetual motion scheme, an impossible idea."[27]

But although he had no clear idea what he could do to resolve the problem, Tesla was not put off. "When I undertook the task it was not with a resolve such as men often make. With me it was a sacred vow, a question of life and death. I knew that I would perish if I failed. . . . A thousand secrets of nature which I might have stumbled upon accidentally I would have given for that one which I wrested from her against all odds and at the peril of my existence."[28] Yet, as he later admitted: "All my remaining term in Gratz was passed in intense but fruitless efforts of this kind, and I almost came to the conclusion that the problem was insoluble."[29]

But then, in the way of all true romantics, a few years later, inspiration, that crucial 1 percent of genius, came to him during a walk in the park with a friend. The solution "came like a flash of lightning and in an instant the truth was revealed. I drew with a stick in the sand the diagrams and my companion understood them perfectly. The images I saw were wonderfully sharp and clear and had the solidity of metal and stone."[30] Within a decade Tesla had built the world's first AC induction motor, the machine his professor had deemed impossible.

Nikola Tesla was a highly cultured man. By the time he graduated from the University of Prague, he had mastered twelve languages, was widely read, and demonstrated an enviable versatility, a man equally comfortable discussing Chaucer, managing complex mathematical equations, or handling the practical problems presented by electrical engineering. But he was also a strangely misanthropic character, very different from his great scientific rival, Thomas Edison.

Tesla was uncomfortable with many of the people he met and often created the impression of being arrogant and superior when in fact he was merely unable to properly express himself. There is no record of him enjoying any form of close personal relationship. When he became a successful scientist and was for a while a wealthy and celebrated New Yorker, he grew friendly with some of the luminaries of the time, in particular Mark Twain, but he never became emotionally involved with anyone and remained a bachelor. He had grown up in a large family, but clearly the loss of his brother and the emotional detachment of his parents (in particular, his father) affected the development of his personality enormously. For all his intellectual power, his versatility as a thinker and a man with an extraordinary imagination, his character was flawed, a fact that was to play a painfully influential role in his life and to overshadow his scientific brilliance.

Tesla's first job was in Budapest, working for a friend of his recently deceased father who was installing the first telephone system in the city. This job lasted less than a year, and by 1882 he found himself in Paris working as a troubleshooter for the European wing of Edison's young company, the Continental Edison Company.

Tesla adored Paris. He wrote of it later: "I never can forget the deep impression that magic city produced on my mind. For several days after I arrived I roamed through the streets in utter bewilderment of the new spectacle."[31] He worked diligently, scraping by on a very modest income, prompting him to comment to a family friend: "When Mr. Puskas asked me how I was getting along in the new sphere, I described the situation accurately in the statement that: 'the last twenty-nine days of the month are the toughest!'"[32] Yet all the while he was thinking of ways to build his AC motor. He had no resources, no laboratory, no support and had merely to hope that one day someone could set him up with the equipment and financing he needed.

And to reduce his chances further, Tesla had none of the entrepreneurial skills of Edison. He imagined how a device should work, thought it through with amazing intellectual dexterity, and only put his thoughts down on paper after the concept was fully formulated. He was quite aware he would have to build models for his designs and then a fully functioning prototype, but he had no idea how to find backing.

Edison was very different. He came up with an idea, developed it to a far less detailed level than Tesla would have done, sought the financial support he needed, then employed people to help him to realize the concept. When the project reached fruition he took a percentage of the profit it accrued and reinvested this money into other schemes and projects.

In the very different ways these two men worked we may see a reflection of the worlds in which they lived. Edison epitomized the hardworking, no-nonsense, practical, go-getting, commercially minded young America, thrusting forward to forge new, bold dreams. Tesla was far gentler, quiet, unassuming, cerebral. Buffeted by the new commercialism of the 1880s, he was as fragile and vulnerable as a child. And to make matters worse, Tesla was absurdly slow to learn the lessons offered by the world of the late nineteenth century. He was very intelligent and did not hide his brilliance, but he did not know how to properly take advantage of the attention he attracted.

But if a man such as Tesla appears on the scene at the right moment, even his extraordinary lack of commercial instinct cannot keep his talents obscured for long. The first person to notice Tesla and to make a huge impact upon his career was an Englishman named Charles Batchellor, a confidant of Edison and head of his European company. Batchellor was quick to recognize Tesla as a gifted electrical engineer whose ideas were attracting considerable attention in Edison's Paris office. He also realized Tesla might be just the man needed at the moment to help him resolve a problem that was causing the company severe embarrassment.

In 1883 the Continental Edison Company had completed one of its biggest projects, the construction of a power station and lighting for the Strasbourg railway station. The grand opening for the new system was attended by no less a figure than Emperor William I, who, with great ceremony, threw the switch to light up the station. Unfortunately, the workmen who had installed the system had wired it incorrectly, and the entire network short-circuited, causing an explosion that seriously damaged the station and nearly killed the emperor.

This was a particularly unfortunate accident because Edison's company was based in Paris and the station in Prussia; the Franco-Prussian War had ended only a few years earlier, and relations between the two states were still fragile. The reputation of Edison's European company hung in the balance. The Prussian authorities demanded an inquiry and only allowed the company access and the chance to put right the problems on the condition their staff were closely supervised by Prussian officials.

Tesla would have heard all about the fiasco in Strasbourg, and soon after it happened Batchellor approached him with an offer to travel there as the company troubleshooter. Batchellor knew Tesla was a brilliant engineer and also realized he was extremely naïve. Knowing that Tesla was seeking funds with which to establish his own laboratory, Batchellor proposed that a bonus of twenty-five thousand dollars might be paid if the situation in Strasbourg could be salvaged and the system there repaired.

So, early in 1883 Tesla dutifully traveled to Strasbourg to begin work on the project. He quickly assessed what was needed and thought the job would take him a few weeks, after which he could return to Paris to claim his reward. But unfortunately, concerned authorities in Strasbourg insisted they check and recheck everything Tesla did, and it took a year of frustrating diplomacy and compromise to complete the job.

Returning triumphant to his old job in Paris, Tesla immediately set about claiming his bonus. But of course he was to be disappointed. First he approached Batchellor. Batchellor passed him on to an underling, who referred Tesla to another manager. And so it continued until Tesla ended up back in Bachellor's office. Then Tesla finally realized he had made a mistake in not having any agreement in writing and accepted that he had worked in Strasbourg for nothing.

Surprisingly perhaps, Tesla did not resign. He probably believed that by staying with the Continental Edison Company he would remain at the center of activity in his field and that this might at some point give him the opportunities he needed to find a backer and set up on his own. But before any such opportunity arose, the boss intervened again. Having heard of Tesla's idea for an AC generator, Batchellor concluded this Croatian, with his wild ideas and indefinable genius for electronic engineering, might just be on to something important, something that could represent a threat to Edison's plans. If Tesla's motor could actually be made to work, it could challenge Edison's DC system, and the entire company would be placed in a vulnerable position. Batchellor decided the best way to deal with this potential problem was for Tesla to meet Edison, for him to explain his ideas, and for the two men to work together; in that way Tesla's maverick genius could be contained and exploited.

In June 1884 Nikola Tesla found himself a penniless passenger on the ocean liner *Saturnia* bound for New York. Before leaving Calais he had been robbed of all his luggage, money, and tickets and had only managed to take his cabin aboard ship thanks to the gift of a photographic memory—at the gangplank he had recited his ticket number. His cabin and his meals had been paid for in advance, but he had to make the Atlantic crossing in the clothes he stood in and with just four cents in his pocket.

In 1884 New York was a young, vibrant city. Vanderbilt and Morgan had built grandiose mansions in the city, but Manhattan was also home to some of the worst criminals in America and gangs who ran a network of rancid brothels and gambling dens. Tesla, fey and self-conscious, arrived at the rat-infested dock smelling and dirty and wondering where his first meal in America would come from. The only scrap of security he could cling to was folded in his pocket, a letter of introduction to the famous Thomas Edison written by Tesla's former boss, Charles Batchellor.

Edison had recently moved back to New York from Menlo Park and was living in a large, fashionable house near Broadway, Number 65, Fifth Avenue. Tesla and Edison met in a cluttered office at the back of the house. Edison had heard a great deal about Tesla from Batchellor and received him warmly. He took the letter of introduction and read Batchellor's glowing recommendation: "I know two great men and you are one of them, the other is this young man."

Tesla and Edison made an odd pair, and it is easy to see how their association did not last long. Edison was a rough-and-ready anti-intellectual, and Tesla prided himself on being a bookish, multilingual, intuitive thinker.[33] Even their outward images were diametrically opposite. Edison was stout and dressed messily, his hair disheveled, food stains on his waistcoat; Tesla was tall and thin, his manner reserved. Edison would almost certainly have perceived him as an arrogant dandy.

And aside from these contrasts, they conducted the business of invention in entirely different ways. Edison worked ferociously hard, but because he had little mathematical ability and failed to grasp much of theoretical science, he was often forced to do things by trial and error. This meant he had to spend far longer on a problem than was strictly necessary. By contrast, Tesla was a formidable theoretician who was also masterfully practical and able to turn his visions into nuts and bolts through an intuitive and honed grasp of fundamental principles.

Tesla once said of Edison that "if he had a needle to find in a haystack he would proceed at once with the diligence of the bee to examine straw after straw until he found the object of his search. I was a sorry witness of such doings, knowing that a little theory and calculation would have saved him 90 percent of his labor."[34] Although this came from the mouth of a wronged man, it was not far from the truth.

But Edison knew Tesla would be important to him. Although he had enjoyed huge success with his phonograph and adaptations of Bell's telephone and was forging ahead with his development of electric light, his plans for electrifying the world were built on shaky foundations, and he could no longer ignore the obstacles he was faced with. It is not surprising then that Edison immediately offered Tesla a job.

Tesla was in the mood to take on anything, and Edison used this to his advantage. His company had struck another problem with an installation. The generators for the electric lighting aboard a new ocean liner, the SS *Oregon*, had failed, casting the entire ship into darkness. The staff of the Edison company who had surveyed the damaged generators claimed they could only be repaired by transporting them back to the factory; but they were too large to be taken out of the ship, which had been built around them.

The day after meeting Edison, Tesla visited the ship at the New York docks and appraised the situation. Identifying the problem as damaged wiring in the main coils, he worked for twenty hours straight and had the lights operating by the following day. Edison was obviously delighted.

And for some time Tesla felt comfortable being employed by Edison. He was doing work he enjoyed, and he felt he had the genuine respect of those around him as well as the approval of Edison himself. As he settled into this new life, he still thought about the AC generator but knew it would take time for him to find the financial support he needed. He may have considered approaching Edison with the idea, but there is no record of this. But then a series of events destroyed forever any possibility of Tesla and Edison working together and led inexorably to the conflict central to the lives of both men.

Edison still had no alternative to a system in which each electrified house required a generator. Coupled with this he had found that, individually, his DC generators were not powerful enough to provide electricity to anything larger than a house. For any greater demands he had to use a collection of generators, each powering a part of the lighting system. The problem with this was that the generators had to be linked so they could work in parallel. But this was also fraught with difficulties. The worst problem was how to synchronize the generators. Because Edison had only a nebulous grasp of theory, he did not understand how generators could be linked so that the pulses of current from each generator were synchronized. In his large installations the pulses were sometimes out of phase, lights would come on and go off at random, and short circuits were almost comically frequent.

Batchellor of course knew Tesla had the theoretical knowledge to find a solution. Indeed, the Englishman had seen a design Tesla had produced for an automatic dynamo regulator—precisely the device needed to synchronize two or more generators or dynamos. Batchellor talked with Edison, and they orchestrated a plan to further exploit Tesla's skills without any unnecessary commitment on their part.

Edison explained his synchronization problem to Tesla in the Fifth Avenue office and told him that a reward of fifty thousand dollars would be offered if he could produce a practical solution. Again, excited by the challenge and too naïve or impatient to insist upon a written agreement, Tesla threw himself into the task of adapting his design for a regulator as a means to link the enormous generators Edison was now using to power four-hundred bulb theaters and offices.

Working long hours and putting aside all other tasks, Tesla committed himself to the project for a year, during which he produced no fewer than twenty-four designs for a vast range of applications, along with a totally original control system. He tested them all thoroughly and demonstrated them to Edison. Delighted, Edison applied for patents on the devices and immediately put them into production.

Having fulfilled his side of the bargain, as the applications for patents were being analyzed and the workshops geared up, Tesla felt it was time for his reward, and he made an appointment to discuss the matter with Edison. The money had been promised; now, Tesla reasoned, all he had to do was to sign a few documents and have the funds transferred. At last he would be able to set up his own laboratory and develop the machines that had been filling his mind for so long. But of course Tesla had been duped again. When he made his formal request, Edison simply smiled and said: "Tesla, you don't understand our American humor."[35]

Tesla was stunned and furious. He had now been conned twice by Edison and his company. He turned on his heel and walked out, never to exchange a civil word with Edison again. The Battle of the Currents had begun.

A S WE HAVE SEEN, by the early 1880s Thomas Edison had become an iconic figure. The press adored him, and in the public perception he was a man who could do no wrong, the greatest genius of the modern age. But at least one man now thought otherwise.

Justifiably upset by the way he had been treated by Edison, Tesla was in no mood when he walked out of the Edison Electric Light Company to begin working again as an electronic engineer, even though he had a respected position in his field and was eminently employable. Instead, even though he was almost penniless, living in a cheap apartment, and still dreaming of finding the financial support he needed, he chose to work as a laborer for two dollars a day.

It is difficult to imagine Tesla working on a construction site. He was certainly fit but wiry; he was still young, but his energies came from his mind not his muscles. But aside from the physical demands of this new job, emotionally he was quite ill suited for it. Fastidious to an extreme, Tesla hated dirt and grime and took an almost obsessive interest in his personal hygiene, washing his hands repeatedly and keeping his room spotlessly clean. But imagine him we must—Tesla, the misanthropic foreigner slaving on a frozen New York construction site, unable to communicate properly with his fellow workers, who probably thought him a lunatic ranting about Edison and his own ideas for crazy inventions. But Tesla was resilient. He was later seen by many as an abrasive character who failed to make the most of opportunities because of his awkwardness, but this trait had also insulated and protected him since boyhood.

Amazingly, he dug ditches and carried bricks for almost a year until his salvation came in the form of the work foreman, who became interested in Tesla's seemingly bizarre talk of how to light up the world with electricity in a far better way than Edison could have imagined.

The foreman happened to be a friend of an executive at the Western Union Telegraph Company, A. K. Brown, who was very interested in the commercial possibilities of a rival system to Edison's. Funding was arranged, and almost immediately Tesla found himself a 50 percent shareholder in the Tesla Electricity Company with a casting vote and Mr. Brown holding the remaining shares. Early in 1888 he left the construction site, was installed in a small laboratory and apartment, and began to build the dynamo he had been visualizing for almost six years.

Six months later the prototype was completed. It was sent to Cornell University to undergo tests, and patents were applied for and given. Then, on May 16, 1888, less than a year after swapping a shovel for a voltmeter, Tesla was invited to deliver a lecture describing his invention to the American Institute of Electrical Engineers, it was perhaps the most important event of his life.

He described in detail the way he could build an AC dynamo or generator and the associated machine, an AC motor. Professor Poeschl, his teacher at the Karlstadt Gymnasium, had claimed it would be impossible to build a working AC motor because as the current changed direction the armature of the motor would swing back in the opposite direction. Indeed, this had been shown to be the case ever since Faraday had created his first dynamo and tried to use alternating current to produce a motor.

Tesla's elegant solution was to use two alternating currents that were out-of-step with each other, so as one current changed direction and the armature was about to stop spinning and reverse direction, the other current, perfectly out of phase with the first, would send the armature on again in the original direction. As the effect of this second current began to diminish, the first current would replace it, so the armature would have a constant motion, creating a practical, continuously running motor.

In itself, the meeting may not have borne any great commercial benefits; Tesla's ideas were on the very edge of contemporary theory, and most of those attending the talk would have remained skeptical about the application of his small-scale machine on a large industrial or domestic scale. However, Brown had invited along one of the few men who could help Tesla realize his dreams.

George Westinghouse was something of an inventor himself and had grown up in a wealthy family whose money came from the lucrative business of designing and building rolling stock for the growing railway network. As a young man Westinghouse had invented an air brake for trains and had gone on to develop a keen interest in the application of electricity. He had mastered some of the theory of electricity and had designed the broad outline of an AC system to rival Edison's plans. A short time before meeting Tesla, he had paid fifty thousand dollars for an English patent for a transformer that could be used to step down the high voltages of an AC generator to a safer level so that the current could be used in domestic situations. He went on to find the other pieces of equipment that could form an integrated lighting system.

And Westinghouse was no friend of Edison. Realizing that all the elements had to be in place in order to capture the market for an entire electric light system and that one company had to control all the major elements, he had applied what were clearly unethical business practices to produce a lightbulb almost identical to the one Edison had patented. He had bought a patent for a rival to the Edison filament, one that was actually inferior but close in design, and had modified it (illegally, as time was to prove) so that it was almost identical to Edison's original. He had then incorporated this filament into his design for a complete AC electrical system. Even so, Westinghouse still lacked some of the key elements to produce his system. Most important, he knew no way of creating an AC generator or dynamo, the precise device Tesla demonstrated at the institute meeting.

Fortunately for the development of modern electronic engineering, Tesla and Westinghouse enjoyed an immediate rapport. The day after the meeting, Westinghouse visited Tesla's new laboratory, where the inventor demonstrated a series of devices and the two men talked of how to create a delivery system that would be far superior to Edison's.

Knowing that the inventions Tesla had begun to develop were the missing elements he needed, Westinghouse immediately offered Tesla a $1 million advance for his machines and a royalty of one dollar per horsepower of electricity delivered by his system.*

To get this offer into some sort of perspective, we may disregard the $1 million and simply look at the royalty. Within a decade of this deal, electricity was being consumed in the industrialized world at a rate of millions of horsepower per month, and today the world uses millions of horsepower of electricity per second. If the deal had been watertight and powerful business interests had not excluded those most responsible for making the entire proposal possible, Tesla would have become the richest man in history. But that was not to be.

As Tesla was creating his AC generator, Edison had been far from idle.

---

*There is some disagreement about the real value of this agreement. According to a recent biography of Tesla, he was offered only $60,000 in cash and Westinghouse stock but a royalty of $2.50 per horsepower. This would of course have meant that in the long run Tesla could have become even richer than by the usually accepted figures (Margaret Cheney, *Tesla: Man Out of Time* [New York: Dorset, 1989], p. 40).

Pushing ahead with his plans for direct current, he had been installing independent power stations and stand-alone electric lighting systems. In a propaganda war against his then great rivals, the gas supply companies, Edison's staff wrote an endless stream of pamphlets attacking the safety record of gas lighting. Through Edison's extensive media contacts, he ensured article after article was published in which every aspect of the gas industry was criticized and any reports of explosions given disproportionate coverage. An example of this hype comes from a pamphlet produced by the Edison Light Company describing a recently installed electrical system that had replaced the old gas equipment in a New York wholesale grocers. "In this room fifty clerks do clerical work all day," it began. "The heat from the gas has proved injurious to health, and the gaslight has proved injurious to eyesight. This room is now lighted by one of our isolated plants and the injurious effects are completely removed." In the same pamphlet the writer proclaimed as fact the false finding that gas light caused shortsightedness.[36]

Edison certainly had not forgotten about Tesla, and he had almost certainly heard rumors of his partnership with Westinghouse, because the world of electrical engineering on the East Coast of the United States was then very small. But rumor was probably all he did hear; Westinghouse, Tesla, and their associates were doing their utmost to keep their plans secret.

Tesla would have had little time to talk to anyone about his efforts; he was now working tirelessly with Westinghouse. For a year he lived in Pittsburgh, the center of Westinghouse's operations, and he tried to adapt his ideas to those of the other designers and scientists who worked there. But such was not Tesla's way; he functioned best on his own, applying his idiosyncratic methods. Eventually, Westinghouse realized Tesla's move to Pittsburgh had been a mistake, made arrangements for him to return to New York, and installed him in his own laboratory.

Back in Manhattan, Tesla began to enjoy life a little. He was always misanthropic, but he was now wealthy and relatively independent, doing the work that inspired him. He began to mix with the wealthy and famous and was invited to dine regularly at Delmonico's and other exclusive restaurants, where his talk of strange experiments and the almost mystical power of electricity excited bored socialites across the white linen tablecloth.

This sudden attention had an odd effect upon Tesla. Unused to admiration perhaps, he started to revel in his success and to believe his own press a little too readily. During this period he loved nothing more than working long hours in his laboratory and then, donning evening dress, making his way in a cab to a glittering dinner at a fine hotel or joining a group of wealthy friends at one of New York's fashionable restaurants. After the meal he would invite his companions back to his laboratory, where he would demonstrate his inventions and thrill his guests with tricks and stunning visual displays.

Tesla understood electricity; he could live as one with it, control it, make it submit to his will. Before his awed and probably terrified friends, he would insulate himself and then pass what would in other hands have proved to be deadly quantities of electricity through his body. He would darken the room and allow lightning to jump from the tips of his fingers, and using his prototype designs for induction coils and generators, he could turn on electric lights without wires, merely by pointing a strange-looking device at the bulbs strung out like Christmas tree lights at one end of his laboratory. According to an eyewitness account: "With invisible fingers [Tesla] set objects whirling, caused globes and tubes of various shapes to glow resplendently in unfamiliar colors as if a section of a distant sun were suddenly transplanted into the darkened room, and crackling fire and hissing of sheets of flame to issue from monster coils to the accompaniment of sulfurous fumes of ozone produced by the electrical discharges that suggested this magician's chamber was connected directly with the seething vaults of hell. Nor was this illusion dispelled when Tesla would permit hundreds of thousands of volts of electricity to pass through his body and light a lamp or melt a wire which he held."[37]

These were Tesla's halcyon days and the high point of his material comfort. But as he was enjoying himself, Westinghouse, claiming the superiority of Tesla's revolutionary devices, began to clash openly in the press with Edison and his supporters.

Most evangelical as an advocate for Edison's cause was Harold Brown, who had been a senior engineer at Menlo Park. Brown was a cruel and obsessive character desperate to maintain Edison's favor and happiest when basking in the great inventor's reflected glory. Sadly, he also appealed to Edison's worst instincts, and the two men fed off each other's overambition and readiness to regard as worthless anything or anyone standing in their path.

Brown and Edison knew instinctively that the best chance of having their own DC system accepted over AC was to generate public anxiety about the safety of alternating current. Edison had recently experienced just such a challenge when he had begun his successful campaign to turn the public against gas lighting by exaggerating the risks of explosion and gassing. But both men realized that, because the difference between the two forms of electricity was subtle, the distinction would be hard to publicize and something stronger than pamphlets and articles in the press would be needed.

Which of the two men first devised a campaign to try to destroy Westinghouse's efforts and Tesla's work is unknown. But whether he volunteered or was pushed into the role of proselytizer, for a short time Brown became famous as the public face of Edison's attempts to impress DC upon the world. And in his efforts to do this, he employed some extremely unpleasant methods.

Brown first burst into press reports covering the Battle of the Currents after he organized a private meeting at the Columbia School of Mines in New York on July 30, 1888. He had invited a large group of scientists, reporters, and members of the Electrical Control Board to a series of talks and demonstrations that he and Edison hoped would alert the audience to their claims of dangers linked with AC.

Brown took the podium and began the meeting with a diatribe against alternating current. "I've not become involved in this controversy because of any connection to financial or commercial interests," he lied. "I am here only because of my sense of right. There are three classes of current in commercial use: the continuous, the intermittent, and the alternating. Investigation has shown that the first and second could by proper safeguards be made harmless to the general public while the third is by its very nature hopelessly deadly." Pausing to survey the audience, he went on. "I have applied a current of 1,410 volts continuous current to a dog without fatal result. And I have repeatedly sent to eternity dogs with as little as 500 volts alternating current. Those advocates of the alternating system who claim to have withstood shocks of 1,000 volts without injury must have been wearing lightning rods. Surviving such a shock is impossible, and those who state otherwise are compelled either by ignorance or guileful commercial interest.

"To demonstrate the veracity of my beliefs," he went on, "I've asked you gentlemen here today to witness the experimental application of electricity to a number of brutes."

Brown then placed a large dog in a cage, muzzled the animal, and bound it with strong leather straps. Wires were applied to the dog's skin and connected to a DC generator. "We will first apply the continuous current at a pressure of 300 volts," Brown announced. As the electricity flowed, the hapless dog growled angrily and struggled against the leather restraints. "Observe, gentlemen, that our subject, though discomfited, is still quite healthy," Brown reported.

He then increased the voltage to 400 volts. This time the dog drooled, saliva sprayed across the stage, and as the poor beast struggled, growls turned to whines. Brown stepped up the voltage to 700 volts. The dog shook and shivered, its eyes rolled, and it began to struggle for breath. "Finally, we increase the current to 1,000 volts," Brown announced, his voice raised above the agonies of the dog.

Some in the stunned audience were now protesting, and a few stood up; one or two walked out in disgust. But Brown continued, seemingly unaffected by either the suffering of the animal or the anger of some of the gathering. He threw the switch once more. The dog gave a pitiful yelp and collapsed against the straps; the smell of burned flesh and singed hair filled the auditorium.

A few minutes later the leads from the cage were wired up to an AC generator, 300 volts AC was delivered to the dog, and it died almost instantly. Brown was about to lead another dog onto the stage when one of the audience stood up, identified himself as a representative of the Society for the Prevention of Cruelty to Animals, and demanded the display be terminated there and then. Others joined in the pretests, claiming quite rightly that Brown's barbaric "experiment" was a sham. The dog, they pointed out, had been almost dead before the AC was applied, and there were no relays in the wiring for the AC, so electricity had been flowing far longer than seemed the case from the throwing of the switch onstage.[38]

Brown was forced to stop, but he was far from finished with his crude attempts to demonstrate the dangers of AC. Having overcome the legal restrictions of the SPCA, within three days he was back in the same lecture theater before a similar gathering. This time he killed three dogs with less than 400 volts AC before showing how another could suffer 1,000 volts DC before it died, a shivering, charred mess.

During the following three months, Brown and his helpers conducted public electrocutions of a horse, a calf, and several more dogs. Meanwhile, at Menlo Park, Brown and a group of what Edison referred to as "muckers" carried out hundreds of electrocutions of animals. The experiments were conducted at night to reduce the local disturbance caused by the noise of sacrificial animals.

During one session, Edison's right-hand man, Charles Batchellor, was almost killed. Struggling to force a particularly lively puppy into a makeshift electric chair, Batchellor received the full force of several hundred volts. "I had the awful memory of body and soul being wrenched asunder . . . the sensation of an immense file being thrust through the quivering fibers of my body," he wrote, before moving on with further experiments.[39]

In a further crude attempt to turn public opinion against the proposed use of AC, photographs of electrocuted dogs from Edison's experiments found their way anonymously into newspapers, some even appeared in posters plastered to walls on the streets of New York. And as Brown and Edison made increasingly desperate attempts to sway public opinion, they also expended great effort in trying to convince the authorities to introduce a law against AC.

Brown and Edison completely failed to do this, but they did succeed in making the argument over which form of current to use in the home a high-profile public issue. From the lecture theater at the Columbia School of Mines, the rows and vitriol spread to the pages of *The New York Times*. On December 13, 1888, the newspaper printed a letter from Westinghouse in which he claimed Brown to be "in the pay of Edison Electric Light Company; that the Edison Company's business can be vitally injured if the alternating current apparatus continues to be as successfully introduced and operated as it has heretofore been; and that the Edison representatives from a business point of view consider themselves justified in resorting to any expedient to prevent the extension of the system." He then pointed out that his company had sold 48,000 lights in the month of October 1888, whereas, by Edison's own figures, Edison Electric Light had taken orders for only 44,000 lightbulbs the entire year. "I have no hesitation in charging that the object of these experiments," he added, referring to Brown and Edison's tests on animals, "is not in the interest of science or safety, but to endeavor to create in the minds of the public a prejudice against the use of the alternating currents."[40]

Brown's response was published five days later. Naturally, he refuted all charges leveled by Westinghouse and accused him of manipulating the scientific facts to suit his argument. He then concluded childishly: "I therefore challenge Mr. Westinghouse to meet me in the presence of competent electrical experts and take through his body the alternating current while I take through mine a continuous current. . . . We will commence with 100 volts, and will gradually increase the pressure 50 volts at a time, I leading with each increase, until either one or the other has cried enough, and publicly admits his error."[41]

When Westinghouse sensibly refused to rise to this ridiculous suggestion of a "duel," Brown claimed cowardice and accused Westinghouse of being happy to let his public suffer the risks he would not be willing to take himself. Clearly, by this stage of the battle Brown and Edison had allowed their commercial ambitions to completely subsume any sense they once had of scientific impartiality.

In 1888, Edison was only forty-one. However, he was no longer a go-getting risk taker but a man past his prime as an inventor. His reputation was sustaining him, and in the public imagination he remained a grand figure, but his famous vigor and adventurousness were beginning to fail him. The historian Harold Passer highlighted this in his account of the early electrical industry and wrote: "In 1879, Edison was a brave and courageous inventor. In 1889, he was a cautious and conservative defender of the status quo."[42]

Even some of Edison's own senior employees were growing disillusioned with the famous inventor and his half-crazed collaborator and implored him to embrace AC rather than waste his time or further damage the company's reputation with this fruitless fight. Modern evidence has also highlighted the fact that some of Edison's more independent-minded researchers were conducting research into AC at Menlo Park even as Edison was at his most active attacking its application.[43] But Brown's evangelism could not easily be contained, and rather than listen to his advisers and censor his helper, Edison merely encouraged him.

Brown wrote a book financed by Edison titled *The Comparative Danger to Life of the Alternating and Continuous Current*. It was little more than a collec-

tion of newspaper articles, speeches, and descriptions of his animal demonstrations, to which he had appended reports from a collection of largely spu-
rious sources condemning the domestic use of AC. In early 1889 he
produced a pamphlet, again funded by Edison, that was posted to every
mayor, politician, insurance agent, and prominent businessman in any
American city with a population larger than five thousand.

"I address you in a matter of LIFE AND DEATH, which may personally
concern you at any moment," Brown's declaration began. He went on to
disparage Westinghouse and Tesla's methods, claimed they were merely
interested in commercial gain, and reported without verification, the grue-
some deaths of innocent users of AC. Calling AC "this EXECUTIONER'S-
CURRENT," Brown concluded his diatribe with a plea that his readers do all
they could to ban the use of any current above three hundred volts in their
towns and cities, a move that would have prevented Westinghouse's system
from operating because the step-down transformers had to be close to populated areas.[44]

In October 1889 Westinghouse responded to these attacks with his own
book, *Safety of the Alternating System of Electrical Distribution*, in which he
defended his cause using clear explanations and showing that, if the correct
guidelines were followed, AC would be a harmless means of powering
homes, offices, theaters, and factories.

So the vitriolic exchange of booklets, newspaper articles written by
biased journalists, and dramatic public performances went on, with each
side becoming increasingly personal in its attacks and each drawing upon a
growing body of supporters. With the grisly botched death of the ax murderer William Kremmler in August 1890, the press coverage of the arguments reached fever pitch.

Through Brown, Edison had pushed hard to initiate the use of the electric chair because he saw it as a golden opportunity to create a dramatic
demonstration for his claims that AC was a deadly form of electricity.
Together Brown and Edison had lobbied politicians, conducted public
demonstrations worthy of a macabre Barnum, and manipulated the press as
fully as they could.

In 1889 Edison and his cohorts practically took over the autumn issues
of *North American Review*. In September, one of Edison's most vociferous
supporters in the fight against AC, Elbridge Gerry, wrote an article describ-

ing the horrors of hanging and the relative merits of using electricity to kill a criminal. The same issue carried a piece titled "The Danger of Electric Lighting," in which Edison readily adopted the role of an Olympian rising above the economic considerations and tawdry squabbles of mere electronic engineers and attacked Westinghouse mercilessly. In the November 1889 issue of the same magazine, Brown waxed lyrical in a piece titled "The New Instrument of Execution," in which he described his idealized vision of the new method: "The deputy-sheriff closes the switch. Respiration and heart-action instantly cease," he wrote with undiluted optimism. "There is stiffening of the muscles, which gradually relax after five seconds have passed; but there is no struggle and no sound. The majesty of the law has been vindicated, but no physical pain has been caused."[45]

By this time Brown at least seems to have lost all analytical sense and completely failed to realize the hypocrisy of attempting to claim AC was painless as a means of execution but caused horrible agonies if misused by the public. This man, who had executed countless animals, should have known better.

Edison and Brown did eventually succeed in persuading Austin Lathrop, the state of New York superintendent of prisons, to gain approval for an experimental execution by means of electricity, and a date and subject for this showcase event were chosen. *The New York Times* wrote of the decision: "In addition to a Westinghouse dynamo, each prison will require an 'exciter' to be used as an auxiliary to the dynamo; a strong oaken chair, in which the convict is to sit and be killed; an electrical cap and electrical shoes."[46]

The documents authorizing the use of the process had been signed and the dates set, but throughout the media all talk centered on what the new means of execution should be called. In his article in the September 1889 edition of the *North American Review*, Gerry had referred to "electrolethe," but this was a clumsy term. *Scientific American* added little with its technical terms—"thanelectrize," "fulmenvoltacuss," and "electropoenize." Edison seriously proposed that the New York authorities follow the example set by Dr. Guillotin and name the method Westinghouse. *The New York Sun* countered by suggesting the method be named after the true father of the electric chair, Harold Brown. But just before the execution of Kremmler, *electrocution* had started to gain general acceptance as a word to describe death from an electric shock, so this name was adopted both officially and by the media.

Meanwhile, as the half-serious business of naming the means of execution occupied journalists, Brown and Edison were faced with a genuine problem. They had fired up the authorities to use AC to kill a man in the electric chair, but they had no means of supplying the necessary equipment. The quiet experimenting into alternating current carried out by a few of Edison's rebellious workers at Menlo Park had achieved little, and Westinghouse was certainly not going to supply the generators and other equipment they needed.

But of course Brown was undeterred and quickly succeeded in finding a clandestine route to his goal. Through an intermediary he heard of an academic working at Johns Hopkins University who wanted an AC generator. Edison covered the eight-thousand-dollar price tag, and through a long chain of buyers and sellers, Brown succeeded in fooling Westinghouse into parting with all the machinery needed by Auburn Prison. Westinghouse knew nothing of the deception until the story of Kremmler's horrible death appeared in the newspapers.

This was perhaps the low point of the dispute between those who supported AC and those who abhorred it, but ironically the vivid press reports of Kremmler slowly burning in the electric chair did nothing to destroy the onward march of AC. Those who opposed the system had harnessed massive publicity for their cause and fronted it with the execution at Auburn Prison confident such a spectacle would forever tarnish the words AC, Tesla, and Westinghouse, but they failed because greater impulses were moving public opinion, and those interested in having electricity in their homes were convinced by practicalities, not melodrama.

First, notwithstanding their nefarious efforts, Edison and Brown had failed to educate the public concerning the differences between AC and DC. After years of publicity, still very few people could distinguish between the two systems, so they were merely guided by the press and by practical demonstrations. Brown and Edison also failed to realize that there would be a backlash against their methods and that the public was interested only in having electricity delivered efficiently. The dangers Brown tried to highlight were obscured by need, and visions of a future powered by electricity had been enthusiastically promoted by influential sections of the press.

Within a year of Kremmler's execution, AC was by far the preferred system for domestic power, approved by the wealthy individuals and the growing number of businesses already switching to electricity. Direct current required the construction of an individual power station and could operate only a restricted number of appliances. Alternating current was led from a distant power supply, was many times cheaper, yet was conspicuously more versatile and powerful.

Two high-profile events sealed the fate of the two delivery systems. In May 1893 Westinghouse beat Edison in a bid to power the spectacular lighting at the Chicago Fair. Crowds flocked to the fair, where, using Tesla generators, the Westinghouse company illuminated no fewer than 96,620 incandescent bulbs, a feat that Edison could only have matched using hundreds of separate and extremely costly power stations. Then, a little over two years later, Westinghouse teamed up with General Electric to win the bid to construct a power station at Niagara Falls to supply electricity to the city of Buffalo and much of the surrounding region. Again Tesla's method was used, each element of the system constructed according to his original patents.

The public had accepted Tesla's methods over Edison's. The underdog, financed by the fabulous wealth of Westinghouse and his business partners, had overcome the dirty tricks, the callousness, and the infamous public spectacle of one of the most famous men in America.

ALTHOUGH TESLA'S GENIUS quite literally electrified the world, he gained little from it personally. By accepting Westinghouse's money and working with him on the creation of the first practical AC dynamo, he had become his servant; the application of his ideas and inventions were controlled completely by George Westinghouse. Tesla's acceptance of a royalty payment had allowed Westinghouse to own the patents, so the inventor was swept along, powerless to control the fate of his own creations.

Yet Westinghouse did not mistreat Tesla. They had entered into a business agreement in good faith, but the flux of big business quickly ruined Tesla's chances of gaining a personal fortune from his work. In 1890, just as AC was about to sweep all before it, Westinghouse faced financial difficulties; his business was expanding too quickly, and he was forced to find a cash-rich partner. Taking the advice of bankers, Westinghouse Electric Company merged with the U.S. Electric Company and the Consolidated

Electric Light Company to form the Westinghouse Electric and Manufacturing Company.

This saved Westinghouse and could have turned Tesla into a billionaire, but, crucially, a nonnegotiable condition of the merger stipulated that Westinghouse had to pay off Tesla and cancel his future royalties. Never an astute businessman, but perhaps knowing the planned conglomerate would further damage Edison, Tesla agreed to a one-time payment of $216,000.

This was of course an insignificant fraction of what he would have earned from the royalty he relinquished, but it was a handsome sum that could have provided Tesla material comfort for the rest of his life. Sadly, within a few years he had spent it all and borrowed more to continue his research. This investment led to ideas and discoveries that were completely ignored or tied up by other businessmen, who took Tesla's creations and placed them beyond his control.

To compound Tesla's problems, when the Westinghouse Electric and Manufacturing Company was established, it immediately ensured that all Tesla's patents were so carefully protected from abuse that no other inventor in the field could begin to make headway with any idea that even came close to encroaching upon their products. This generated huge resentment from Tesla's fellow electrical engineers and inventors, many of whom unfairly turned their resentment toward Tesla himself rather than the financiers who had created the problem.

William Blake wrote: "To be in Error and to be Cast out is a part of God's design." And so it was that Tesla, a man obsessed with discovery and invention, the man who did more to develop the technological age than almost anyone else, gained neither financial reward nor the kudos owed him. Instead of being remembered as a great inventor, his name became associated with what grew to be seen as unfair patenting law.

Typically, Tesla did not blame Westinghouse for his misfortunes and said of him: "George Westinghouse was in my opinion, the only man on this globe who could take my alternating current system under the circumstances then existing and win the battle against prejudice and money power. He was a pioneer of imposing stature, one of the world's true noblemen, of whom Americans may well be proud and to whom humanity owes an immense debt of gratitude."[47]

Tesla was an inveterate dreamer, a man whose imagination was his strongest asset and his most profound fault. He could not resist writing and talking publicly about any new idea that came to him and would pontificate on any subject, no matter how outlandish. And this served to damage his reputation still more. By exaggerating Tesla's weakness for attention and his often misjudged enthusiasm, his enemies made sure he was never again taken seriously, in spite of the fact that his work was responsible for massive technological advance.

Ironically, many of Tesla's enthusiasms are now readily accepted. He wrote about the possibility of life on other planets, missions to Mars, and the production of death rays, global communications, and strange military devices. But with an insatiable fondness for offering his ideas to any journalist who would guarantee him a splash in his paper and pay a reasonable fee, Tesla could be encouraged to say almost anything. Naturally, this lowered his capital and damaged his image, and those who despised him exaggerated his statements and turned them back upon him, even when some of those same enemies knew that often what Tesla was saying was not so far-fetched.

Tesla was certainly a difficult man to know and had only a vague understanding of how to behave in company. He was famed for ridiculously long speeches, in which he would bore audiences with stories from his childhood or descriptions of his working methods. He was obsessed with his work and the promotion of his own image to a point that he could talk about little else, in spite of having been graced with a wonderfully diverse education.

A particular illuminating example of this was when in 1915 a *New York Times* reporter heard a rumor that Edison and Tesla were to be jointly awarded the Nobel Prize. The journalist approached Edison for an interview but was turned away. Tesla, by contrast, was ebullient, and the newspaper carried a long interview with him describing how he believed his new ideas would at last receive the attention they deserved. Sadly, the rumor was nothing more than that, and the award that year was given to the father-and-son team of W. H. and W. L. Bragg for their pioneering work on X-ray diffraction. Edison knew all along that Nobel Prize winners were contacted privately before any announcement was made to the press.

But, to pile irony upon irony, Edison himself, although more circumspect than Tesla when it came to Nobel Prizes, was no stranger to exaggerated claims for the future of technology. He wrote articles and spoke publicly on a vast range of ideas at the very fringes of science as it then was.

His distrust of the scientific establishment, the college professors, and those he sneeringly called "the-o-ret-i-cal" scientists, never diminished. He was often ridiculed by those same academics for his lack of technique and his readiness to jump to bizarre explanations for physical effects. A good example was his insistence that electromagnetic waves responsible for his famous Edison effect were a new "etheric force" rather than a readily explainable aspect of conventional science.

Edison was also eccentric in his spiritual leanings. In 1878, just as he was gaining fame as an inventor, he became an active member of the Theosophy movement, a maverick pseudointellectual gathering of mystics and occultists that had been established in New York in 1875 by Madame Helena Blavatsky. Members of the movement believed in occult forces and etheric beings who were supposed to guide humanity through a predetermined course and proposed that the human race was standing upon the threshold of God-like status.

Edison also espoused some distinctly odd ideas about mind and spirit and wrote in his private papers about spiritual entities that control our minds and bodies. "They fight out their differences, and the stronger group takes charge," he declared. "If the minority is to be disciplined and to conform, there is harmony. But the minorities sometimes say 'to hell with this place, let's get out of here.' They refuse to do the appointed work in the man's body, he sickens and dies, and the minority gets out, as does too of course, the majority. They are all set free to seek new experience somewhere else."[48]

Toward the end of his life, Edison even talked of a machine he claimed to have designed that could allow the living to hear the voices of the dead. "I have been at work for some time on building an apparatus to see if it is possible for personalities which have left this earth to communicate with us," he told one journalist.[49]

Tesla hated talk of the occult, and although he put his name to endless articles about ideas often decades ahead of their time, he shunned anything that was not scientifically valid. Yet even today Edison is viewed as the empiricist, the grand old man of American science, while Tesla, if remembered at all, is too often identified with pseudoscience, his legacy unjustly sullied.

Both men lived to reach old age. Edison maintained an honored position in the world of science. He was responsible for inventing the first film projector and worked with George Eastman to produce the earliest motion pictures. He remained financially comfortable, and his undoubted business acumen served him well. He died in 1931, at the age of eighty-four.

Tesla's reputation peaked with the introduction of AC, then plummeted to a point where he was almost completely ignored by the scientific community. In 1943, at the age of eighty-seven, he died in poverty in a hotel room shared with dozens of pigeons he habitually coaxed in from his window ledge.

*E*LECTRICITY IS SUCH a fundamental aspect of our lives it is almost impossible to imagine a world in which almost anyone could not flick a switch and see light flood a room. Electricity, an invisible, untouchable, hard-to-define, nebulous thing, is now so much a part of the fabric of civilization it is difficult to visualize how different life might have been without the rivalry between Tesla and Edison.

Tesla was certainly the more able theoretician, but he was also a match for Edison's practicality. Through a combination of hard work and inventiveness, Edison gave us the electric lightbulb—a practical way to create artificial light—but his system for delivering the electricity to power the lightbulb was entirely misguided. Tesla's great contribution was his understanding of a far superior method, AC, a form of electricity he knew how to harness and control.

It might be argued that others may have developed a working lightbulb not long after Edison; indeed, inferior versions were designed before his, and his method of trial and error combined with an enviable practical ability could have been applied by someone else. But Tesla's contribution was more sophisticated and reliant upon a very rare talent. Tesla combined a powerful imagination with a personal drive founded in the peculiar circumstances of his childhood. These were then blended with a peerless grasp of the theory of electricity.

Perhaps Tesla and Edison would have achieved more if they had been able to cooperate, but their personalities were quite incompatible. Rather than trying to understand Tesla, Edison, driven by a genuine need to make money from his science, abused Tesla and repelled him.

The rivalry between the two men was played out under the shadow of high finance. George Westinghouse was the man who took Tesla's ideas and applied them to the marketplace, and Harold Brown was Edison's disciple who tried to push upon the public a clearly inferior system so that Edison's company might earn a fortune. The often brutal conflict between these factions certainly drove Westinghouse to persuade the rich and powerful to take AC and discard Edison and Brown's overtures. From this platform Tesla's devices eventually fed electricity to every home and workplace.

Without the challenge of the DC enthusiasts, Westinghouse may not have worked so hard or so swiftly to develop the technology that precipitated his eventual victory. It might even be argued that because the race to electrify was accelerated by the conflict over AC and DC, it moved ahead so rapidly and captured the public imagination so swiftly that the enemy of both groups—the gas companies—had no chance to slow public acceptance of electric lighting.

This was the final grand scientific conflict of the nineteenth century, played out almost exactly one hundred years after the phlogistonists had fought the modernizers of chemistry and lost. And in some ways this battle might be seen as the first of the modern age, a conflict in which money played a key role in both the initiation and the development of the ideas as well as fueling the aggression that led to a particular scientific advancement.

And as Edison and his colleagues were forced to accept defeat and the electrons that give a semblance of form to electricity flooded the world through wires and filaments, the human race was set to enter a new age, one governed and ruled by powerful new business interests, ripped apart by global conflict, but, perhaps most significant, in thrall to those electrons and the subatomic particles associated with them.

The Atomic Age was approaching fast.

# FIVE

## Of Atom Bombs and Human Beings

Uncertainty in the presence of vivid hopes and fears,
is painful, but must be endured if we wish to live
without the support of comforting fairy tales.
—*Bertrand Russell*

*Zurich, December 18, 1944*

*O*N THE WALL of the small seminar room at the University of Zurich,
the clock shows 4:15 P.M., and the twenty people in the room settle themselves in preparation for a talk delivered by their esteemed guest, the Nobel
Prize–winning German physicist Werner Heisenberg. Heisenberg is tall
and thin, with light brown hair and large blue eyes. He moves across the
front of the room, talking quickly and scribbling equations on the black-

board. After a few minutes the board is covered with, a muddle of equations.

Most of those present can follow what Heisenberg has been saying; they are physicists, specialists who work in the same field as Professor Heisenberg, and they are here to learn the latest ideas of the man who led the earliest development of quantum mechanics, some two decades earlier.

But after only a few minutes, one man in the room finds himself lost. In spite of spending many hours preparing for this seminar and reading all he could about nuclear physics, the speed of Heisenberg's delivery and the increasingly elaborate mathematics he is using quickly become too much. This listener's concentration wandering for the first time, he puts his hand into his jacket pocket and fingers the loaded automatic nestled there.

The man is Morris Berg, a former baseball champion, now an undercover agent for the Office of Strategic Services (OSS, an American intelligence agency and forerunner to the CIA). His mission is to determine from Heisenberg's talk whether or not the German scientist has cracked the secrets of constructing an atomic weapon, and if Berg believes the Germans are on the brink of building such a device, in front of the gathered scientists, he will remove his pistol and shoot Heisenberg between the eyes.

But for Berg the talk has become a cascade of meaningless words, the writing on the board akin to blurred hieroglyphics. And, he begins to feel nervous. What is Heisenberg saying? What do the symbols mean? Berg feels the palms of his hands grow clammy. Can he really shoot this man in cold blood? If he did he would surely be caught and killed himself, and for what? Maybe the man is innocent. Although Berg's knowledge of physics is extremely limited, he is a superb linguist, yet most of what he is hearing makes no sense; they could be the words of the guilty or those of the blameless.

Berg gradually calms down. Slowly, he removes his hand from his pocket. He cannot go through with a murder, he has no evidence, no real reason, and he cannot bring himself to take the life of an innocent man without definite proof. As Heisenberg concludes his lecture, Berg rises with the others, exchanges a few words with those standing beside him, and leaves quietly.

Although it sounds like a scene from a novel, this cloak-and-dagger operation to assassinate the leading German scientist of the day really did happen. Morris Berg, a man ill prepared to know if Heisenberg was indeed discussing ideas instrumental to the construction of the atomic bomb, was nevertheless convinced the man at the front of the lecture hall was not withholding any monumental secrets that could alter the course of the war. And so, unmolested, a few days after his talk Heisenberg returned home to Germany, where he could continue his researches oblivious to the fact that he had almost met a violent death in Zurich seminar room.

By this time the Manhattan Project, the Allied effort to create an atomic bomb, was close to completion, the Allies were advancing through Europe, and Hitler had just months to live. But rumors had long been circulating that the Führer had some form of "wonder weapon," some secret device that could yet change the course of the war. Determined to thwart Nazi plans, two physicists working on the Manhattan Project, Hans Bethe and Victor Weisskopf, had suggested a kidnap or assassination attempt upon Heisenberg and had even volunteered for the job. But, fortunately for Heisenberg, when Berg found himself in the seminar room in Zurich and in a position in which he could have killed the German, he had become convinced that if a superweapon really did exist, Heisenberg was not the man behind it.

However, this was only Berg's opinion. Most of Heisenberg's former colleagues, who were by this time working in Allied countries, felt that he certainly was capable of building an atomic bomb, just as they were attempting to do at Los Alamos. None of these scientists could be sure, but while there was any form of doubt, they knew they had to do everything they could to ensure the Allies produced a bomb first. The alternative was too frightening to contemplate.

THE SECRETS OF the atom have long been an intellectual Holy Grail for humanity. And just as ancient myth claims anyone not pure of spirit drinking from the Lord's chalice would instantly perish, the Holy Grail of the atom offers humanity the chance to use it for great good and pure evil.

The notion of atoms is found in some of humankind's most ancient writings. As we have seen, Aristotle preferred the doctrine of the four elements—fire, earth, air, and water. Another Greek, Democritus, who had taught two generations before Aristotle, had presented a lucid argument for the existence of fundamental particles including his famous statement "Color exists by convention, sweet by convention, bitter by convention, in reality nothing exists but atoms and the void." But Aristotle's voice was the louder, and Democritus was almost entirely forgotten for two thousand years.

So the world was cast into scientific darkness, a darkness the fires of the alchemists could do little to brighten. Alchemists were totally ignorant of the true nature of matter. They took Aristotle's misconception of the four elements and sprinkled in random experiment, poetry, contrived magic, and wishful thinking that produced little of lasting value.* The thousands of alchemists who toiled for centuries and gave their very lives to transmute base metal into gold (a rearrangement of atomic structure, no less) were acting like blind men with palettes of color attempting to produce a Rembrandt.

But after centuries of chaos, things did begin to change for the better. Millennia after his death, the ideas of Democritus were rediscovered by a few, including the early-seventeenth-century natural philosophers Pierre Gassendi and René Descartes. Through them the ideas of atomism were revived just as Aristotle's dogma was coming under fierce and continued attack by Robert Boyle, Thomas Hobbes, and later, most crucially, Isaac Newton.

The Enlightenment of the late eighteenth century was a time illuminated by the mechanics of Newton, and to men like Antoine Lavoisier and, later, the English chemist John Dalton, the Russian pioneer Dmitri Mendeleyev, and others, the universe was clearly constituted of matter made from elementary "chunks of material," or atoms. Little more than a decade after Lavoisier's death in 1794, chemistry had progressed to the point that it was accepted that molecules consisted of simple blends of atoms (a term Dalton had resurrected from Democritus), atoms that were believed to be indivisible and indestructible. Almost exactly a century after Dalton's death in 1844, both of these last premises were shown to be wrong, and in dramatic style. By the 1930s it was known that atoms are indeed divisible and far from indestructible; it was from these qualities that the horrors of Hiroshima and Nagasaki emerged and the atomic power that today fuels perhaps as much as a quarter of all human industry is produced.

Of course, scientific progress is all about stepping-stones, or standing

---

*Except, as we have seen, a set of useful practical techniques and a body of work that (perhaps perversely) inspired Isaac Newton.

on giants' shoulders, as Newton would have it, and this may be seen most clearly in the development of atomic theory. Each contribution, from the early thoughts of Democritus though the advances of Boyle and Newton, the revolutionary ideas of Lavoisier, and the concepts presented by Dalton has brightened the inner world of atoms and molecules and drawn us on. But it was not until the turn of the nineteenth century that scientists began to delve into the world imagined by the alchemists and to use science and mathematics to make real the fantasies of the occultists.

Between 1890 and 1920, the description of the atomic world was completely transformed by the work of a small group of towering scientific figures. Wilhelm Röntgen discovered X rays in 1895; a year later Antoine Becquerel noticed the phenomenon of radioactivity, and during the following few years, the husband and wife Marie and Pierre Curie elucidated the meaning of this discovery—what it was about atoms that created the effect. The Curies were the first to realize that radioactivity was produced when certain unstable atoms were changed into more stable forms. Marie Curie went on to describe how both uranium and thorium transform in this way and are hence radioactive. Soon after this Curie discovered two new elements; the first she named polonium after her native country (Poland), the other was given the name radium. For these efforts, the Curies and Becquerel shared the Nobel Prize in 1903.

In 1897, less than twelve months after Becquerel announced his findings, a Cambridge don, J. J. Thomson, discovered what he thought was a new particle he called an electron, which appeared to be quite independent of atoms. Within another year the New Zealander Ernest Rutherford, studying under Thomson at Cambridge, showed that Becquerel's radioactivity consisted of at least two forms—alpha and beta radiation. Two years later he discovered a third form of radiation, since named gamma rays.

But Rutherford took his ideas and experiments further still. In 1907, after he had moved to Manchester and was running his own laboratory, one of his students noticed that if alpha particles were fired at a sheet of platinum foil, about 1 in 8,000 of them was deflected straight back, almost along the line of fire. This was a stunning revelation because alpha particles were fast-moving, and it was known that atoms are incredibly small, so an alpha particle coming right back at the experimenter seemed improbable. Rutherford described the result as similar to firing a fifteen-inch shell at a sheet of tissue paper and having it come back and hit you. He immediately realized this was an important discovery, describing it as "quite the most incredible event that has ever happened to me in my life."[1] Soon he had succeeded in deducing what was happening.

He knew the alpha particles are positively charged, so the only explanation for what he had seen was to suppose that atoms are largely empty space but that at the center of each lies a tiny, positively charged region that can repel the alpha particle if it is fired at a precise angle (that's why only 1 in 8,000 of the alpha particles bounced back). He went on to show that in order to balance this supposed positive charge at the center of the atom there must be a countercharge. Enter Thomson's electron.

During the course of little more than a decade our understanding of atoms had gone from seeing fundamental particles as indivisible, discrete units that combined to make molecules (basically what had been known as far back as the early 1800s) to a vision of the inner working of atoms—the idea that they are made of smaller particles. Rutherford's model described an atom with a positive nucleus surrounded by an appropriate number of electrons to counter the size of the positive charge of the nucleus. Accordingly, the nucleus, made of positively charged protons, held the negatively charged electrons in place by electrostatic attraction. From this it was clear each element in the periodic table—a carefully elucidated table of all the known elements in the universe—had a different number of positive charges in the nucleus and therefore a different balancing charge provided by the electrons. Furthermore, the model could explain the difference between purely chemical processes, such as those investigated by Lavoisier more than a century earlier, and the newly discovered "nuclear processes," such as the radioactive emissions studied by the Curies. Chemical reactions, it was now realized, involve only the electrons Rutherford described as orbiting the nucleus like planets around a star, and nuclear transformations (the dream of the alchemists) are dependent solely upon changes wrought within the nucleus of the atom.

But not content with these explanations, Rutherford went on to a truly epoch-making discovery, one that was to plant the seed that led to the creation of the Manhattan Project and with it the Atomic Age. He reasoned that if certain elements, like uranium or thorium, broke up naturally to produce smaller elements and emit radiation, then perhaps they could be artificially induced to do this under controlled conditions.

Once again Rutherford used alpha particles, this time he fired them at atoms of nitrogen. To his surprise, he found protons were knocked off of some of the atoms. In other words, he had replicated the process of radioactive decay of atomic nuclei seen in nature. The press of the day dubbed him "the first man to split the atom."

Although these were massive advances and the results of remarkable insight and imagination, Rutherford's was a ridiculously simplistic model, and during the following decade it underwent considerable revision at quite breathtaking speed. This model became a template from which more and more sophisticated models were devised, an effort that occupied many scientists across Europe working together to perfect the picture of the atom.

The first idea to change was that the atom is made up of only two types of particles. If this were true, all atoms would be extremely unstable; to explain why most atoms remain stable over long time periods, physicists had to employ a third particle in their descriptions. This particle, it was assumed, possessed no charge but acted a gluing agent or stabilizer. Using a mathematical description of the structure of the atom, it was shown that this particle could exist only within the nucleus of the atom. Remarkably, in 1932, almost a quarter of a century after it popped out of the equations, the English physicist James Chadwick discovered this particle—the neutron—in the nucleus of the atom.

But soon Rutherford's model began to face more serious criticism. Most important, it was found that as a consequence of his arrangement electrons orbiting the nucleus would emit energy continuously. This would send them spiraling into the nucleus. It was calculated that for the most stable element known, hydrogen, this destruction would take only about one-billionth of a second. This obviously does not happen, so, as it stood, Rutherford's model had to be fundamentally flawed.

At its heart, the problem with Rutherford's model was that it was based upon Newtonian mechanics and treated the nuclear world as a microcosmic version of the everyday macrocosm. In other words, it was a *classical* model that treated atoms as balls of matter. Electrons and protons were visualized as lumps of material not dissimilar to the chunks of matter with which we are most familiar, large conglomerates that are affected by gravitational forces. The forces within the atom, however, are not gravitational but something altogether different, what we now know to be the weak nuclear and strong nuclear forces. And, to explain how the atom is really put together and for a better description of why the world is the way it is, physicists had to begin devising a wholly new way of looking at the atom. This required an approach called quantum mechanics.

Niels Bohr was one of the first to venture ideas that formed the bedrock of this new science. He proposed that atoms spend most of their time in what is called a ground state. The ground state is the lowest possible energy state of the atom, and using this idea Bohr was able to circumvent the problem of electrons spiraling into the nucleus and the atom decaying. In Rutherford's model, electrons could exist anywhere outside the nucleus, (some described this as the "plum pudding" model), but Bohr proposed that in all atoms electrons have to be arranged in definite locations, and he created a set of rules to describe the configuration of electrons that explained some of the chemical characteristics we now know are true for the elements.

But even Bohr's model worked with any accuracy only for the simplest element, hydrogen (which contains just one electron). Gradually, through the work of many theorists (including further contributions from Bohr, who was seen as a father figure in the physics community of the 1930s and 1940s), the first sketch of a quantum model was greatly elaborated to produce a working form of quantum mechanics that is still evolving today.

One of the clearest statements from the earliest quantum model is that we can assign only *probabilities* to the position of electrons. According to this theory, at one extreme it is possible the electron could reside in the nucleus; at the other it could be found almost infinitely far from there. But the chances of either of these is so remote they would almost certainly not occur in the lifetime of the universe. Instead, an electron is *most likely* to be found in a position similar to one of those suggested by Bohr. However, this new way of looking at the atom leads to some very strange, counterintuitive effects.*

Because we can assign only probabilities to the positions of electrons in atoms, nothing in the material world can be said to be "certain." This led to Werner Heisenberg's great contribution to quantum mechanics, the uncertainty principle. He arrived at this principle after intense discussions with both Bohr and Albert Einstein and went on to explain the concept in his most important book, *The Physical Principles of Quantum Theory*, published in 1928, for which he was awarded the Nobel Prize in physics four years later.

The uncertainty principle shows that there are absolute limits to the accuracy with which pairs of physical quantities may be measured. For example, if the momentum of a particle is known accurately, then its position remains a mystery. So, from the fact that the quantum world can be observed only on the basis of probability, we are led to the steely truth of the uncertainty principle. From here it requires only another step to demonstrate that the physical world of which we are all a part is a nebulous thing indeed. Yet, from this uncertainty, unimaginable forces may rage and the very flames of Hell burst forth.

ALTHOUGH EINSTEIN DID much to influence the thinking of Heisenberg, Bohr, and other luminaries at the birth of this astonishing new science, he loathed quantum mechanics, once famously declaring: "God does not play dice." By this he meant he could not accept the idea of a universe in which the behavior of matter and energy seemed to be entirely random. Yet, ironically, Einstein's erudition and lateral thinking did greatly influence his colleagues during the 1920s and 1930s, when they met at conferences and informal gatherings across Europe to discuss their descriptions of the atomic world. Furthermore, his own work outside particle physics generated vast and unexpected change.

---

*We should remember that quantum mechanics is not a fantasy but an approach that has proved to be one of the most successful scientific ideas in history. If you find this hard to accept, think for a moment of what has come from quantum mechanics. The list includes television, microchips, lasers, satellite communications, and fiber optics, to name but five.

When Einstein published his most famous work, the theory of relativity (the special theory in 1905 and the general theory in 1916), he could never have suspected that an idea tucked away in his equations would one day provide the impetus for a radical leap in atomic theory and lead a few years later to the atomic bomb. The most famous equation in history, $E = mc^2$, comes from Einstein's special theory of relativity and tells us how much energy we can expect to obtain from the nuclear process that causes an atomic explosion.

Back in 1897 the Curies had shown how certain substances could give off radioactivity when their atoms changed from less stable to more stable forms, but by the late 1930s physicists were moving closer to an understanding of what was happening at an atomic level when a large and unstable atom produced radioactivity. Toward the end of 1938, two German radiochemists, Otto Hahn and Fritz Strassmann, and their Austrian colleague, the physicist Lise Meitner, created an experiment in which they fired neutrons at a sample of uranium. Thinking they could make the neutrons stick to the uranium nuclei, they intended to create an atom that was larger than uranium. Instead they produced the opposite effect.

To their astonishment Hahn, Strassmann, and Meitner discovered that uranium atoms bombarded with neutrons could be made to break up into at least two smaller atoms. They found that one of these atoms was barium, which is about half the size of a uranium atom. They were actually the first to produce artificial nuclear fission. Unable to believe what thy had done, the three scientists repeated the experiment again and again and studied the fragments from the process. When they measured the masses at the beginning and the end of the experiment, they discovered that the total mass of the fragments produced by the process was less than the mass of uranium atoms they had started with. This could mean only one thing—some of the original mass had been converted into energy.

Galvanized by this experimental evidence, Hahn, Strassmann, and Meitner immediately set about explaining their discovery in a paper and calculating the sorts of energies involved. Meitner concluded that the two fragments from a simple fission of uranium would be repelled from each other at about one-thirtieth the speed of light, an astonishingly high velocity. Unfortunately, she had no idea how to calculate the energy this represented until she recalled a lecture of Einstein's she had attended thirty years earlier in

Salzburg. It had been one of the first times Einstein had mentioned the formula that would forever be attached to his name, the relationship $E = mc^2$.

The equation states that the energy we can obtain from a given amount of matter is equal to the mass of the matter multiplied by the speed of light and then multiplied by the speed of light again. In other words, because $c^2$ ($c$ multiplied by $c$) is such a huge number, a tiny amount of matter can be converted into a vast amount of energy. In fact, the fission of one pound of uranium provides 3 million kilowatt hours, the energy produced by 3 million one-bar heaters burning for an hour. In an atomic explosion this vast amount of energy is released in a microsecond.

The paper describing the fission experiments, written by Hahn and Strassmann, appeared in the German journal *Naturwissenschaften* during the first week of 1939. It had been sent to the offices of the journal in Berlin before Meitner had come up with her calculations. The journal published the findings without Meitner's mathematics, but these and other calculations from Bohr appeared in subsequent issues and helped to clarify the discovery.

Hahn, Strassmann, and Meitner's findings were received with utter astonishment by the physics community. Yet they did not lead immediately to ideas of constructing superweapons. Many scientists, including Heisenberg in Germany, realized this discovery might one day initiate the development of a bomb, but almost no one could believe the technical problems associated with producing an atomic weapon could be overcome in the foreseeable future.

However, as scientists around the world delved into the equations governing nuclear fission, a few physicists who had fled Nazi Europe could already visualize the horrifying potential of what Hahn, Strassmann, and Meitner had found. Most farsighted and therefore most worried was a Jewish Hungarian physicist named Leo Szilard, who was then a junior lecturer, or *Privatdozent*, at the University of Berlin. Although still in his twenties when quantum mechanics was first postulated and only a relatively junior researcher, Szilard had met and gotten to know some of the great names of the physics community. He was ambitious and versatile, knowledgeable about politics, and, unlike many of his contemporaries, he was acutely conscious of current affairs. In 1933, when the Nazis gained genuine political power, Szilard prudently left Germany and set up his home in England.

Szilard had long been fascinated with the work of the English science fiction writer and social commentator H. G. Wells and had spent considerable time and energy developing what might, if not for the intervention of the war, have become a political party or social movement based upon some of Wells's ideas for a new world order. Szilard called this political endeavor *Der Bund*. This would be, he declared: "a closely knit group of people whose inner bond is pervaded by a religious and scientific spirit. If we possessed a magical spell with which to recognize the 'best' individuals of the rising generation at an early age . . . then we would be able to train them to independent thinking, and through education in close association we could create a spiritual leadership class with inner cohesion which would renew itself on its own."[2]

In essence, this was a concept that had been fermenting in the minds of many intellectuals of the period. It is not far from Einstein's own utopian ideals, which became the foundation of his crusading antiwar, united world government stance of the early 1950s, and it is reminiscent of the intellectual purity of the Nobel Prize–winning author Hermann Hesse's novel *Magister Ludi*, published in 1943.

Central to Szilard's philosophy (and another prophetic concept gleaned from Wells) was the idea of a weapon created by using what Wells called the "secret power" of the atom. Although he then had no idea how this energy could be extracted, let alone harnessed, as early as the 1920s Szilard had pondered what power lay locked in the heart of the atomic nucleus. When he read Strassmann and Hahn's paper in January 1939, he knew that reality had caught up with his and Wells's imaginations and that he must take action.

By the time Meitner, Strassmann, and Hahn made their discovery, Europe was in complete turmoil, and Nazi oppression had precipitated a mass exodus of scientists from Fascist states. Within a few years of working together on the most abstruse and innovative puzzles in Nature's magical maze, the scientific community had splintered along national lines. The great majority of those engaged in nuclear research elected to work for the Allies. Many ended up at Los Alamos, while the German Werner Heisenberg and a small team chose to remain at the Kaiser Wilhelm Institute in Berlin, where they were later to begin development of the Nazi bomb.

Many émigrés, including Albert Einstein, John von Neumann, and Hans Bethe, traveled first to England, but most continued west and found senior academic positions in American universities or were recruited by the new intellectual powerhouse, the Institute for Advanced Study at Princeton. By 1938 Szilard had joined the flow to America, knowing that if he was to make a difference it would be through galvanizing the scientific community and the political forces in the United States.

In popular myth, news of nuclear fission caused an immediate reaction among politicians as well as scientists, but this is simply not true. As we have seen, even within the scientific community there were many who saw the link between the process and the potential to produce a weapon but believed the technical problems involved would be insurmountable. Those politicians who were aware of the revolutionary discovery, and even those very few who understood it, relied on their technical advisers, so they too were slow to accept the possibilities, and the incumbent dangers.

Szilard, however, knew how politicians thought, and he predicted that they would see nothing in the discovery of artificial fission unless they were prompted. As soon as the discovery was announced (some eight months before Germany and Britain were at war), he set about trying to generate interest in building a nuclear weapon in America.

Beyond a small cadre of European scientists, Szilard was totally unknown. He had no political influence but was on good terms with many scientists who did. It was therefore natural that his first efforts should be to persuade the most important scientist in his immediate circle to accept that the construction of a bomb was indeed feasible.

A fellow émigré, Enrico Fermi, had won the 1938 Nobel Prize in physics principally for his work on slow-moving neutrons and their properties, and soon after this he had emigrated to the United States with his wife, Laura. He could not have been closer to the center of research on the process that led to fission and chain reactions, and his status as a Nobel laureate meant that he could open doors. Fermi was Szilard's first point of contact.

Szilard had known Fermi for some years and they were friends, but even so he found him resistant to his ideas. Fermi was far from convinced of the practicality of constructing an atomic bomb and was not easily drawn into any plans for alerting the authorities to the dangers Szilard was visualizing. Such resistance did not come from any lack of patriotism or sympathy with the Nazis but because at first Fermi simply doubted that the leap from laboratory curiosity to atomic bomb would be possible for many years.

But after lengthy discussions during January and February 1939, Szilard managed to convince Fermi that quick action was essential. After a succession of false starts, they succeeded in arranging a meeting with a group of high-ranking officials in Washington. The officials listened patiently to Fermi's discourse. They sat silently watching the slides he presented; then they dismissed outright the idea of any form of "superweapon" that could be based upon Hahn, Strassmann, and Meitner's discovery.

Szilard was shocked and frustrated, but he knew he was right, and he was not going to give up his campaign so easily. His political judgment told him that he was already racing against the clock, that the Germans would not be slow to start their own atomic bomb project. But his experience as a scientist also made him aware that with each passing month there would be more evidence from experiment to confirm the feasibility of building a bomb, and that gradually some of the deeper questions and the problems they offered would find solutions.

Szilard did not have to wait long. Within a few weeks of the disastrous meeting between Fermi and the U.S. Navy, another step in the theory behind the bomb was revealed. On April 22, 1939, paper was published in *Nature* that added further evidence to Szilard's claims that a sustainable chain reaction could be achieved and a bomb built on the principle. In a series of experiments conducted at Marie Curie's Radium Institute in Paris, Frédéric Joliot-Curie concluded that a typical fission of uranium nuclei produced an average of 3.5 neutrons. This meant that each time a uranium atom was hit by a neutron and made to undergo fission, it emitted between 2 and 4 neutrons (giving an average of 3.5 because of statistical skew). "The chain reaction," he wrote, "will perpetuate itself and break up only after reaching the walls limiting the medium. Our experimental results show that this condition will most probably be satisfied."[3] In other words, uranium did not just undergo a limited degree of fission but could be made to undergo a chain reaction.

This was the sort of progress Szilard had hoped for. But others were at least as quick to realize the potential of this finding and the way it supported Hahn, Strassmann, and Meitner's results. The very day of Joliot-Curie's announcement, and within weeks of Fermi's meeting with the U.S. Navy two scientists in different parts of Germany read the Joliot-Curie paper and independently alerted the authorities.

The first of these scientists was a physicist working in Göttingen who anonymously contacted the Reich Ministry of Education. Unlike the navy officials in Washington, the German ministry took the news seriously, so seriously a special conference of the nation's top physicists was called for April 29, all exports of uranium were banned, and the army in newly occupied Czechoslovakia was ordered to send to Germany supplies of radium extracted from the mines at Joachimsthal.

The other scientist to react to the news was a young German physicist named Paul Harteck, who wrote to an assistant at the German War Office: "We take the liberty of calling to your attention the newest development in nuclear physics, which, in our opinion, will probably make it possible to produce an explosive many orders of magnitude more powerful than the conventional ones. . . . That country which first makes use of it has an unsurpassable advantage over the others."[4] Within days the War Office consulted the eminent nuclear physicist Hans Geiger, who proposed the matter be investigated immediately.

Meanwhile, in the United States, the scientists who could see the danger brewing were held in a political vacuum. The media even considered talk of bombs from atoms mildly amusing. A *New York Times* article datelined April 29—the very day of the high-level conference in Germany—described a meeting of émigré and American physicists to explore the possibilities of the newly verified chain reaction: "Tempers and temperatures increased visibly today among members of the American Physical Society as they closed their Spring meeting with arguments over the probability of some scientist blowing up a sizeable portion of the earth with a tiny bit of uranium, the element which produces radium."[5]

Such reactions could only have made Szilard feel more isolated and concerned, but he was not a man to give up; he realized he had to hit harder and push with greater force. He would need, he now understood, the support of a bigger name than Fermi, and in 1939 there was no bigger name than Albert Einstein. Szilard knew Einstein from his time at the University of Berlin, and they had often spoken at the physics conferences both men had attended during the 1930s.

In July 1939, Albert Einstein, who had been at the Princeton Institute for

Advanced Study since 1933, was vacationing on Long Island. Szilard found the sixty-year-old physicist in his study drawing on a pipe and wearing his usual tatty clothes—shoes with no socks and a moth-eaten sweater. Einstein listened and asked astute questions. By this time he was almost completely isolated from the world, absorbed with his own fantastic mental explorations, and he had not even heard about Hahn, Meitner, and Strassmann's developments. But, as Szilard explained the idea of a chain reaction and brought Einstein up to date with the Joliot-Curie figures as well as his own extrapolations, Einstein quickly grasped the potential of the process.

In many ways Einstein was a very naïve man who spent his last years sincerely believing an elite group of intellectuals could make the world a better place by force of reason alone. But he was no stranger to the darkness in the souls of many of the world's powerful figures, and he knew the fire at the heart of the atom could be used to devastate and destroy.

Before his visit Szilard had formulated the idea that Einstein should talk to the president, man to man, but when he discussed this with Einstein, he quickly realized it would be impractical. Einstein was charismatic and charming, he was a global celebrity and commanded huge respect, but he would not be good at conveying such an important message as Szilard's to a politician like Franklin Roosevelt. Instead, it was agreed that Einstein should write a carefully constructed letter to the president in which he could explain the danger and the need for urgent planning of an atomic weapons program in the United States. Einstein's famous letter began:

> Sir,
> Some recent work by E. Fermi and L. Szilard, which has been communicated to me in manuscript, leads me to expect that the element uranium may be turned into a new and important source of energy in the immediate future. Certain aspects of the situation seem to call for watchfulness and, if necessary, quick action on the part of the administration. I believe, therefore, that it is my duty to bring to your attention the following facts and recommendations.
> In the course of the last four months it has been made possible—through the work of Joliot in France as well as Fermi and Szi-

lard in America—that it may be possible to set up nuclear chain reactions in a large mass of uranium, by which vast amounts of power and large quantities of new radium-type elements would be generated. Now it appears almost certain that this could be achieved in the immediate future.

This new phenomenon would lead to the construction of bombs, and it is conceivable—though much less certain—that extremely powerful bombs of a new type may be thus constructed.

He then remarked: "In view of this situation you may think it desirable to have some contact maintained between the administration and the group of physicists working on the chain reaction in America."[6]

Soon after Einstein wrote this letter, Fermi found a direct route to the president via an old friend of Roosevelt, the economist Alexander Sachs. Roosevelt trusted Sachs's judgment, and Szilard was convinced he would be a good man to act as a go-between. However, such were the demands upon Roosevelt's time that even the influential Sachs could not arrange a meeting with the president until October that year. Szilard had to contain his impatience and hope the wait would be worthwhile.

According to reports, when Sachs discussed the matter and showed him Einstein's letter, Roosevelt quickly understood the danger. Absorbing Einstein's comments, he apparently declared: "This requires action."[7]

But for the moment at least, these proved to be entirely empty words. In October 1939 the United States was still over two years from military involvement in World War II, and Britain had just declared war on Germany. Apart from the remaining doubts about the scientific potential of this research, there was the practical problem of justifying the allocation of funds to such a task at this stage of the conflict. Roosevelt realized immediately that vast sums of money and large numbers of workers would be required to develop a weapon as radical as an atomic bomb, and siphoning off a substantial budget for it would take time to arrange.

Mired in doubt and skepticism, the U.S. government actually did nothing. Interested scientists continued to find answers to the many complex theoretical issues associated with the idea of an atomic bomb entirely in their spare time, and unpaid. The military and the government perceived Szilard's and Fermi's efforts as hysteria and even suspected them of ulterior motives. A military intelligence report written in 1940 declared: "Enrico Fermi is supposed to have left Italy because of the fact that his wife is Jewish.... He is undoubtedly a Fascist.... Employment of this person on secret work is not recommended. Mr. Szelard [sic] is a Jewish refugee from Hungary. It is understood that his family were wealthy merchants in Hungary and were able to come to the United States with most of their money.... He is stated to be pro-German.... Employment of this person on secret work is not recommended."[8]

Even Einstein was not trusted. In spite of his fame and global respect, many politicians and high-ranking military figures in the United States considered Einstein an enemy alien with no understanding of the real world and the methods of politics.

But indifference toward the potential of atomic energy was not an attitude shared by the Germans.

*W*ERNER HEISENBERG WAS born in 1901 in Bavaria. His father had been an eminent scholar, a professor of Greek at the University of Munich, so the young boy grew up in an academic, bookish environment. Apparently Heisenberg was introduced to the world of science through Plato's *Timaeus*, which he read when he was sixteen. There is very little of what we would today call "real" science in this important text; Plato writes of a universal flux guided by his beloved pure mathematics. But it seems that something in the prose inspired the boy to visualize atoms, and it led him to a lifelong fascination with atomic theory.

Heisenberg was something of a golden boy. He had classic Teutonic looks, blond hair and blue eyes, and he was a brilliant student through school and university. When he was eighteen he was taken under the wing of the great Bavarian classical physicist Arnold Sommerfeld, a contemporary of Rutherford. The roster of Sommerfeld's students include luminaries such as Hans Bethe and Wolfgang Pauli (both of whom left Germany for the United States just before the war), but of all those guided by Sommerfeld, Heisenberg was said to be the most naturally gifted and inventive.

Sommerfeld introduced Heisenberg to Niels Bohr in 1922. At that time Bohr was seen as the most innovative particle physicist in the world, an equal of Einstein, and the man who bridged the gap between the classical model of physics left over from the nineteenth century and the modern nonclassical or quantum view. Bohr, then in his late thirties, took Heisenberg as his assistant in Copenhagen, and for the next three years the German Ph.D. student spent much of his time in Denmark, with occasional trips back home to study under other great names in Europe. He spent some time in Göttingen in 1925 and became friends with Robert Oppenheimer, and he knew the British mathematician Paul Dirac, who was later made Lucasian professor at Cambridge. Along with other young students of that vintage who would go on to become iconic figures in the world of physics, Heisenberg, Oppenheimer, and Dirac had gone to Göttingen to sit at the feet of Max Born, a contemporary of Bohr and guru of quantum mechanics. All three men would eventually make enormous contributions to the development of the new physics. Oppenheimer and Heisenberg would also change the world in more immediate ways.

In 1927 Heisenberg returned to Germany to take up a professorship at the University of Leipzig, the institution where a little over two hundred and fifty years earlier Gottfried Leibniz had been refused a doctorate. It was here that Heisenberg made his most important contributions to the still embryonic science of quantum theory.

The uncertainty principle of 1927 made Heisenberg's name within the physics community, and by the end of the decade he was viewed as one of the most important scientists in the world, working at the cutting edge of quantum theory and writing some of its basic tenets. He had grown close to Niels Bohr during their work together in Copenhagen, and their relationship had survived Heisenberg's return to Germany. They collaborated often, wrote to each other frequently, and became the central figures (along with Einstein and Born) at the conferences where physicists came together to discuss the latest discoveries and to meld their ideas concerning quantum mechanics.

By the late 1930s, as the world shuffled to the brink of global catastrophe, Heisenberg was considered an important figure in Germany, and although he was never a Nazi sympathizer, he became a scientific figurehead for the new state. When he made a lecture tour in America during the summer of 1939, his many friends who had fled Europe tried in vain to persuade him to stay and to help them unravel the mysteries of the atom using the best facilities in the world; any post would have been his for the taking. But he refused.

Fermi, who had been a close friend for many years, pressed Heisenberg hard. "Whatever makes you stay on in Germany?" he asked. "You can't possibly prevent the war. . . . In Italy I was a great man; here I am once again a young physicist, and that is incomparably more exciting. Why don't you cast off the ballast too, and start anew? In America you can play a part in the great advance of science."[9]

There is no record of Szilard making any direct attempts to dissuade Heisenberg from returning to Germany. They were not close, and during Heisenberg's visit Szilard was preoccupied with his efforts to educate the higher echelons of government. All he could think about was the need to persuade the administration to share his vision, to step up several gears and create a dedicated program to develop a nuclear weapon. Besides, Szilard's attention probably would have had no effect upon the German; Heisenberg, it seems, was bound by duty to his country, to his roots, not to the ideology that had overtaken the country with the rise of Hitler. Like many Germans, he responded to an older form of patriotism and appears to have believed that he could play a role in moderating the energies of the Third Reich and at the same time serve his country the best way he knew how.

During Heisenberg's visit to the United States, the German War Office was already making advances in organizing serious research into nuclear fission. How much of this was known to Heisenberg when he was telling friends he had to return home is unclear, but by September 1939, a few days after war was declared between Germany and Britain, Heisenberg was called to give his expert opinion on the feasibility of constructing a bomb.

At this stage the Germans were certainly some way ahead of the Allies in the development of a nuclear weapon. Resources and finances were in plentiful supply, and experimentation was generally encouraged. More important, although the United States was sending material aid to Britain, the Allies were not then allies in the true sense of the word. Mainland Europe was about to be overrun, and the United States would not enter the war for over two years. In 1939 and into 1940, the Germans forged ahead with their atomic bomb program, and by 1941 Heisenberg had been appointed head of a research group exploring all aspects of atomic energy based at the Kaiser Wilhelm Institute in Berlin.

During the two years between the start of the war in Europe and the bombing of Pearl Harbor, Szilard passed through frustration and out the other side, but he never despaired. In spite of Roosevelt's bold claims to his friend Sachs, new objections were constantly raised to prevent funding the development of an atomic bomb in America. Coupled with this, Szilard's cause was hampered by the continued skepticism of many of his colleagues in the physics community. Some believed the production of a bomb was quite impossible, and those less pessimistic still saw it as, at best, a remote possibility, realizable in perhaps a decade or two and costing unimaginable sums.

In some ways this pessimism was justified. Although the Germans had shown themselves to be more open to new ideas and the race had begun with their willingness to experiment, both sides soon encountered a succession of what seemed at first to be impenetrable technical (rather than theoretical) barriers. It was the need to solve these problems that presented the greatest challenge to the enthusiasts and required some of the most determined efforts from both engineers and theoretical physicists.

$\mathcal{N}$UCLEAR FISSION IS the breaking of a large, unstable atomic nucleus into smaller pieces by bombardment using neutrons. The fission of the large nuclei into smaller pieces also liberates more neutrons, which may then go on to split other atoms in the material, creating a chain reaction. Each time fission occurs, energy is released. In itself, the release of energy from a single fission is tiny, but when we consider the vast number of atoms of a substance in a sample of an unstable material such as uranium and the fact that the energy from each fission is released almost simultaneously, it is easy to see that the burst of energy is huge, about 25 million times that produced by the combustion of the same quantity of fossil fuel. However, even after Hahn, Strassmann, and Meitner had conducted a small-scale laboratory demonstration of this process, turning nuclear fission into an effective weapon was a problem of a quite different order.

The first and perhaps greatest problem came from the need for isotopic purity within the fissile material. An isotope is a variety of any substance that has a number of neutrons different from the parent element. Take the example of carbon. Most atoms of carbon contain six protons and six neutrons, giving it an atomic weight (the combination of protons and neutrons) of twelve; this is called carbon 12. But there is an isotope of carbon that is relatively less stable and radioactive, carbon that contains eight neutrons (and six protons) in its nucleus, giving it an atomic weight of fourteen, this is carbon 14.*

Uranium has two common isotopes: the most abundant are uranium 238, which contains 146 neutrons and 92 protons; and uranium 235, with 143 neutrons and 92 protons. Within any sample of uranium, the ratio of these two substances is about 140:1; in other words, for every atom of uranium 235 there are on average 140 atoms of uranium 238.

The problem for the bomb makers was that only the uranium 235 isotope readily undergoes fission by bombardment with neutrons. Also, the neutrons that are produced from this fission process are "fast" neutrons that have the potential to cause further fission in other atoms. But these "secondary" neutrons are made less potent as fissile neutrons because some of their energy is absorbed by the uranium 238 atoms in any natural sample: This means that any attempt to bring about a chain reaction in a typical piece of uranium containing both isotopes would fail because neutrons produced by the relatively tiny amounts of uranium 235 would be made ineffective by the much larger quantities of uranium 238.

As research into the uses of fission for the construction of a weapon got under way, a simultaneous effort was started that concentrated on the possible peaceful uses of a chain reaction, the idea of using fission as an energy source. Enrico Fermi built the world's first nuclear reactor in a squash court at the University of Chicago, and it was tested successfully on December 2, 1942. To tame the chain reaction that could lead to a nuclear explosion, Fermi used graphite rods inserted into a mass of uranium. This slowed neutrons produced by the uranium 235 in the core but not so dra-

---

*This is the isotope used in carbon dating.

matically as the uranium 238 would do, so he was able to create what was effectively a slow-burning nuclear explosion or controlled nuclear fission. Based upon this success, scientists could show that, in theory, the energy from fission could be used to power steam turbines and supply the nation with electricity.

But the scientists working on using a chain reaction to create a nuclear explosion did not want the neutrons to be slowed at all. They wanted a situation in which the energy produced by billions of atoms undergoing fission could be released simultaneously. For them, the only way to achieve this seemed to be to purify the uranium, to separate the uranium 235 from the uranium 238, and to allow an uncontrolled chain reaction to begin in the pure uranium 235.

However, purification (or enrichment of uranium) proved to be extremely difficult. After a number of schemes were considered and rejected by researchers in the United States, a particularly complex technique, known as gaseous diffusion, was attempted. This involved putting gaseous uranium through a superfine porous barrier. The barrier prevented the passage of the larger atoms of uranium 238 but allowed through the smaller atoms of uranium 235.

This was an ingenious solution in theory, but in practice it proved extremely difficult. Most significant, in its natural state uranium is not a gas but a metal. So for this process to work, uranium had to be converted to a suitable gaseous compound. Unfortunately, the only uranium compound that could be separated by the gas diffusion process was uranium fluoride, which is one of the most corrosive substances known. At the time no known material could contain uranium fluoride for long, and there seemed no way to make usable laboratory apparatus for working with it. A whole new set of industrial processes had to be created even before the substance was used in the diffusion process. New materials had to be produced that could be molded into suitable containers, and pieces of dedicated apparatus, including hundreds of taps, miles of pipes, and a vast array of holding vessels, were designed and manufactured.

Before the team of scientists had gathered at Los Alamos, a vast uranium enrichment plant was built from scratch at a fifty-two-thousand-acre site at remote Oak Ridge in Tennessee. Its construction involved forty-five thousand workers, and when it was completed, twenty-five thousand technicians and scientists worked there. When it was operating at full capacity, it consumed more electricity than Pittsburgh, and its construction had taken a considerable chunk of the total budget of $2 billion assigned to the production of the atomic bomb.

Yet, as far as the war effort was concerned, most of this turned out to be time and money wasted. As the massive plants were being built and the minds of twenty-five thousand technicians and scientists were bent to the task of enriching uranium to produce pure uranium 235, in at least three laboratories around the world, physicists had almost simultaneously stumbled upon a fact that would make the process of uranium enrichment redundant.*

What the theoreticians had noticed was that when the common isotope of uranium, uranium 238, was bombarded with neutrons, some of the atoms captured a neutron and formed a different chemical species, uranium 239, which has 92 protons and 147 neutrons in its nucleus. This is a very unstable and short-lived atom, and after one of the neutrons in uranium 239 changes to a proton, it decays to neptunium 239 (with 93 protons and 146 neutrons). This atom then rapidly decays to plutonium 239, which has 94 protons and 145 neutrons.

Plutonium 239 is very important for two reasons. First, because it is a completely different substance than uranium 238, it has entirely different chemical properties and may be separated easily from any sample of uranium 238 that has been bombarded with neutrons. Second, and most crucial, plutonium undergoes fission.

This breakthrough occurred in the summer of 1941, just as the United States was beginning to take seriously the growing conflict overseas. And the timing was to prove hugely auspicious. By never easing their efforts to attract the attention of the government and the military, Szilard and his colleagues had finally begun to convince politicians of their arguments, and at the same time the scientific hurdles were falling. Within months of Pearl Harbor, the United States was compelled to undertake what would turn

---

*Although later it would play a crucial role in the peaceful uses of atomic energy.

into the most ambitious scientific project in human history. During the summer of 1942, the Manhattan Project was born.

But, across the Atlantic, the Germans were also hot on the nuclear trail.

Aʜ s FERMI WAS PREPARING his reactor in a squash court in Chicago in the summer of 1940 Werner Heisenberg and the physicist Carl von Weiszäcker mirrored the early stages of this work in a newly constructed wooden building at the Kaiser Wilhelm Institute. To discourage the curious, they called the building the Virus House and set up preliminary experiments with the intention of building a nuclear reactor. As a result of these experiments, by early 1941 the Germans had concluded that two possible moderators could slow down neutrons. One of these was graphite, the material chosen by Fermi in Chicago; the other was heavy water (water that contains a heavy isotope of hydrogen called deuterium).

Late in 1940 Walther Bothe, a colleague of Heisenberg at the University of Heidelberg, had concluded a series of experiments to determine graphite's ability to absorb neutrons (a property called the cross section). He had obtained a value for the cross section of graphite that was very different from those found by Fermi's team. Bothe's figures suggested graphite would make a poor moderator. Unfortunately for the Germans, his values were inaccurate, but in a hurry to move forward, Heisenberg accepted them without verification and went with the less effective alternative, heavy water.

Notwithstanding this mistake, by early 1941, while scientists and bomb activists in the United States were still mired in bureaucracy, the Germans were incredibly well placed to build an atomic bomb. They possessed the scientific talent (even though they now had fewer first-rate scientists than either the British or the Americans); they had thousands of tons of uranium ore from occupied Belgium and the Belgian Congo; and they had access to the world's only heavy water factory, at Vemork in German-controlled Norway. They also had financial backing and support from the higher echelons of the Nazi Party. But, during the course of the next six months, everything was to change.

By choosing heavy water over graphite, Heisenberg and his colleagues had inadvertently stacked the odds against their success. As Fermi had shown, graphite was the best moderator and Bothe's experiments had been seriously misleading. However, this choice in itself was not fatal to their effort. And Germany had plenty of heavy water (at least until the British destroyed the Vemork plant in February 1943).

A more serious problem was Heisenberg's failure to see how uranium could be converted to plutonium and then separated before its use as a fissile material. This meant the German researchers were pursuing the only possible method of obtaining uranium 235, using uranium fluoride. And, as mighty as the German war machine had grown by early 1941, the resources needed to produce enough pure uranium 235 for a bomb were simply not available.

By the summer of 1941 Heisenberg had become aware of these errors, but it is also clear from surviving documents that he was increasingly concerned about the moral and political implications of his work. In September 1941 he obtained special permission to travel to occupied Denmark to talk to his old mentor Niels Bohr. Officially the trip was to discuss physics, and the German authorities probably sanctioned it because they believed Heisenberg might obtain important information. But he was following his own agenda.

By this time many scientists living in occupied Europe no longer trusted Heisenberg; to them his refusal to leave Germany when he could have was tantamount to agreeing to support the Nazi cause: For those suffering daily privation in occupied lands, there was no middle ground. Bohr's assistant in Copenhagen, the Polish physicist Stefan Rozenthal, would not even talk to Heisenberg when he arrived.

But Bohr did agree to meet him; to the annoyance of his colleagues, he invited Heisenberg to dinner at his home. However, from interviews with Heisenberg conducted after the war and from recollections of what Bohr told friends immediately after the meeting, it is clear that from the start the Danish scientist treated Heisenberg frostily and remained tight-lipped about his own work.*

After dinner the two men talked, and Heisenberg apparently tried to break through Bohr's resistance. We must remember that, to Heisenberg,

---

*This meeting is the subject of Michael Frayn's play Copenhagen, which has won plaudits for performances in London and on Broadway.

Bohr was a magus, and he would have been deeply distressed by his old tutor's cold attitude. Bohr, however, now perceived Heisenberg as a German first and a scientist second and therefore felt he could not be trusted. Yet Heisenberg had gone to great lengths to make this visit and had a clear objective, this was obvious even to the oversensitive Bohr.

During their talk Heisenberg became uncomfortable and depressed. For some time he had been developing what can only be described as a siege mentality. While the German state enjoyed its period of greatest expansion, Heisenberg was feeling increasingly isolated from the rest of the scientific world and sensing the resentment and reproach of his former colleagues. Their perception only made what he had to say appear more abrupt and therefore open to misinterpretation. Furthermore, Bohr was so distrustful he would not have been inclined to give Heisenberg any benefit of doubt.

They started badly by talking about the Germans' war efforts. Heisenberg offered the opinion that Germany would soon be in control of Russia and that this was a good thing. This angered Bohr, who interpreted the comment as a clear sign of Heisenberg's support for the Nazi cause. Then Heisenberg moved awkwardly on to the subject he really wanted to discuss.

After the war, Bohr's assistant Victor Weisskopf recalled: "Heisenberg wanted to know if Bohr knew anything about the nuclear programme of the Allies. He wanted to propose a scientist's decision not to work on the bomb, and he wanted to invite Bohr to come to Germany to establish better relations."[10] Of course Bohr was stunned by talk of atomic weapons and immediately went on the defensive. He assumed Heisenberg was trying to lead him into divulging anything he might know of an Allied project to build a bomb.

Meeting only stony silence, Heisenberg went on to ask Bohr if he thought it morally right that scientists should engage in any effort to build an atomic bomb. Bohr was again startled and replied with honesty that he could not imagine it ever being possible to build such a device.

What Heisenberg said next depends on whose testimony one believes. According to those Bohr talked to after the meeting, the German made it very clear that he was working on a project to develop an atomic device, what he called a "uranium machine." But, according to Heisenberg, he simply said: "I know that this is in principle possible, but it would require a terrific technical effort, which, one can only hope, cannot be realized in this war."[11]

By now, Bohr was furious. And as Heisenberg tried to explain himself more clearly, he merely made matters worse. The next day Bohr reported the conversation to his family and a few close colleagues at work. "Either Heisenberg is not being honest, or he is being used by the Nazi government," he declared to one witness.[12]

Further depressed and still more isolated, Heisenberg returned to Germany and did not speak to Bohr for many years. He had taken a huge personal risk in approaching Bohr. He did not know if his own government trusted him, and from reliable sources it appears that for some time he had been actively discouraging Nazi interest in his research. The Jewish physicist Fritz Reiche, who had escaped from Germany and reached the United States in March 1941, had brought with him a message for the physics community. "Heisenberg," he reported, "will not be able to withstand any longer the pressure from the government to go earnestly and very seriously into the making of the bomb, despite the fact that Heisenberg tries to delay work as much as possible."[13]

Heisenberg must have known that Bohr would immediately alert his colleagues and try to get news of their conversation to the Allies. He must have worked through the consequences and realized that if the Allies took the idea seriously, they would begin a rival project that would certainly outstrip anything he was doing.

From the evidence available, it seems clear that late in 1939, as war was breaking out across Europe, Heisenberg had thrown himself into researching atomic fission with the enthusiasm one would expect of a particle physicist then at the height of his powers. But by 1941 he had come to realize that the interest of his government (in particular men with technical backgrounds, like Albert Speer) might lead to his purely scientific intentions becoming perverted to military ends. To a man like Speer, there would be no clear distinction between experiment and application, and he would push hard for the latter.

Heisenberg had wanted to experiment, but when it became clear the Nazis would demand much more, he was faced with a difficult moral dilemma. On the one hand, he was a patriot, and like many Germans he was particularly fearful of the Russians (hence his comments to Bohr), but he also knew that, given sufficient resources, in the wrong hands the realization of his research would have far-reaching and possibly devastating consequences.

If we assume this to be true, it seems likely that in his rather clumsy way Heisenberg was actually trying to defuse the situation by talking to Bohr. Whether or not he really believed it possible to create a band of scientists (a *Bund* not dissimilar to Szilard's ideal perhaps) that could influence governments and act independently of tyrants and traditional democrats is impossible to judge. Perhaps such thoughts are academic anyway, for such a scheme would have stood almost no chance of success at that volatile time.

Whatever Heisenberg's intentions that September evening in Copenhagen, his thoughts and plans were quite insignificant. Forces far greater than one scientist's ideology had already been unleashed, and these would make it impossible for Heisenberg to build a bomb even if he had wanted to.

Hitler himself probably would not have cared what Heisenberg might have achieved. When advisers tried to explain the principle of the atomic bomb to him, he had shown almost no interest. Ironically, the most dramatic factor that halted the German effort to build an atomic weapon was the military decision Heisenberg had claimed to admire during his talk with Bohr. When Hitler decided to open the Eastern Front in the summer of 1941, all resources viewed as nonessential, including Heisenberg's experiments at the Kaiser Wilhelm Institute, were diverted to this cause. Just as Heisenberg was beginning to pursue the most expensive and slowest option in the development of an atomic bomb, his budget was slashed and his men relocated to other projects.

From late in 1941, just as the first serious moves were being made to set up the Manhattan Project, the German effort to build a bomb was in tatters.* But, crucially, almost no one outside Germany knew this. Those few who did were to have the greatest influence upon the course of the war and indeed the political structure of the postwar world.

---

*Since the opening of the U.S. National Archives material covering this period of the war, there have been many theories about how advanced the German atomic bomb program was and the culpability of scientific figures (not just Heisenberg). There has also been plenty of speculation about the involvement of the other Axis powers. A recent report (Philip Henshall, *The Nuclear Axis: Germany, Japan and the Atom Bomb Race, 1939–1945* [London: Sutton, 2000]) proposes that the Japanese atomic program was actually more advanced than the Germans', but there is only circumstantial evidence to support this notion.

· · ·

SZILARD AND HIS SUPPORTERS had been working for two years to establish a serious bomb project, but the decision to move ahead was made only at the most politically auspicious moment. Throughout 1940 and 1941, the British government had been pushing the United States into committing itself to nuclear weapons production. Winston Churchill had seen the potential of atomic bombs immediately and had put his weight behind the idea. British physicists had been at the cutting edge of atomic research for decades, and many of the most respected scientists in the field were British. Britain also owned large stocks of uranium ore saved from mainland Europe as the Nazis had advanced, as well as supplies from the colonies. What they did not have was money and manpower. By 1941 Britain was almost bankrupt, battered by two years of war and on the point of exhaustion. America had the money, the manpower, and the material resources to make an atomic weapon. Clearly, the two nations were compelled to work together on this huge undertaking.

But Churchill's powers of persuasion were effective only once the United States was actually at war with Germany. By coincidence, late 1941 was also the point at which sufficient numbers of scientists were beginning to believe that an atomic weapon was theoretically and technically feasible. Everyone involved, scientists and politicians, knew it would be an astonishingly expensive undertaking, but the fear that Hitler's Reich might soon have its own atomic bomb presented a powerful incentive.

Roosevelt, greatly influenced by his scientific advisers, was without doubt slow to react to the claims of Szilard, Fermi, Einstein, and others, but when he did realize the potential of atomic weapons, he made sure things happened very quickly and all necessary resources were made available.

The supply of materials was a problem solvable with financial commitment; managing human relations presented a thornier challenge. It was soon realized the Manhattan Project could work only by bringing together very strange bedfellows—the military and the scientific community. Such a union had never been attempted on this scale before, and parties on both sides were skeptical the two could work together at all. To administer this coalition, Roosevelt appointed Vannevar Bush, who throughout the three years of the Manhattan Project was answerable only to the president. Bush was ultimately responsible for the organization of hundreds of thousands of staff, the establishment of Los Alamos itself, and a budget of $2 billion (equivalent to $30 billion today), an effort as demanding in its time as the construction of the pyramids had been for the ancient Egyptians and a project that by the end of World War II had grown to be larger than the entire American car industry.

$\mathcal{T}$ODAY LOS ALAMOS, situated in the New Mexico desert, is a city with a population of twenty thousand. It has schools, movie theaters, hospitals, and stores, but until 1942 it did not exist. The area, thirty-five miles northwest of Santa Fe, is dominated by a seventy-two-hundred-foot mesa, and it had been the site of the Los Alamos Ranch School, established in 1917.

The military commander of the newly created Manhattan Project, Bush's second, General Leslie Groves, chose Los Alamos after inspecting several sites around the country during the summer of 1942. The site was so remote it had no electricity or gas supplies and possessed only one Forest Service telephone. Satisfied this was the ideal location, Groves paid $440,000 for the land, the buildings, sixty horses, two trucks, and two tractors. In the autumn of 1942, contractors were brought in to begin building the research establishment that would soon house, as he put it: "the biggest collection of eggheads ever."

The Manhattan Project got its name because much of the early work on the creation of the bomb was done by scientists from Europe who had gathered in New York after the start of the war. From the beginning it was a turbulent marriage of scientific and military sensibilities, but fortunately the two men appointed by Bush to run it, General Groves and the lab director, Robert Oppenheimer, enjoyed an immediate rapport and held a deep and lasting respect for each other.

Groves was born into a traditional middle-class family. His father had been a small-time lawyer who became a successful minister. He had encouraged Leslie to work hard at school and watched proudly as his son graduated from the University of Washington and West Point. Leslie Groves then served with distinction in Europe and Central America, and when he decided to make the army his career, he rose through the ranks quickly. By the time of the Manhattan Project, he was deputy chief of construction for the entire U.S. Army.

Weighing in at 250 pounds, Groves was a big man. He was loud and outspoken and exuded an infectious confidence and determination. Many of those he worked with disliked him personally but could not fail to admire his considerable talents. One of his closest assistants, Lieutenant Colonel Kenneth Nichols, who served with Groves for the second half of the war, called him "the biggest sonovabitch I've ever met in my life, but also one of the most capable. He had an ego second to none, he had tireless energy—he was a big man, a heavy man but he never seemed to tire. He had absolute confidence in his decisions and he was absolutely ruthless in how he approached a problem to get it done. . . . I hated his guts and so did everybody else but we had our form of understanding."[14]

Groves was usually a good judge of character, and once his loyalty was won, he never wavered. He immediately took to Oppenheimer, once declaring. "He's a genius, a real genius. While Lawrence [the physicist Ernest Lawrence] is very bright he's not a genius, just a good, hard worker. Why, Oppenheimer knows about everything. He can talk to you about anything you bring up. Well, not exactly. I guess there are a few things he doesn't know about. He doesn't know anything about sports."[15]

J. Robert Oppenheimer came from a background very different from Groves's. Born into privilege, he was cushioned from many of life's vicissitudes, but as he grew older his family's wealth began to disturb him. Although he appreciated and enjoyed the finer things in life (including the van Gogh he inherited from his parents) and benefited enormously from the best education available, he was also troubled by what he saw as an overabundance of advantages bestowed upon him by chance of birth.

His Jewish émigré father, Julius, had arrived in America during the 1880s, quickly established a successful business, and in 1903 married into a wealthy, academic family. Robert's mother, Ella, had been a talented artist in spite of the fact that she had been born without a right hand. She greatly influenced Robert, and it is from her that he acquired a strong artistic streak, an aspect of his personality that not only rounded him intellectually but also informed his approach to science. Oppenheimer became an early exponent of the quantum theory, and throughout his career, he placed profound emphasis upon uniting an artistic and a scientific vision of the interpretation of this new science.

During the 1920s the Oppenheimer family lived on the Upper West Side of Manhattan in a vast apartment overlooking Central Park. Robert and his brother, Frank, eight years his junior, enjoyed frequent foreign travel, a grand summer home on Long Island and the trappings of wealth and privilege.

Robert showed academic brilliance from an early age and was as gifted in the arts as he was in mathematics and science. He was enthusiastic about learning but was seen by others of his age as a bit of a sissy; at camp he was vulnerable and bullied mercilessly. He became a frail, sickly, intellectually intense youth. He was constantly worried by his own easy path through academia and the immediacy with which he grasped concepts, describing it in a letter to a friend as "the awful fact of excellence."[16] This was no pose but a genuine feeling that he had only to reach out and anything he wanted, materially or intellectually, would be his.

A family friend took him on an extended trip to the wilderness of New Mexico when he was eighteen. This not only initiated his love affair with the Southwest of America (a place he would of course return to some quarter century later) but also toughened him. By the time he arrived at Harvard as a chemistry undergraduate in 1922, he had matured remarkably.

Oppenheimer thrived at Harvard and merged perfectly with the other students, many of whom came from backgrounds as privileged as his own. He formed a close friendship with a fellow science student, Francis Fergusson, who remained his most trusted male companion until Oppenheimer's death in 1967. They traveled through Europe together during summer vacations and wrote long letters to each other after Fergusson left for Oxford in 1924. In these Oppenheimer enthused about the latest literature and painting and the physics that was beginning to dominate his thinking. Along with these accounts he included poems and, once in a while, a short story mixed in with complex discourses on philosophy he had unearthed from arcane texts.

After graduating from Harvard summa cum laude in 1925, Oppenheimer left chemistry behind and decided to pursue his budding interest in physics. His tutor placed him with the world-famous Ernest Rutherford at his laboratory, the Cavendish, in Cambridge.

In 1925 Cambridge was one of the focal points of the "new physics." Einstein's general theory of relativity was only nine years old, and the first paper to describe elements of quantum mechanics, written by the French physicist Louis de Broglie, was published the very month Oppenheimer arrived in England.

Ernest Rutherford was a classical physicist, but at the time he was seen as probably the most important figure in physics, a man who had built his reputation upon his gifts as both a theoretical and an experimental physicist. Unfortunately, Oppenheimer was neither interested in experimental work nor especially good at it, so Rutherford took little interest in him. This rejection came at the very moment Oppenheimer was discovering his fascination with theoretical work, in particular the innovative findings of men such as Niels Bohr, Max Born, and Erwin Schrödinger.

Nevertheless, Rutherford's cold shoulder, blended with Oppenheimer's own growing anxieties about his future career, precipitated what we now know was a nervous breakdown. On vacation with Francis Fergusson during his first Christmas in England, Oppenheimer's inner rage and frustration burst into the open; without provocation he physically attacked his friend. He later apologized and tried to explain his odd behavior, but back in college others had grown concerned by what they saw as his increasingly eccentric manner. Then, the following Easter, Oppenheimer was again on holiday with Fergusson and a group of college friends when he suddenly announced he was returning to England that very evening because he had "left a poisoned apple on a colleague's desk in Cambridge."[17]

It was not long before Oppenheimer's parents heard of these disturbing episodes, and after they visited him at Cambridge he was sent for psychiatric treatment at a Harley Street practice in London. Here he was diagnosed as suffering from dementia praecox, but Oppenheimer took, little notice, telling a friend that he thought the doctor too stupid to follow him and that he knew more about his troubles than any psychiatrist could.[18]

But after he completed his first miserable year at Cambridge, everything changed for Oppenheimer. He left England and began to study under the great Max Born at Göttingen, where he met and became friends with the English mathematical physicist Paul Dirac. It was during this twelve-month stay that Oppenheimer discovered his vocation and began truly to come to grips with quantum mechanics. Before retuning to the United States he completed his Ph.D. and started devising (with Dirac) a theory to describe the existence of an antiparticle, the positron.*

By the early 1930s Oppenheimer had established himself as one of the brightest young physicists in the United States and had the luxury of a dual

---

*The existence of this particle was later verified experimentally by C. D. Anderson.

appointment as research fellow at Cal Tech and the University of California, Berkeley. Then as now, Cal Tech was at the cutting edge of modern physics, and Oppenheimer's star ascended as the young faculty (founded in 1921) grew to become a center of excellence.

During the decade leading up to the Manhattan Project, Oppenheimer enjoyed the privileges of his academic position as well as his inherited wealth (his mother died in 1931 and his father in 1937). He published a succession of acclaimed papers on theoretical physics, which concentrated on the interactions between subatomic particles, and these earned him the respect of his colleagues. He had remained close to his younger brother, Frank, who was also a physicist, and between college terms he traveled extensively. He also continued to write poetry and acquired a new and profound interest in music.

Not surprisingly perhaps, Oppenheimer had many friends and a succession of female companions. On one date he crashed a newly purchased sports car while racing the coast train near Los Angeles. With him was Natalie Raymond, a society beauty he had met at a party in Pasadena a few years earlier. Neither Oppenheimer nor his passenger was seriously injured, but Miss Raymond was knocked unconscious. Oppenheimer thought he had killed her and was later so full of remorse over the near disaster he gave her a Cézanne drawing.

As Fascism began to dominate the political scene in Europe, Oppenheimer followed an example set by many other intellectuals of the time and acquired a newly fashionable interest in Communism. This took him to meetings held by activists within the Californian universities, but he never became a member of the party. Even so, this involvement plagued Oppenheimer for the rest of his life.

Around this time he met and fell in love with a militant Communist named Jean Tatlock. Before long Oppenheimer and Tatlock had began a torrid affair, but while he wrote her love poems and was utterly devoted to her, she would take off on long, lost weekends, returning with tales of her sexual exploits. Oppenheimer was torn apart by jealousy and close to another breakdown before he managed to disentangle himself from her.

Then, in 1939, he met a twice-divorced reformed alcoholic and another Communist Party member, Katherine (Kitty) Harrison, who was newly married to an English physician Stewart Harrison. Within a year of her marriage to Harrison, Kitty and Oppenheimer had fallen in love, and by the autumn of 1940 she was pregnant by him. She succeeded in divorcing Stewart Harrison, and Oppenheimer married her on November 1, 1940, six months before their first son, Peter, was born.

By 1942, as plans for the Manhattan Project were progressing, Oppenheimer was comfortable and content with his academic life. He was working on a small set of theoretical problems that absorbed him, and he had established himself as a fine teacher. Hans Bethe, who worked with Oppenheimer at Los Alamos, said of his teaching: "Probably the most important ingredient he brought to his teaching was his exquisite taste. He always knew what were the important problems, as shown by his choice of subjects. He truly lived with these problems, struggling for a solution, and he communicated his concern to his group.... He was interested in everything, and in one afternoon [he and his students] might discuss electrodynamics, cosmic rays, electron pair production, and nuclear physics."[19]

It is in this statement that we may see why Oppenheimer, rather than any other scientist, was chosen to lead the Los Alamos team. Although he was a brilliant thinker, and by 1942 had already made original and important contributions to at least two branches of physics (particle physics and astronomy), he was not a theoretician in the class of Bohr or Heisenberg. He could also be arrogant and elitist, but his greatest talent was his ability to manage scientists and to organize a complex technical operation with grace and delicacy.

Groves saw these qualities in Oppenheimer immediately, and their blend with Oppenheimer's obvious intellectual power impressed the general enormously. The last thing Groves wanted was a yes-man, and he knew Oppenheimer would never be that. He also clearly admired his scientific versatility and evident skills as a communicator.

Yet, most crucially, it was the very fact that Oppenheimer was not a specialist that made him the perfect scientific administrator for the Manhattan Project. Uniquely, he possessed an exceptional talent for understanding everything that was going on. This enabled him to converse with physicists, mathematicians, chemists, and engineers on their own level, whatever their specialty. Oppenheimer quickly became a consummate leader, administrator, and facilitator. Oppie, as he was soon nicknamed, could talk to the blinkered military men, appease the government, and deal with the many prima donnas whose delicate intellectual powers he knew had to be nourished and coaxed.

Life at Los Alamos was not easy for anyone working there, or for the families who had set up home at the site. Built on a mesa, the laboratory complex was soon christened the Hill. During the first stage of its development, in the early months of 1943, it was little more than a huddle of ramshackle buildings hastily erected on the muddy plateau. The planned site was meant to house a maximum of one hundred scientists, but the population soon reached three times this figure. Most of the scientists and engineers were young—their average age, twenty-five—and many were newly married. Not surprisingly, during the first year at Los Alamos, resources were stretched still further by a mini–baby boom.

They had been promised an up-to-date working environment, with comfortable housing and modern facilities, libraries, a movie theater, restaurants, and a laundry service, but when the team arrived they were stunned to find a shantytown almost medieval in its squalor. Water had to be pumped from Guaje Canyon, seven miles away, and it carried a strong taste of chlorine and leaf mold. Almost everyone who moved to the Hill spent the first three days vomiting and suffering stomach pains from a bug they dubbed Hillitis, and before anyone could run a bath, a gauge had to be consulted to see if there was any water available.

The electricity supply was equally unreliable. There was no electricity company in the region, and four old generators found in an abandoned local mine had to be refurbished to supply the needs of the project. These broke down frequently and often at the most inopportune moments.

The Los Alamos site was little more than a military camp; some even thought of it as looking and feeling a little like a concentration camp. Armed guards patrolled a high metal fence that skirted the Hill, mail was censored, and the movements of all were closely scrutinized. During the first year only Oppenheimer and the post commander were allowed telephones, and most communication with the outside world was conducted via Teletype. Richard Feynman famously took great pleasure in pitting his wits against the security measures by inventing codes so he could get letters past the censors and staging escapes under the noses of the guards.

After the damp and the mud, security was the most hated aspect of working at Los Alamos, and the general feeling among the scientists was that it was overdone. But, unknown to most of the Los Alamos researchers, high security was the price for a bargain Oppenheimer had struck with Bush and Groves during planning of the establishment.

Because security is in the blood of military men and government officials, they cannot function without it, neither can they visualize a world in which individuals are free to discuss matters deemed sensitive. And there could be nothing more sensitive than military secrets during a time of global war, especially if the raison d'être of the military establishment was to win a race to build something as important as an atomic bomb.

But equally, science cannot work properly in an atmosphere of secrecy, which goes completely against the mind-set of most scientists, for whom the free exchange of ideas is the wellspring of all invention. Although Newton was obsessed with secrecy and Darwin was scared to publish, both these cases involved personal reasons and do not reflect the general attitude of scientists toward the communication and eventual acceptance of their discoveries. Science is, in the grand sense, a communal activity in which some individuals contribute vastly more than others.

Scientific journals publish any new discovery deemed original and verifiable; scientists offer their thoughts for their colleagues to develop. Somehow, science progresses through rivalry and competitiveness as well as a sharing of fundamental information. As progress is driven by human lust for glory and personal recognition, it also thrives upon the free exchange of ideas. This ethos has always been incredibly strong within the scientific community, and the notion that an outside agent—in this case the military—should interfere and break down such a tradition was resented enormously by scientists involved with the Los Alamos Project.

Oppenheimer knew the atomic bomb could be made reality only if all the scientists working on the project were able at all times to exchange ideas and to know what other scientists were doing, just as they had always done. Yet the military wanted to establish groups of scientists working in isolation, on a "need to know" basis. Obviously a compromise had to be reached. But before the first scientists arrived on the Hill, Oppenheimer had convinced Groves the military approach would fail utterly. Groves persuaded Bush, and he eventually gained clearance from President Roosevelt. But such a break in protocol came at a price; the Hill itself had to be a secure unit, the entire population treated as suspect, all mail read and if necessary censored.

Everyone was issued a pass, and the scientists and their families were allowed off the site only for specific, officially sanctioned reasons. Many of the more important figures were under constant surveillance. When they left the mesa they were trailed by FBI agents, who filed reports on their activities.

Throughout the three years of the Manhattan Project, no one was suspected more than Oppenheimer himself. Vannevar Bush was convinced Oppie was a spy and tried repeatedly to have him forced out, but Groves always fought back and won. This paranoia was fueled by Oppenheimer's past involvement with Communist sympathizers. The ghost of his relationship with Jean Tatlock haunted him still. She had become increasingly unstable, drank heavily, and was often suicidal. Even after he had married Kitty, Oppenheimer could not resist responding to Tatlock's pleas for help. By the time she committed suicide in January 1944, the FBI had a fat file on the couple, including surveillance records and intimate photographs. And a decade later, when Oppenheimer fell under Joseph McCarthy's suspicion, this relationship was again dragged up.

In one of the most ironic twists of the saga, security problems also haunted Leo Szilard. The man who more than anyone had forced through the need to develop an atomic bomb was never invited to Los Alamos because General Groves (displaying a rare lapse of judgment) viewed him as a subversive. After the war he called Szilard "the kind of man that any employer would have fired as a troublemaker."[20]

More than anyone, Szilard opposed the idea of segregating scientists and forcing them to work in an atmosphere of secrecy. He understood the need for security and the risk from foreign agents but also knew that science cannot function within fabricated military guidelines. When the authorities realized Szilard was talking freely with his colleagues across the world about the processes involved in the creation of an atomic bomb (just as he had been doing since 1939), Groves, Bush, and the security services immediately concluded he was a spy and attempted to have him arrested. This was forestalled only by the intervention of Szilard's influential friends, and a job was arranged for him working with Fermi and Arthur Compton at the atomic pile in Chicago. There Szilard stayed for the duration of the war, under constant surveillance.

During the early planning stages of the Manhattan Project, Albert Einstein had been suggested as a key figure to work on the Hill. But Einstein had nothing to do with the project. His expertise had little direct relevance to making bombs, and by the 1940s he was entrenched at the Institute for Advanced Study and could not have been persuaded to work at Los Alamos. Yet, even if he had wanted to work there, government prejudice would have prevented him.

Again, Bush was at the heart of this paranoia. "I'm not at all sure that if I placed Einstein in entire contact with his subject he would not discuss it in a way that it should not be discussed," he wrote in a memorandum to the director of the Institute for Advanced Study early in the war. "I wish very much that I could place the whole thing before him . . . but this is utterly impossible in view of the attitude of the people here in Washington who have studied into his whole history."[21]

Yet perhaps most ironic is that, for all their misguided ideas about who should be under strict surveillance and who should not, the authorities completely missed the real spy—the Los Alamos researcher Klaus Fuchs—who, throughout the final two years of the war, was busily passing on technical information to the Russians.

Aside from the security precautions, the pressure to deliver brilliant physics and flawless engineering in a race against the Nazis was immensely stressful for everyone on the Hill, and this pressure initiated some destructive disputes among members of the team. One of the most serious came during the summer of 1943 and originated with a practical problem that stretched the abilities of the Los Alamos scientists more than any other. Having made the decision to switch from isotope separation to plutonium production, the next problem they faced was to find the best design for the bomb itself. Key to this was the construction of a mechanism to control the fissile material until the moment the bomb was to be released.

The scientists at Los Alamos knew that in order for a chain reaction to be sustained, a certain *minimum* mass of fissile material was needed. This is called the critical mass. To produce a bomb, two noncritical masses must be kept apart to prevent premature chain reactions, but they must be brought together very rapidly to create the critical mass and initiate an almost instantaneous release of energy.

To accomplish this, the team designed a bomb in which a small piece of uranium would be fired into the center of a large piece using a "gun" detonator. In theory, the two pieces would then reach a critical mass and the bomb would explode. The problem with this method was that the "bullet" had to travel extremely fast, at around 3,000 feet per second. If it traveled any slower, neutrons from the smaller mass might reach the larger piece of uranium early and set off small explosions. Unfortunately, in 1943, when the calculations were complete and the system proposed, the army had no gun capable of coming close to propelling a mass at such high speed. So an alternative method had to be found.

A young physicist named Seth Neddermeyer proposed that a mass of uranium could be packed into a tube and the tube surrounded with high explosives. When the explosives were detonated, they would force the uranium into a more dense form and create a critical mass and a subsequent chain reaction.

His idea was approved by Oppenheimer, and Neddermeyer established a group to carry out tests. However, it was soon realized his theory was not practical because the explosives had to be packed in such a way as to produce a perfectly uniform implosion. But, to the growing frustration of others at Los Alamos, Neddermeyer spent months testing various methods totally without success. After hearing Neddermeyer's latest theories at one meeting, his friend the outspoken Richard Feynman told the assembly, "It stinks."[22]

Another team who had been working on the bullet method, led by the military-trained Captain Deke Parsons, argued that Neddermeyer was wasting time and valuable resources and ought to be stopped. "I question Dr. Neddermeyer's seriousness," Parsons declared. "To my mind, he is gradually working up to what I shall refer to as the Beer-Can Experiment. As soon as he gets his explosives properly organized, we will see this done. The point to watch for is whether he can blow in a beer can without splattering the beer."[23]

From this opening shot, the arguments between the two team leaders raged on for almost a year. Parsons demonstrated time and again that Neddermeyer was going about things the wrong way; he was convinced the only way Neddermeyer's method could work would be to pack the explosives around a perfect sphere of uranium rather than a tube. But Neddermeyer claimed this was impossible in practice because any imperfection in the sphere created "shock waves" that produced a nonuniform explosion and ruined the bomb.

Eventually, Oppenheimer was forced to intervene. He listened carefully to both sides, then overruled Parsons and agreed to give Neddermeyer more time. But still the method could not be made to work. By the autumn of 1943 Oppenheimer was forced to change his mind about Neddermeyer's idea, and he created a team consisting of both sets of researchers, with Parsons in overall control and Neddermeyer acting as an adviser. It was a delicate compromise and a bitter pill for all, especially Seth Neddermeyer, to swallow, but each believed the project had to come before personal grievances, and for a time both men worked hard to cooperate.

The peace between them did not, however, last long. By early 1944 Parsons and Neddermeyer were once more clashing openly. The Ukrainian chemist George Kistiakowsky, who arrived at Los Alamos in early 1944 and was assigned to help Neddermeyer, recalled how personal conflicts were very much an everyday element of life on the Hill and how the Neddermeyer-Parsons arguments became almost unmanageable. "Everything in books [about Los Alamos] looks so simple, so easy, and everyone was friends with everybody.... After a few weeks I found that my position was untenable because I was essentially in the middle trying to make sense of two men who were at each other's throats. One was Captain Parsons who tried to run his division the way he had done in military establishments—very conservative. The other was of course Seth Neddermeyer, who was the exact opposite of Parsons, working away in a little corner. The two never agreed about anything and they certainly didn't want me interfering."[24]

Eventually, it took a fresh approach to solve the problem of implosion and to create the mechanism that would be used in the bombs dropped on Hiroshima and Nagasaki. The Hungarian mathematician John von Neumann in collaboration with Richard Feynman found the solution in a series of diagrams scribbled on coffee-stained scraps of paper during an intense lunchtime discussion. Attempts had been made to use a spherical container for the implosion method, but all efforts had failed to create a regular shock wave that would produce an even distribution of force. Within the sphere, eddies were formed, which led to an irregular implosion. From purely mathematical considerations, and without conducting a single experiment,

Neumann and Feynman showed that to compensate for the eddies and establish a regular implosion, the sphere had to be made from wedges and contain different types of explosives.

But although Oppenheimer was a talented diplomat and a calming influence, he also had his own battles to fight. The most prolonged of these and potentially the most damaging was his clash with one of the most senior scientists on the Hill, Edward Teller.

Teller, a friend of Leo Szilard and a fellow Hungarian émigré, had arrived in the United States in 1935 and was quickly appointed professor of physics at George Washington University. He had worked closely with Enrico Fermi on the first nuclear reactor at the University of Chicago and arrived at Los Alamos in 1943. But it was clear from the moment he settled into his lab that he was never going to be much of a team player and was really only there to pursue his own ideas about atomic weapons.

As early as 1941, when he had been working in Chicago, Teller had become obsessed with the idea that a far more powerful weapon than a simple nuclear fission bomb could be created. The process that interested him was a nuclear *fusion*. Fusion relies on the fact that two small nuclei, usually the heavier isotopes of hydrogen—deuterium (composed of one proton and one neutron) and tritium (composed of one proton and two neutrons)—may be fused together to produce a helium atom. This fusion creates neutrons that then help further fusion. Most crucially, this process releases vast amounts of energy, far more than that generated by an equivalent nuclear fission reaction.

Teller was a great theoretician and was respected by all who knew him, but he was also extremely headstrong and career-minded. Like all the scientists working at Los Alamos, he understood the need for urgency, and he feared the Nazis as much as anyone, but he was so absorbed with creating a fusion bomb, which he quite wrongly believed would be easier to manufacture than a fission bomb, he could not bring himself to focus on what the others were heading toward.

Oppenheimer could of course see the potential of the fusion process, but anyone not obsessed with it could also instantly realize that, for the purposes of defeating Germany, the atomic bomb was the obvious choice and could be produced far sooner than the hydrogen bomb.

Teller was alone in proposing that the fusion bomb take center stage, and when, inevitably, his wishes were overruled and repeatedly opposed by every other scientist at Los Alamos and the military commanders, he withdrew his commitment to the project.

His most vehement critic was the German American Hans Bethe, a man every bit Teller's scientific equal and officially his superior at Los Alamos. "He declined to take charge of the group that would perfect the detailed calculations on the implosion [based upon Neumann and Feynman's creative mathematics]," Bethe recalled almost forty years later. "Since the theoretical division was very short-handed, it was necessary to bring in new scientists to do the work that Teller declined to do."[25] Bethe was infuriated by Teller's intransigence at a time when everyone at Los Alamos was racing to meet one deadline after another and many of the teams were working sixteen-hour shifts. He took the matter to Oppenheimer.

Oppenheimer understood Bethe's frustration, but he also wanted Teller at Los Alamos producing the great work everyone knew he was capable of. Using his considerable powers of diplomacy and persuasion, Oppenheimer tried hard to reach a solution to the problem. However, after Teller twice more refused to help with fission calculations, Oppenheimer was forced to write to Groves suggesting Teller be replaced by the German British physicist Rudolf Peierls, who had just arrived at Los Alamos with a large group of British scientists.

Referring to the problems Teller had caused, Oppenheimer wrote: "These calculations were originally under the supervision of Teller who is, in my opinion and Bethe's quite unsuited for this responsibility. Bethe feels that he needs a man under him to handle the implosion program." He added: "This is of the greatest urgency."[26]

Teller's threats to leave had almost certainly been a bluff. For all his independence of thought, he, like every other scientist on the Hill, valued the energy and intellectual stimulation such a unique gathering of minds provided. At Los Alamos he also had access to the best facilities available anywhere. But Oppenheimer was also quite aware of Teller's value. He removed him from Bethe's team, replaced him with Peierls, and allowed Teller free rein to theorize about the fusion bomb—but at a price. Teller's access to facilities on the Hill was greatly restricted, and he was obliged to attend a weekly one-hour brainstorming session with Oppenheimer. In exchange, Teller could watch the others work as he thought through his own plans.*

Even though Teller had been given more than resources would normally allow, he continued to see himself as the injured party. For their part, the government and the army were careful to indulge Teller because he was producing something that most military men and scientists agreed would one day constitute an important arms development. However, if he had not been so obsessed with his own efforts, he could have been of far greater help to the Manhattan Project.

The feeling at Los Alamos, and an opinion still held by many today, was that Teller held a deep and abiding bitterness toward Oppenheimer and this provoked his contentious behavior. According to one modern historian, writing during the late 1990s, "Edward Teller ... is still deeply resented by many in Los Alamos."[27] Teller was envious of Oppie's power, but more than this, he could never really forgive him for not appointing him head of the Theoretical Division, the position given to Bethe. Naturally, there was no love lost between Bethe and Teller. Indeed, at a reunion of Los Alamos veterans in 1995, fifty years after the end of the war, organizers had been "forced to take extreme measures to keep Teller and Hans Bethe from running into each other, such was their bitterness towards each other."[28] But at the time, instead of expressing his anger toward Bethe, Teller turned it on the top man on the Hill, and his resentment never faded.

They had been such close friends during Oppenheimer's Berkeley days, a colleague described their relationship as a "mental love affair,"[29] and Teller had always respected Oppenheimer's abilities. But after 1943 they could no longer communicate on a personal level. The morning after Oppenheimer made the terrible blunder of forgetting to invite Rudolf Peierls to a party soon after his arrival at Los Alamos, he apologized profusely, then added with a wry smile: "But there is an element of relief in this situation: it might

---

*According to other senior researchers at Los Alamos, Teller threatened to resign. They believe that Oppenheimer was so keen to keep Teller aboard he went to great lengths to appease the Hungarian. Apparently Groves did not want Teller to leave either, and Oppenheimer eventually gave him almost everything he wanted (Hans Bethe, "Comments on the History of the H-bomb," *Los Alamos Science* [Fall 1982], p. 3).

have happened with Edward Teller."[30] And to see Teller's vindictiveness at its worst we need only consider how his testimony before the House Committee on Un-American Activities in 1954 helped decimate Oppenheimer's reputation.

In 1947, as head of an investigative committee that had proposed a slowdown in development of the H-bomb, Oppenheimer had attracted the wrath of two powerful men. One of these was Teller himself, who since Los Alamos had devoted himself to developing the hydrogen bomb. The other was a self-made millionaire businessman, Lewis L. Strauss, who had worked his way so far into the government that he had acquired the plum job of heading the newly created Atomic Energy Commission. Strauss was also a vehement supporter of hydrogen bomb development and a close associate of Teller.

Seven years later, in 1954, paranoia concerning Communism had reached a peak of hysteria, fueled by the delusional Senator Joseph McCarthy. The Rosenbergs had just been executed, and the temperature of the Cold War was dropping fast. At Strauss's instigation, Oppenheimer, the man who had played such a crucial role in the Allies' victory, was hauled before the House committee. During a series of public hearings, Oppenheimer responded to charges that he had Communist sympathies, but many thought he was lying. One member of the committee commented off the record that he believed J. Robert Oppenheimer was more probably than not a Russian spy.[31]

Toward the end of the proceedings, Strauss called Teller to give testimony. When asked: "Do you or do you not believe Dr. Oppenheimer is a security risk?" Teller replied: "In a great many cases I have seen Dr. Oppenheimer act—I understand that Dr. Oppenheimer has acted—in a way that is for me exceedingly hard to understand. I thoroughly disagreed with him in numerous issues and his actions frankly appeared to me confused and complicated. To this extent I feel that I would like to see the vital interests of this country in hands which I understand better and therefore trust more. In this very limited sense I would like to express a feeling that I would feel personally more secure if public matters would rest in other hands."[32]

This carefully worded and self-serving testimony was crucial in sealing Oppenheimer's fate. And after the hearing, Teller could not resist twisting the knife. "I knew I would have to fight you, Robert," he said quietly, "and now I've won."[33]

Within days Oppenheimer had been stripped of his security clearance, removed from all official positions linked to government or the military, and made a pariah in Washington; only a strongly worded letter of protest signed by such luminaries as Albert Einstein and Kurt Gödel saved his job as head of the Institute for Advanced Study in Princeton.

During 1943 and 1944, as Oppenheimer struggled to head the project to make the first atomic bomb and to help rivals and personal enemies work together, his own career nadir lay a decade in the future. Throughout the three years of the Manhattan Project, he could not allow any animosities to fester for long. Because of fears for security, no one was dismissed from the Manhattan Project, so every effort was made to smooth problems and find solutions to satisfy everyone. It is a shame Teller failed to realize this, or perhaps he did.

Edward Teller was a great thinker and a visionary scientist. But Los Alamos was never short of great minds, and the absence of cooperation from one temperamental scientist did not destroy the project. Indeed, we need only look at the extraordinary list of researchers who worked and lived there to see that Teller was by no means unique. Of the three dozen or so scientific staff at the start of the project, the really big names included Teller, Bethe, and Fermi. Later came a young genius named Richard Feynman and a visionary mathematician, John von Neumann, then at the peak of his powers. At the end of 1943, a party from Britain led by Peierls and the discoverer of the neutron, James Chadwick, arrived, and a few weeks later Niels Bohr, one of the founders of modern atomic research, managed to escape Denmark and spent the final eighteen months of the war on the Hill.

The task of building a bomb was possible only because the project brought together the best minds and the greatest resources then available. Heisenberg, working six thousand miles away, was an equal of anyone at Los Alamos, and some of his team, especially Otto Hahn and the elderly Nobel laureate Max von Laue, were world-class physicists, but some of the elements that led to Allied success were missing in Germany, and the Nazi-funded project never really had a chance.

In July 1945 the first atomic bomb was exploded at the Trinity test site in the New Mexico desert a little over 200 miles south of Los Alamos, 120 miles south of Albuquerque. It was deep desert and part of an existing army test area, the Alamogordo Bombing Range. Contractors had built a tower ten feet high to support the bomb at ground zero, a base camp about 20 miles away, and a series of shelters at different distances from the bomb, as well as miles of tarmac roads crisscrossing the area.

At 5:30 A.M. on July 16, 1945, the bomb was exploded at ground zero. The morning sky was rent by a pure white light watched from behind goggles or through pieces of treated glass to protect the eyes. Moments later an enormous ball of fire swelled from the ground. Next came a wave of sound booming against eardrums. Slowly, ominously, the mushroom cloud formed and ballooned miles into the atmosphere. Conservative estimates before the test suggested the payload would be around 5,000 tons of TNT. It actually produced a blast equivalent to 18,600 tons.

Edward Teller described it as "like opening the heavy curtains of a darkened room to a flood of sunlight."[34] Another of the observers compared it to "opening a hot oven with the sun coming out like a sunrise."[35] "Most experiences in life," said another, "can be comprehended by prior experiences, but the atom bomb did not fit into any preconceptions possessed by anybody."[36]

Three weeks later, at 8:16 A.M. on August 6, 1945, a U.S. Army Air Force B-29 bomber, *Enola Gay*, dropped an atomic bomb called Little Boy. As it exploded fifteen hundred feet above the ground, a wave of heat and light swept outward from the epicenter of the blast. People were vaporized where they stood; all that was left of them were their shadows cast upon walls. In mothers' arms babies dissolved, and old men pushing carts through the streets vanished along with their animals, carts, and the goods they carried.

In an instant, the world's first atomic bomb dropped on an enemy target, a device about two-thirds the power of the Trinity test weapon vaporized a region some one thousand yards in diameter. Within the circle were an estimated 20,000 human beings. Within a few minutes a further 50,000 civilians were dead from the blast. A conservative estimate produced after the war placed the death toll from injuries and the effects of radiation at 140,000. Three days after this first bomb, another, Fat Man, was dropped on Nagasaki: 100,000 people died.

As news broke of the Hiroshima atomic bomb, a world accustomed to pain upon pain, death upon death, was stunned. At Los Alamos the members of the team who had worked for almost three years to achieve this each reacted to the news in his own way. Some were emotionally cauterized by what they had done. Oppenheimer was deeply moved but remained convinced they had done the right thing. Hearing of the success of the bomb, he is said to have whispered words from the Bhagavad Gita: "I am become Death. The destroyer of Worlds."

By August 6, 1945, the war with Germany had been over for almost three months. Heisenberg and his team were then interned at Farm Hall near Cambridge. The first they heard of the explosion was on a six o'clock news bulletin from the BBC, and they were so staggered by the report they could barely focus on the details.

Bugs hidden in all the rooms of Farm Hall recorded their reactions. Heisenberg at first simply refused to accept the news, and a colleague, Walther Gerlach, felt devastated the Allies had beaten them to it. But the physical chemist Otto Hahn, who had the previous year received the Nobel Prize for chemistry, produced the most telling reaction. It was Hahn who had discovered the phenomenon of nuclear fission some seven years earlier, and it had been his work that had initiated the race for the bomb. Almost inconsolable, most believe because of the realization of what he had unleashed, he turned on Heisenberg, calling him a "second-rater."

The other scientists believed Hahn was irrational and on the verge of suicide that night. One of the officers guarding the internee's later claimed: "Their first reaction, which I think was genuine, was an expression of horror that we should have used this invention for destruction." During another taped conversation the following day, Hahn asked Heisenberg why he thought Gerlach had taken the news the way he had. Heisenberg replied: "Because Gerlach was the only of them who had really wanted a German victory."[37]

For Werner Heisenberg, the outcome of the war and the success of the Los Alamos scientists produced a whirlpool of emotions. He was first and foremost a scientist, and it was in his nature to want to reach a scientific goal before anyone else. But like Hahn and many of the others working with him, Heisenberg had had no love for the Nazis and no sympathy for their ideologies. It is largely from evidence gathered at Farm Hall that most historians are now sure Heisenberg deliberately kept his discoveries and developments from Nazi hands.

Hans Bethe, who read the transcripts of the Farm Hall recordings almost fifty years after they were made, declared it obvious that Heisenberg knew a great deal more about making an atomic bomb than any of the others on his team and that he must have deliberately withheld much of what he knew.[38] By then Bethe was glad his plans to kill Heisenberg had come to nothing.*

Heisenberg himself claimed constantly that he had never been a Nazi sympathizer, and as we have seen, after the war he made it clear that he had made deliberate attempts to withhold nuclear secrets, that his meeting with Bohr had been an effort to communicate his progress, and that he was genuinely shocked by the success of the Allies in August 1945.

Even so, confusion remains over what Heisenberg was trying to do when he met Bohr in 1941. Did he think the Allies had not considered a project of their own and wished to plant the idea to encourage them? Was he betraying Germany? We will probably never know. As Victor Weisskopf, who knew Heisenberg well, has mused: "We never talked [about Heisenberg's moral stance during the war]. I blame myself most for this. I never went to him and said, 'I have as long as you like—tell me what happened.'"[39]

However, a far more intriguing and ultimately more important question is how much the Allies really knew about the German efforts to construct an atomic weapon. It is obvious that during the first years of the war, many scientists were afraid that the German physicists led by Heisenberg would be capable of constructing a bomb. Many years after the war, one of the most creative scientists working at Los Alamos, Eugene Wigner, commented: "[I] was quite conscious of an immoral element in my action. But I was far more concerned with the moral failings of a man across the ocean: Adolf Hitler."[40]

Fritz Reiche's testimony early in 1941 that Heisenberg was on the point of capitulating to Nazi demands to build a bomb must have resonated deeply. It is likely that, even as late as 1945, the scientific community still believed it was engaged in a genuine race. The truth was revealed only after

---

*Again, like so much to do with this story, there is controversy surrounding the Farm Hall recordings. The originals have been lost, and research is based entirely upon transcripts. Some historians believe that the German scientists were fully aware they were being taped and faked their responses. But of course this is only conjecture.

the war, when Heisenberg and the others on his team could report some aspects of their experiences firsthand.

Strikingly, throughout the early part of the war there appears to have been almost no intelligence concerning German progress. What little there was came from the British spy network, but this was almost destroyed when the Germans made incredibly rapid advances through Europe during the summer of 1940. The resistance movements in France and the Low Countries were preoccupied with more immediate concerns, and at that time U.S. Intelligence simply was not listening. Consequently, the Allies appeared to know almost nothing of Heisenberg's progress.

Yet when we consider the vast resources that were later diverted to the Manhattan Project, genuine, near-total Allied ignorance seems truly remarkable. Is it unreasonable to assume that before committing so fervently to such a mammoth task, the British and the Americans would first have checked and double-checked Heisenberg's progress to see how serious the threat might be?

What is certain is that such ignorance ended in early 1944. In January of that year, British spies learned that the Germans had achieved little with their atomic research program. An intelligence report of the time informed Churchill's government: "All the evidence available to us leads us to the conclusion that the Germans are not in fact carrying out large-scale work on any aspect of TA [tube alloys, the wartime code name for atomic bomb research]. We believe that ... the German work is now confined to academic and small-scale research."[41]

However, Groves chose to ignore this. In his memoir, *Now It Can Be Told*, he decided he would tell almost nothing on the subject of espionage and sidestepped the entire issue of why the Allies had virtually no intelligence on the German bomb during the early part of the war, laying it at the door of "the unfortunate relations that had grown up among [the intelligence agencies]."[42] He also made no reference to the intelligence from Britain in 1944, about which he was almost certainly informed.

After the war Groves claimed that if anyone with knowledge of atomic physics had been sent to Germany as a spy, he would have placed at risk the secrecy of the Manhattan Project. So instead he did nothing until the war with Germany had been won. Official history relates that after the Normandy landings and as the Allies advanced across Europe, Groves established a special unit to penetrate the German lines. Code-named Alsos, the unit was created late in 1943 and was headed by a former teacher with little scientific knowledge, a man named Boris Pash. In April 1945 Pash discovered Heisenberg and his men. Most of them had escaped Allied bombing of their research facility and were holed up in the Black Forest in southern Germany.

By this time the Manhattan Project had progressed so far and had cost so much, it was quite unstoppable. After the defeat of Germany and before the bombing of Japan, Leo Szilard and other scientists in the United States tried to prevent further development of the bomb by presenting to politicians and military leaders cogent arguments against nuclear weapons, but to no avail. Szilard pointed out that other nations would soon possess the technology, but that because of its newfound wealth, the United States could always outproduce conventional weapons and hold a premier position in world politics, so the instigation of what he saw as the inevitable arms race would not be in the nation's longer-term interest. In response to this, one senior politician, James Byrnes, nicknamed by Roosevelt Assistant President, put the matter bluntly: "We have spent two billion dollars on developing the bomb," he stormed, "and Congress would want to know what we had got for the money."[43]

Such an attitude must prompt the question Was it realized early on in the war that atomic weapons would allow the United States to dominate the world stage after the conflict was won?

Immediately after the war, there was certainly a great deal of confusion surrounding the ethical and practical aspects of atomic weapons, and a mélange of opinions circulated in the upper echelons of the scientific, military, and political communities. Niels Bohr was arguing for general release of information about the bomb so the Soviets could build their own weapons. He believed that this would stop secrets leaking out to create future conflict, and that if the United States gave the secrets of the bomb to the Russians, Stalin would show greater trust of the West. In 1945 the secretary of war under President Harry Truman, Henry Stimson, held the view that the bomb could be used to encourage democracy in Russia, arguing: "It would be inadvisable to put it [the pooling of scientific research into atomic weapons] into full force yet, until we had gotten all we could in Russia in the way of liberalization in exchange."[44]

At a meeting between Roosevelt and Churchill in 1943, the British and U.S. governments had agreed to a free exchange of atomic secrets and had drawn up what became known as the Quebec Agreement. But with Roosevelt's death in April 1945 and Truman's administration holding a different set of priorities, some within the United States even began to question whether Britain should be excluded from secrets surrounding the atomic bomb, a weapon it had played a major role in creating. A senior member of the administration, James Conant, who had been involved with the Manhattan Project from the beginning, believed the British were trying to "con" secrets about the bomb from the United States in exchange for the deposits of uranium they had supplied for the project.[45] His views were overruled by Truman and others.

Since the earliest days of its development, commentators and researchers have accepted that each participant in the race for the bomb had very limited knowledge of what the other side had achieved. For the Allies, it was a race motivated by blind fear that the Germans might explode a bomb any day.

Intelligence gathering is, of course, an inexact science, so perhaps the U.S. and British governments simply assumed the worst, never allowed for the possibility that the Germans had achieved little, and opted not to place their nations at risk of nuclear attack. Under the circumstances, these would seem admirable sentiments.

But there is an alternative reason the Allies continued building the bomb and used it against a different target from the one originally intended. Remembering the stance of some of the key politicians involved with the way atomic secrets were handled toward the end of the war and the many misguided suggestions about the potential political value of the bomb, we might find it conceivable that Roosevelt and his most trusted advisers knew the Germans were far from developing their own weapon and went ahead with the Manhattan Project primarily for its postwar value.

It is almost certain no one at Los Alamos knew of this. Groves may have suspected it, but the scientists working on the bomb were probably too busy and far too politically naïve for such an idea to occur to them. For them the bomb was a science project first and a weapon second. Szilard, Einstein, and the others involved in the initiation of an Allied project to build an atomic bomb were certainly acting out of genuine fear created by the German threat. Each of them knew Heisenberg and was sincere in his conviction that he and his team had the technical ability to develop atomic weapons. It was others who may have turned this fear into a project with dual potential: a way to develop a weapon to win the war and a means by which they could hold the trump cards in the postwar world, a world in which politics would, they knew, be rapidly and completely overhauled.

Perhaps we should do well to remember the words of Paul Harteck in his letter to the German War Office in 1939 quoted earlier: "That country which first makes use of it has an unsurpassable advantage over the others."

THE USE OF atomic weapons and the effect their creation has had upon the world have been debated and discussed in great detail since 1945. Advocates of nuclear arms point to the fact that there has been no global war since 1945. Opponents claim this has been thanks more to luck than anything governments have done. The simple truth is that nuclear weapons are now a part of our lives and can never be "uninvented."

More subtle than the arguments about the morality of atomic bombs are those concerning the role of the scientist in this revolutionary change in human development. Many curse the physicists involved in the Manhattan Project for giving the world objects of potential annihilation; some even go so far as to blame science in general for pushing humankind into the Atomic Age. These people see the creation of nuclear weapons as another in a long list of misguided scientific developments.

But it is not science that is misguided. Science is used by humans, and in particular by powerful humans who almost always work to their own political and personal agendas, people who are often motivated by selfishness and greed. Politicians have used and misused the science that allows us to understand the atomic world. Science itself is chaste: It is politicians who transformed equations and theories into charred human remains, and to a degree it is politicians who manage social systems in which the energy liberated from the atom might one day end all life on Earth.

In spite of the determined efforts of a few government officials to make sure the United States would be the sole possessor of atomic secrets, details of the bomb could not remain hermetically sealed for long. Largely through the efforts of spies such as Klaus Fuchs and the Rosenbergs, the Soviet government built a bomb in 1949. As the poor relation in the development of the atomic bomb, Britain had to wait until 1952 to test its first nuclear weapon, and the French exploded their first test device in 1960. By 1964 the People's Republic of China also possessed atomic weapons.

The new world order created in 1945 gestated from utter confusion. But it is ironic that work derived from pure intellectual effort and a productive competitiveness during the early part of the twentieth century led, during World War II, to a fantastically expensive race to build a bomb, which then facilitated a standoff between the West and Communist powers that dominated the politics of the second half of the century.

The race for the atomic bomb was a rivalry like no other witnessed up to that point in human history. If any should doubt that this struggle advanced our understanding of atomic physics, they need only recall how, as late as 1942, many scientists at the cutting edge of nuclear research thought the bomb an impossibility, or at best achievable decades in the future. As nations were propelled to the battlefield, the minds of many of the world's greatest scientists were concentrated upon one far-reaching goal. Armed with the resources of the richest country in the world and the genuine belief they were in a race to the death with a rival team of equally ingenious physicists, this "biggest collection of eggheads ever," accelerated the natural evolution of scientific progress enormously. Perhaps more than any other scientific endeavor in history, the making of the atomic bomb exemplifies how pure intellect, corrupted by greed and fear, and supercharged by vast material resources, is capable of transforming the course of society.

# SIX

## The Race for the Prize

A good scientist values criticism almost higher
than friendship: no, in science criticism is the height
and measure of friendship.
—*Francis Crick*

*I*T IS AN OVERCAST SPRING MORNING in 1953; a light, misty
rain falls on the wet Cambridge streets as a large chauffeur-driven car pulls
up outside the Cavendish laboratory. The driver opens the back door and
holds an umbrella over the head of Linus Pauling, the most eminent
chemist of his generation. Pauling strides into the main entrance of the

Cavendish, where he is met by an old rival, Sir Lawrence Bragg, Nobel laureate and head of the laboratory.

Bragg is the elder of the two by a decade, but he and Pauling are perfectly matched as scientists. Each specializes in understanding the structure of molecules but from quite different directions. With his father, Bragg had become famous forty years earlier as the cocreator of the science of X-ray crystallography, a technique by which the structure of crystals may be displayed on special photographic plates almost like snapshots of crystal lattices. Pauling, who would go on to win two Nobel Prizes (one for chemistry in 1954, the other for peace in 1962), is a theorist working at Cal Tech on the West Coast of the United States. By 1953 he had done more than anyone to develop an understanding of how atoms bond to form molecules. James Watson, with a hint of insouciance, later described Pauling as "Cal Tech's fabulous chemist."[1]

But on this April morning, neither man is here to discuss his own work. Pauling is en route to a high-level meeting, the Solvay Conference in Brussels, which Bragg will also be attending, and he has dropped into Cambridge to witness for himself a piece of science that is about to make headlines around the world and to alter radically our perception of ourselves and our place within the grand scheme of life.

The two men take the stairs, with Bragg leading the way. At the top they are met by Pauling's son Peter, who has been studying at Cambridge for the past six months. Father and son shake hands and embrace, and together the group enters the room from which Peter Pauling has just emerged.

It is a large room but cluttered. Desks face outward from three of the walls, and upon each are piled papers and books in a random mess, but these are of little interest. Everyone's immediate attention is drawn to a table in the center of the room and a four-foot-tall molecular model made from pieces of wire and metal plates held together with screws and bolts and constructed around a framework of retort stands. In front of the model stand its creators, two young men, James Watson and Francis Crick, who have just elucidated the structure of the chemical of life, DNA.

Both are tall and wiry. Each wears the 1950s academic uniform of shabby dark suit, brogues, white shirt, and plain tie. Crick, the thirty-six-year-old English biologist, stands to the left of the model, hands in pockets, and on the other side stands, the American Watson, more than ten years

Crick's junior. They step forward to greet the party before standing aside to allow the great chemist a chance to study their model.

For the Cambridge pair, this is the final test along the road to acceptance of their work. But they actually already know what they have built is perfect; two days earlier they mailed to *Nature* magazine their definitive paper describing the structure of DNA. Other members of the Cavendish, including Sir Lawrence Bragg, and the team's British challengers in the race for the structure, researchers from King's College, London, Maurice Wilkins and Rosalind Franklin, have all been to study the object in the center of the room, and each has approved it. But Pauling has chased Crick and Watson to the wire. For the past year he and they have been racing to be first to determine the structure of DNA. Pauling had been quite unaware of their efforts until just a few weeks before they succeeded. But for the past eighteen months, Crick and Watson had followed every nuance of Pauling's thinking and theorizing. The Cambridge pair have won the race, and as far as they are concerned, the vanquished have come to confirm their triumph.

Pauling paces the floor in front of the model, first peering closely at a detail of the structure, then standing back to obtain a better overall view. It is not beautiful in any obvious, physical sense. But this model represents not only the culmination of elegant scientific reasoning but a molecule that is fundamentally beautiful because of its meaning for humankind. Not only is Crick and Watson's structure logically consistent with the rules of chemistry but it allows for the most complex task in the living world—the encoding of the information that accounts for the myriad forms of life on Earth.

Pauling appreciates this perhaps more than anyone. After a few silent minutes he stands back, hands on hips, and announces to the room that he needs to check a few figures with the experimenters at King's, who have X-ray diffraction photographs of DNA, but that it looks very much as if Crick and Watson have beaten him to it, that on the table before them they have the correct model for DNA.

A few days later at the Solvay Conference, Pauling, beaten to the most important finding of twentieth-century biology but magnanimous in defeat, announced: "I think that the formation of the structure by Crick and Watson may turn out to be the greatest development in the field of molecular genetics in recent years."[2] It was a prescient comment and, if anything, an understatement.

$\mathcal{D}$URING THE FIRST years of the 1950s, James Watson was something of a scientific wanderer. Coming from a wealthy midwestern family, he had blossomed early intellectually. He was "discovered" by a local TV producer who saw him as a child prodigy and put him on the *Chicago Quiz Kids Show*. Soon after his fifteenth birthday he was enrolled at the University of Chicago to study zoology and went on to complete a Ph.D. at the University of Indiana at Bloomington.

He found college work unfulfilling and was far more interested in ornithology. "I was principally interested in birds," he has written, "and managed to avoid taking any chemistry or physics courses which looked of even medium difficulty. Briefly the Indiana biochemists encouraged me to learn organic chemistry, but after I used a Bunsen burner to warm up some benzene, I was relieved from further true chemistry. It was safer to turn out an uneducated Ph.D. than to risk another explosion."[3]

However, Watson did manage to absorb some chemical and biochemical knowledge, seemingly without effort. His Ph.D. supervisor, Salvador Luria, who apparently "abhorred most chemists," turned him on to the study of bacteriology and microbiology and helped him secure a fellowship with the Merck Foundation to study a class of viruses called phages with the Danish leader in the field, Herman Kalckar in Copenhagen.[4] But Watson found little stimulation in this new position; neither did he take to the harsh climate of Copenhagen. Then, a year into his fellowship, he was offered the opportunity to spend two months at a zoological station in Naples. There, by serendipity, he attended a lecture given by a visiting British biophysicist named Maurice Wilkins, who described his recent X-ray diffraction work on a substance called DNA, deoxyribonucleic acid, a large biochemical that had first been isolated in 1869.

For some time Watson had been pondering how genetic information might pass from generation to generation and how the mechanism worked on a biochemical level. The general consensus among biologists was that proteins carried the information during reproduction, but no one had come up with firm evidence to support or negate this theory. However, some more adventurous researchers were starting to look to DNA rather than protein as the molecule responsible for carrying genetic information.

Watson too had considered the idea and had even wondered if it would be possible to elucidate the chemical structure of the code-carrying substance, whether it was DNA or some other complex biochemical. What had stopped him even thinking about this seriously was the fear that this substance might possess an irregular structure that would resist analysis and modeling. But during his talk in Naples, Wilkins clarified two things that together set Watson off on what would prove to be a very productive train of thought.

First, Wilkins seemed sure that DNA and not a simple protein was responsible for carrying information across the generations. Second, and most startling for Watson, was that Wilkins had taken X-ray diffraction pictures of DNA and, although they were grainy and ill-defined by today's standards, they showed that the substance, which could be purified into a crystalline form, had a regular, ordered structure that could, in theory at least, be studied, analyzed, and perhaps even modeled.

For Watson, these facts were revelatory. As he described it: "Suddenly I was excited about chemistry. Before Maurice's talk I had worried about the possibility that the gene might be fantastically irregular. Now, however, I knew that genes could crystallise; hence they must have a regular structure that could be solved in a straight-forward fashion."[5]

Within weeks of his return to Copenhagen, he had arranged to leave Kalckar's lab and pursue his new interest. Through his mentor, Professor Luria, he managed to secure a position as a research fellow at the world-renowned Cavendish Laboratory in Cambridge, headed by the creator of the X-ray diffraction technique himself, Lawrence Bragg. There could be no better place for him, and, as we shall see, this move was to radically alter the course of what had until then been a rather unorthodox career.

Watson's first impression of Cambridge was not good. The city itself was without doubt beautiful and the academic atmosphere stimulating, but England six years after the end of World War II was not a comfortable place to live. The quality of food and culinary standards left much to be desired, many everyday commodities were still rationed, plumbing was bad, central heating was an eccentric rarity, most houses were damp all year, freezing in the winter, and badly ventilated in summer.

Watson also disliked what he learned of the attitudes and manners of many of the dons. In his best-selling account of the discovery of the structure of DNA, *The Double Helix*, he is scathing about English academics and scientists, referring to the "fuzziness of minds hidden from direct view by the considered, well-spoken manners of the Cambridge colleges." He had little respect for many of his fellow scientists and described them as "cantankerous fools who unfailingly back the wrong horses. One could not be a successful scientist without realising that, in contrast to the popular conception supported by newspapers and mothers of scientists, a goodly number of scientists are not only narrow-minded and dull, but also just stupid."[6]

It is conceivable that Watson may have treated Cambridge as he had other research opportunities and simply passed through, but he was saved by the good fortune of working at the Cavendish and by the fact that Bragg saw immediately that Watson would benefit greatly by being teamed up with one of his own charges, a graduate student named Francis Crick. Although Bragg realized Crick's brilliance, he saw him as something of a maverick who did not fit in well with the regime at the Cavendish. At the same time, he judged Watson to be someone who might develop a mutually supportive relationship with the English biologist. This, Bragg believed, would keep Crick busy, distract him from treading on other people's toes, and it might even lead the pair down some interesting research avenues.

According to Watson's later recollections, although Crick was seen by his colleagues as very bright, a genius perhaps, he was not terribly well liked at the Cavendish. Watson claims this was in part because of Crick's loud, irritating laugh and his penchant for saying what he thought and believed, no matter how controversial his views might be. Watson saw Crick as completely ill suited to living within the staid and stifling academic world of early 1950s Cambridge, a man too bright and too opinionated to be ignored but also, because of these traits, barely tolerated. "Though he was generally polite and considerate of colleagues who did not know the real meaning of their latest experiments, he would never hide this fact from them," Watson says. "Almost immediately he would suggest a rash of new experiments that should confirm his interpretation. Moreover, he would not refrain from subsequently telling all who would listen how his clever new idea might set science ahead."[7]

Crick's recollection of how he first heard of Watson is equally subjective. Apparently, Crick's wife, Odile, had met him first and told her husband: "Max [Perutz, Francis Crick's immediate superior at the Cavendish] was here with a young American he wanted you to meet and—you know what—he had no hair!"[8] What she meant was that Watson had a severe crew cut.

Bragg was particularly scornful of Crick, and because of this Crick was in a relatively vulnerable position at the Cavendish. He was thirty-five and finishing his doctorate when Watson arrived at the lab. Crick was working under Max Perutz, a formidably intelligent and single-minded molecular biologist and one of Bragg's protégés. He and Crick were using X-ray diffraction techniques to try to unravel the structure of proteins. Perutz was a renowned experimenter who had already begun to make some headway with elucidating protein structures (and would go on to deduce the structure of hemoglobin the same year Crick and Watson unraveled the structure of DNA). He was meticulous and methodical, respectful of convention and technique. Crick's mind worked in a very different way; he was primarily a thinker, a theorist who loved to conduct thought experiments and to probe deep intellectual problems. He found it difficult to concentrate on any specific pursuit unless it opened a route to the most fundamental questions and their answers.

Francis Crick was born in the Midlands city of Northampton halfway through World War I, in 1916. His father was a successful factory owner, and Francis won a scholarship to a small private school in Mill Hill on the outskirts of London and went on to study at University College, London. Like Watson, Crick had been unimpressed with the courses offered at the university, believing them dated and narrow. Consequently, he graduated with a second-class degree and a jaded attitude toward the strictures of English academic life.

However, his maverick mind and his determination to do serious scientific work in whatever discipline presented itself kept him within the world of scholarship, and in spite of his poor class of degree, he was accepted as a Ph.D. student. But World War II interrupted his work; he was drafted into the Admiralty as part of a design team working on new mines. He married in 1940 and worked quietly on mines throughout the war. Then, with peace, came a succession of important changes in Crick's life.

In 1946 he was preparing to return to his Ph.D. (working on the design of apparatus for the study of water viscosity) when he became captivated by a book that had been published at the end of the war and is now considered a classic work of scientific literature. The book, called *What Is Life?* was by one of the founders of the quantum theory, the physicist Erwin Schrödinger. In it Schrödinger approaches biology from the perspective of a quantum mechanist and suggests that the mysteries of genetics might be resolved by applying the laws of quantum theory, by building a picture based upon atomic physics and the design of molecular structures.

Crick immediately empathized with this idea and realized the potential it offered. He wanted to investigate Schrödinger's line of reasoning and knew at once that he could not go back to building copper tanks to study water viscosity. The enthusiasm he had unlocked was also, he knew, what lay at the root of his unhappy academic life up to that point. He had always desired to strike at the heart of a problem but had instead been forced to think about matters of everyday science. At last he had stumbled upon an area of science that opened up great opportunities and dealt with some of the most basic and key problems known to humankind.

In 1947 Crick's marriage ended and he moved to Cambridge, where he knew he would find some of the best facilities for the work he hoped to do. But things did not fall into place immediately. For two years the only research work he could find was a rather obtuse study to determine the effect of magnetic fields on the development of cells. Eventually he was taken on by the Cambridge Medical Research Council Unit at the Cavendish, where in 1949 he began working under Perutz.

The day Crick and Watson met at the Cavendish, an instant bond formed, and each knew he had found a kindred spirit. Crick recalls: "Jim and I hit it off immediately, partly because our interests were astonishingly similar and partly, I suspect, because a certain youthful arrogance, a ruthlessness, and an impatience with sloppy thinking came naturally to both of us."[9]

For his part, Watson had felt stifled and oppressed by trivial science and wanted to solve what to him was the greatest mystery—the question of genes and how inheritance worked. Crick, who was interested in everything as long as it delved deep into the fundamental mysteries of Nature, was equally fascinated by what lay behind genetics and how physics could lead to the greatest prize in biology. Like Watson, he rejected the role of proteins as the code carrier and was sure that DNA was the key, that data were somehow inherited using this molecule.

Max Perutz remembers Crick and Watson "[sharing] the same sublime arrogance of men who had rarely met their intellectual equals. Crick was tall, fair, dandyishly dressed, and talked volubly, each phrase in his King's English strongly accented and punctuated by eruptions of jovial laughter that reverberated throughout the laboratory. To emphasise the contrast, Watson went around like a tramp, making a show of not cleaning his shoes for an entire term (an eccentricity in those days), and dropped his sporadic nasal utterances in a low monotone that faded before the end of each sentence and was followed by a snort."[10]

Crick and Watson were clearly both exceptionally brilliant, and when they met that day in 1951—thrown together, so Bragg later claimed, to get Crick out of his hair (he mistakenly thought Watson would act as a foil)— they were both bursting with ideas and energy to follow through. But between them they knew almost nothing about the subject they wished to master. Crick had been a biologist for only two years, but he at least had some experience working with X-ray diffraction. Watson had probably never even seen an X-ray diffraction machine, and he knew almost no chemistry.

Yet somehow, as time would prove, none of this mattered. Crick and Watson were not specialists, they had not refined their intellects to rapier points through years of dedication to a single discipline and then forced Nature to divulge its secrets as Newton or Lavoisier or Einstein had. But in 1951 they were outstandingly intelligent, hungry young men. Watson has made no secret of the fact that he was driven to find the structure of DNA because of an overwhelming desire for fame and fortune; in short, to win a Nobel Prize.

Crick has denied ever seriously thinking such things and claims he was motivated by a pure aesthetic, the desire to reach the root of a problem. He believes Watson had a more bravura approach: "Jim had no doubts about his abilities. He just wanted the answer, and whether he got it by sound methods or flashy ones did not bother him one bit. All he wanted was to get it *as quickly as possible*."[11]

Whatever their motivations, Crick and Watson are unique in the history of science and the only great discoverers in this book who did not conduct any form of experiments but instead took ideas from a collection of disciplines and built an intellectual pyramid from the sandstone blocks of other people's work.* Crick and Watson lay stone upon stone, borrowing ideas from here and from there, combining their thoughts with the theories of others, taking facts from experiments conducted by fellow scientists, and where others missed connections and failed to see lateral links, they sought them out and amalgamated a thousand and one ideas until they reached the apex of the pyramid, the structure of DNA, the "molecule of life."

Within days of their first meeting, Crick and Watson had produced an unwritten manifesto. They were chasing the "Big Idea," seeking the structure of deoxyribonucleic acid, DNA, the molecule they were convinced acted as the vehicle for the inheritance of genetic information. As Watson later put it: "Within a few days after my arrival, we knew what to do: imitate Linus Pauling and beat him at his own game."[12] And what of the great Linus Pauling? Where was he in the chase for the structure?

By 1951 Pauling was considered a master of molecular structure. Born in Portland, Oregon, in 1901, while still a young postdoc he had started making significant contributions to structural chemistry and molecular biology. He received his Ph.D. from the California Institute of Technology (Cal Tech) and went on to teach there for most of his career. His most important contribution to science was the application of quantum theory to chemistry. He had been one of the superbright generation of physical scientists who had been drawn to Europe in the mid-1920s, and he had become friends with many of the figures described in Chapter 5. He was a lifelong friend and supporter of Robert Oppenheimer and was close to Richard Feynman, Paul Dirac, Niels Bohr, and many of the physicists who shaped the nuclear age. But even though he had a masterly command of mathematical physics and the intricacies of quantum theory, Pauling was essentially a chemist. As

---

*This point may need a little clarification. As we have seen, Thomas Edison was very adept at using the ideas of others and relied upon the canon of published work to help him unravel difficulties, but he did experiment and was renowned for his practical skills above his theoretical understanding. By contrast, Crick and Watson were "all theory." The other characters in this book who do no experiments themselves are Bill Gates and Larry Ellison (discussed in Chapter 8), but I would contend that the thousands of men and women who conduct fundamental research at Microsoft and Oracle might be thought of as their teams of experimenters.

Erwin Schrödinger, Max Planck, and the other quantum pioneers applied their mathematical skills to nuclear physics, Pauling became the leading proponent of "quantum chemistry" and reconstructed the science of molecular biology from the ground up, taking the most fundamental premises of quantum theory and applying them to the vastly complex molecules found in Nature.

As well as writing an array of important papers on the subject, mirroring Lavoisier a century and a half earlier, in 1939 Pauling explained his discoveries in a book aimed at chemistry graduates and undergraduates. *The Nature of the Chemical Bond* became one of the most influential science books of the twentieth-century and a classic read by almost all chemists.

Pauling became interested in large biochemicals during the mid-1930s. He first studied the molecule of hemoglobin, responsible for carrying oxygen in the blood to the cells of the body; after the war he began to investigate protein structure. In February 1951, only seven months before Watson's arrival in Cambridge, Pauling published an accomplished paper: "The Structure of Proteins: Two Hydrogen-Bonded Helical Configurations of the Polypeptide Chain," in which he demonstrated the helical structure of proteins and proposed that this form might be common to many large and complex molecules in Nature.

Proteins are a chemical type and possess a simple structural unit known as an amino acid. An amino acid contains an amine group—two hydrogen atoms bonded to a nitrogen atom. Amino acids also contain what is called a carboxylic acid group, which is made up of a carbon atom, an oxygen atom, and a hydroxy group (an atom of oxygen and an atom of hydrogen bonded together). All proteins contain these groups, but the rest of the molecule may be almost any size and can be very complex. Because of this complexity, the range of possible protein structures is almost limitless; a typical animal cell might produce more than fifty thousand proteins. Clearly, then, proteins are one of the essential chemicals of life. They form enzymes (biochemical catalysts) and antibodies and are integral to many huge molecules, such as DNA and RNA (ribonucleic acid).

Pauling's great discovery was that the amino acid units in proteins are arranged in such a way as to give rise to highly ordered, periodic, secondary structures. These include the alpha-helix, which Pauling defined shortly before Watson arrived in Cambridge, and then described in a succession of startlingly inventive papers published in the *Proceedings of the National Academy of Sciences.*

Because proteins can take on so many forms, including more complex structures than the alpha-helix (such as beta-plated sheets and the triple helix), it was understandable that many biochemists believed these molecules were responsible for the transfer of genetic information during reproduction. But Pauling was not one of them, and by 1951, like Watson, Crick, Wilkins, and some others in the field, he had become convinced DNA was the molecule Nature used for the task. So, just as Crick and Watson in Cambridge began to turn their minds toward the first step to elucidate a possible structure for DNA, Pauling was doing the same thing in California.

Pauling was famous for the speed with which he worked, and news that he was now chasing the structure of DNA took little time to reach Cambridge. Even before Crick and Watson had laid the foundations of their structure, they heard Pauling had begun proposing tentative arrangements for the molecule.

Many young scientists in the position Watson and Crick found themselves in at this time would have simply given up the chase. Knowing the world's most eminent chemist, and a man with all the resources of a leading institution at his disposal, was hurrying to solve the same problem would have paralyzed lesser men. But to their eternal credit, Crick and Watson's reaction to such competition had quite the opposite effect. They were so filled with a sense of their own worth and inspired by the dream of the Nobel Prize that instead of retiring from the field they became even more galvanized.

Crick believes one of the reasons for their success was that no others could bring themselves to face up to the task. "Practically nobody else was prepared to make such an intellectual investment," he has written, "since it involved not only studying genetics, biochemistry, chemistry, and physical chemistry (including X-ray diffraction—and who was prepared to learn that?) but also sorting out the essential gold from the dross."[13] Of course they also had little to lose; they were nobodies in the world of science. If they had fallen on their faces, no one would have cared and few would even have known they had tried to take on Pauling in the first place.

Watson and Crick also had two other huge advantages over Pauling. First, the Cal Tech professor could not concentrate upon a single problem for very long. He was only happy when he was working on a dozen projects simultaneously. So, although he was fascinated with DNA and wanted to elucidate the structure, this was far from being the only idea he was chasing at the time. Further, Pauling was preoccupied with many other responsibilities and commitments. He was campaigning against the development of nuclear weapons, lecturing widely, and heading a large research department at Cal Tech.

The other disadvantage for Pauling was that he had at his disposal only one rather poor-quality X-ray diffraction picture of DNA, which had been produced by the English crystallographer W. T. Astbury some five years earlier. Although at the start of their research Crick and Watson also had only this one image to work with, Crick, an old friend of Maurice Wilkins, quickly obtained the latest, far superior images coming from Wilkins's lab at King's College.

But Pauling's problems gave Crick and Watson little reassurance. Such was the man's eminence and reputation for finding fast solutions to the most difficult scientific mysteries, the Cambridge pair could never feel sure he would not surprise everyone and reach the goal first. However, throughout the eighteen months they were struggling to find the solution to the structure of DNA, Pauling was not in fact their most dangerous opponent. Between 1951 and 1953, Crick and Watson remained largely unaware of the advances of another competitor in the race for DNA, a researcher who was every bit as ambitious as they, more knowledgeable, more experienced, and with some of the best data at her fingertips. This rival was the X-ray crystallographer Rosalind Franklin, who in January 1951 had joined the team of researchers at King's College in London.

The King's researchers worked under the banner of the Biophysics Unit headed by John Randall and his deputy, the talented X-ray crystallographer Maurice Wilkins. Wilkins was a highly experienced and versatile scientist. He had worked at Berkeley on isotope separation techniques during the war, and the results of his research had led to important developments in the creation of the earliest nuclear reactors. But with the exploding of the atomic bombs over Japan in 1945, Wilkins had become extremely unhappy with the study of nuclear physics and returned to England intent upon a complete change of career. He soon found a new direction in biophysics and, in particular, X-ray diffraction studies of complex biochemicals. He had enjoyed working at King's, but in January 1951 his contentment was shattered when Randall recruited Rosalind Franklin, an English X-ray crystallographer who had been working for the French government in Paris with the chemist Vittorio Luzatti.

Franklin came from a wealthy, academic Jewish family and had attended the elite St. Paul's Girls' School. At Cambridge she had studied chemistry and gone on to research carbon fibers, work that had led her to the post in Paris. This in turn had taken her to an interest in X-ray diffraction and other crystallographic techniques under the guidance of, Luzatti.

It is unclear why Franklin decided to return to England. She had been extremely happy working with Luzatti, and in choosing King's she made the transition from a very liberal working environment to one that would have been considered conservative and staid even for the time. In 1951 the science faculty of King's College was a largely male preserve, especially in the physical sciences, and the laboratories in the basement of the old building on the Strand running parallel to the Thames were far from salubrious. The offices and labs were cramped and damp, lit entirely by artificial light and painted in dull gray and beige. Franklin had exchanged the Left Bank café society she had loved for a bleak tearoom at the end of a gloomy corridor and the anachronism of a common room reserved for men only.

To make matters worse, she was not a naturally easygoing character. One colleague, Aaron Klug, later said of her: "She was—I can't say she was like a man, not like that at all, but one didn't think of her being particularly like a woman; she wasn't shy, or self-effacing—but she wasn't blustering either. She spoke her opinions firmly, I think people were unaccustomed to dealing with that in a woman. I think they expected women to behave—rather differently, quieter. She expected reason to dominate. She was very much a rationalist."[14]

James Watson was heavily criticized by both reviewers and fellow scientists for his portrayal of Franklin in his famous book *The Double Helix*. Throughout the book he rather flippantly refers to Franklin as Rosy and paints her quite unjustly as a rabid feminist. At one point he says: "The thought could not be avoided that the best home for a feminist was in another person's lab."[15]

Others have given a more balanced and reasoned description of Rosalind Franklin. A co-worker at King's, Mary Fraser, has said: "Dedicated people like Rosalind Franklin (the sex is irrelevant), great artists, scientists, writers, mountaineers, sportsmen, or the Florence Nightingales have an obsession and their fellow human beings are secondary to their overriding passion."[16]

Another female colleague, Sylvia Jackson, has commented: "My impression of Rosalind Franklin was twofold. She was absolutely dedicated, a tremendously hard worker, she strode along rather quickly; she was enormously friendly if you gave her half a chance. But I found her *formidable*."[17] And Franklin's closest colleague at King's, her research assistant Raymond Gosling, recalls: "She was a very intense person, almost eccentrically so. Some people doubtless called her a bluestocking because she was terribly interested in her work. And she was shy, I think, certainly not the person to let her imagination fire up openly."[18]

Some feminist writers have tried to appropriate Rosalind Franklin to their cause, but this is quite irrational, and much that is inaccurate has been written about her in an attempt to burnish the legend of Franklin: the feminist icon with her invaluable role in the discovery of the structure of DNA.

There is no doubt Franklin was immensely unhappy at King's, but this was only in small part because of the old-fashioned style of the place. It has been said for example that she was the only woman working in the department, but this is quite untrue. Of the thirty-one researchers and administrators working under Randall, eight were women. There was indeed sex discrimination concerning the common room, but there was also a women's common room. Even there Franklin did not seem to want to become part of a "club." One of her female colleagues recalls: "Now Rosalind arrived and I suppose we assumed she would fit into the casual role of relaxing amidst the beakers, balances, centrifuges, and petri dishes, but she didn't. Rosalind didn't seem to want to mix. Her manner and speech was rather brusque and everyone automatically switched off, clammed up, and obviously never got to know her."[19] Crick, who after the discovery of the structure of DNA became close to Franklin, has said: "I don't think Rosalind saw herself as a crusader or a pioneer. I think she just wanted to be treated as a serious scientist."[20]

The root of Franklin's discomfort, and ultimately the most important reason she failed to reach the structure of DNA before Crick and Watson, was that she clashed terribly with Maurice Wilkins, a clash that had been initiated by John Randall. Wilkins was lecturing abroad the day Randall interviewed Franklin, and the two of them only met the morning she took up her position on January 1, 1951. Randall had told Franklin she would be working independently at King's and would not be answerable to Wilkins (deputy head of the entire unit). But he had given Wilkins the impression the new recruit would be little more than a glorified research assistant.

It is now believed Randall did this deliberately to keep Franklin and Wilkins as rivals rather than collaborators so that he might step in and share the credit if either was to make a significant discovery. "Randall was an ambitious, ruthless, no-holds-barred researcher himself," one historian has commented. "He undoubtedly recognized in the spring of 1950 [when he first perused pictures taken by Gosling] the possibility that the discovery of DNA structure would far transcend in importance his co-invention of the cavity magnetron."[21]

This is almost certainly true, but others think Wilkins and Franklin may not have worked well together anyway. "They were *both* I think rather difficult people," a contemporary has said.[22] Maurice Wilkins was a quiet, amicable, if rather reserved character. He and Franklin may not have liked each other, but they probably would not have clashed so badly if Randall had not set their relationship off on entirely the wrong footing. Because they were both often taciturn and unable to express their feelings easily, misconception grew, and resentments were left unresolved to fester and obstruct.*

So the rivals to Crick and Watson were as different from each other as could be imagined. In California, Linus Pauling, a loud, ebullient American, an internationally esteemed scientist who could marshal great resources and experience. He cherished building models of molecules using nuts and bolts and pieces of sheet metal twisted into shape, and to any problem he could apply a range of experience and technique. And, in a dingy basement complex beneath King's College in London, the extremely talented and dedicated Rosalind Franklin, teamed unhappily with the quiet, patient, and careful experimenter Maurice Wilkins, two scientists who together could

---

*I was fortunate enough to have been a student of Maurice Wilkins at King's College and remember him from lectures as being a quietly confident, reserved, and self-contained man. He was never cold but displayed little humor or levity, and, much like Rosalind Franklin, he had no time for triviality or small talk.

have produced wonders but instead could barely exchange a civil word. Meanwhile, in Cambridge, the combination of the eccentric maverick Crick and the footloose American genius James Watson had gelled instantly. The pair thought in the same ways and wanted the same things; each was driven and determined, but, compared with their rivals, they were able to muster little more than a patchy knowledge of biochemistry or relevant experience.

An impartial observer, one without the invaluable benefit of hindsight, might have been justified in thinking there could be only one possible victor in the race. But if the clamber to elucidate the structure of deoxyribonucleic acid teaches us one thing about science, it is that research is not often predictable. Indeed, perhaps more than any other example of rivalry in science, this race demonstrates just how important personal relationships can be in determining who is victorious.

ALMOST EVERYTHING we now know about genetics has been discovered within the past century and a half, and the fundamental mechanisms that govern the evolution and reproduction of all living things have been understood only since the 1950s, principally from the discovery made by Crick and Watson.

The Babylonians of 4000 B.C. observed that reproduction depends upon the fertilization of female plants with the pollen from male plants. Aristotle tried to create a general hypothesis concerning reproduction but reached the typically warped conclusion that male seed endowed offspring with "motion" while the female element contributed "matter." Aristotle and other Greek philosophers also developed the curious idea of pangenesis, which proposed that the individual organs of the body each secreted particles that entered the reproductive material and gave it characteristics to be passed on to future generations. This idea was adhered to by most scientists, including Charles Darwin, as late as the nineteenth century.

With the invention of the microscope in the mid-seventeenth century, scientists could discover more about the tiny physical elements that play their role in reproduction. Most accomplished in this field was Robert Hooke, who became an expert microscopist and in 1665 wrote his masterpiece, *Micrographia*, in which he outlined his experiments.

Hooke was the first to describe "cells." Later, sex cells (sperm and ova) were observed, but some experimenters went too far and claimed to see "homunculi," or tiny human forms, within these cells and were led to believe they had discovered a clear and simple mechanism for reproduction. They called this mechanism preformation—the idea that conception simply set in motion the growth of tiny embryos already present within the sperm or female egg—and this idea dominated the thinking of biologists throughout the first half of the eighteenth century. A supporter of the theory, Charles Bonnet, declared it to be "one of the greatest triumphs of the human mind over the senses."[23]

The opponents of preformationism, the epigenesists, led by the French scientist Pierre de Maupertuis, astutely pointed out that there were two serious flaws with the idea. The first, philosophical, noted that the preformationists were merely putting the argument back a stage and their theory still offered no serious mechanism for the formation of the embryo. The second pointed out that the theory took no account of the observed fact that offspring usually possess characteristics of both parents. However, the epigenesists had no verifiable alternative theory and were forced to try to explain reproduction by the introduction of a semimystical "force" that somehow communicated characteristics. And, although performationism went the way of phlogiston theory, so did epigenesis; the entire mechanism of reproduction remained a mystery for another century.

During this time hypotheses came and went. Most were based upon unverified observation and misguided deduction. One of the most popular was the notion that genetic material might be passed from generation to generation via blood, an idea approved by many scientists even after sex cells had been observed. Vestiges of this theory remain with us today in common expressions such as "blood brothers," "blue blood," or, most tellingly, "bloodline." Although the idea that blood was responsible for transferring inherited characteristics might have seemed logical to scientists up to the late nineteenth century, it actually explained nothing. How, for example, did characteristics lie dormant for many generations and then reappear if the medium was blood?

We saw in Chapter 3 how the idea of evolution developed and how Darwin's revolutionary concept of natural selection explained this process, but although the mechanism for change across geological time was understood by applying Darwin's work, the way individual animals might pass on inherited characteristics remained a mystery. Charles Darwin died in 1882 knowing that genetic information must be transferred from generation to generation but quite ignorant of how this could happen. However, unknown to him (and almost every other scientist of the time), in 1866, a priest living in complete isolation, some thousand miles from Down House, had stumbled upon the mechanism of genetics. Darwin never knew of the existence of Gregor Mendel.

Born in 1822 in a tiny Silesian town, then part of the Austro-Hungarian Empire and now within the borders of the Czech Republic, Mendel failed to complete his university education because his parents could not afford to keep him at college. This misfortune led him into the priesthood. Mendel spent most of his life in a monastery in the isolated town of Brno, and in his spare time he continued his scientific education, first by reading any scientific text he could find and then by conducting his own experiments. He was particularly interested in biology and especially the apparently mysterious way the characteristics of individuals were passed on to their offspring.

To investigate this, in 1856 he began breeding experiments using edible pea plants in the garden of the monastery. He concentrated on seven characteristics he knew showed variation between strains of peas. These included the tallness of the plants, the color of their seeds, and the distribution of flowers (comparing those plants that had flowers along their entire length with those that had flowers only at the ends of stems). He conducted hundreds of crosses employing artificial pollination, and during a decade of work he completed an amazing twenty thousand experiments, which he carefully recorded and described in detail. In 1866 he published his findings in the only format available to him; he wrote a paper titled "Experiments with Plant Hybrids" (*Versuche über Pflanzenhybriden*), which was published in the incredibly obscure *Journal of the Brno Natural History Society*.

In this paper Mendel set out his conclusions using three elegant theories that laid the foundations of modern genetics. These have since become known as Mendel's laws. His first law asserts that when sex cells (the egg and the sperm) are formed, pairs of what he described as factors separate. These factors are what we now call genes, so a sperm or an egg will carry one factor (such as that for encoding pea color) but not both. Mendel's second law states that characteristics are inherited independently of each other. For example, the tallness factor may be inherited with any other factor, such as those that determine color or the distribution of flowers in the plant.* The third law says that each inherited characteristic is determined by the interaction of two hereditary factors or genes, one from each parent.

From his experiments Mendel found that one of the factors (or genes) always predominated; for example, he showed that in pea plants tallness always dominated over shortness.† This led to the idea of dominant and recessive genes. The first of these endow a characteristic in an individual even if only one gene for that characteristic is present. Recessive genes need to be present as pairs (one from each parent) to express their characteristics.

As well as writing his paper for the local scientific society, Mendel sent his findings to the most eminent botanist of the day, a German named Karl von Nägeli. Sadly, von Nägeli was unimpressed and wrote back emphasizing the importance of spontaneous generation (an idea he had long cherished); he claimed the hybrids Mendel was writing about were nothing more than "aberrations." Instead of encouraging the priest, Nägeli told Mendel to repeat his experiments or retire.

It is impossible to say with certainty if Mendel was greatly disturbed by von Nägeli, but, in 1868 Mendel was made abbot of his monastery and obliged to give up science in order to fulfill his more orthodox duties. He never returned to experiment and died totally unknown to the larger scientific world in 1884.‡

As revolutionary and brilliant as Mendel's work undoubtedly was, publication in the *Journal of the Brno Natural History Society* guaranteed it

---

*This law was later modified when it was discovered that it is possible to have linkage between genes. This means that sometimes genes are inherited with others positioned close by on the same chromosome.

†The term *gene* was coined by Wilhelm Johanssen in 1909.

‡Bizarrely, during his final years Mendel became a chain smoker. He developed heart disease and a kidney condition along with acute dropsy. Perhaps he was led to the weed through depression.

would be a long time before anyone might notice his theories. Indeed it was not until 1900, sixteen years after Mendel's death, that a group of researchers stumbled upon this seminal paper and hailed its author as the "father of genetics."*

But, as Mendel's work lay ignored in an East European library (and possibly on Darwin's desk), others were developing an understanding of a chemical basis for genetics. The Swiss biochemist Friedrich Miescher was the first to isolate DNA in 1874. Half a decade later Walther Flemming discovered the existence of tiny units within cells, which he called chromosomes. Shortly after, an embryologist named Edoard van Beneden, who was studying a threadworm called *Ascaris megalocephala* found in horses, showed that when an egg was fertilized, an equal number of chromosomes came from each parent.

But at first this discovery caused confusion because Beneden also found that the offspring had the same number of chromosomes as each parent. Faced with this information, Flemming quickly drew the correct conclusion: During reproduction chromosomes split into two identical units, and these then pair up with others in the cell so each individual within a species ends up with the same number of chromosomes.† This process he called meiosis, and it clearly indicated to scientists of the day that some (then unknown) biochemical process lay at the heart of genetics that allowed such processes as the splitting and rearranging of parts of cells.

Early in the twentieth century the American Thomas Hunt Morgan extended Mendel's newly unearthed laws and from experiments with *Drosophila* (fruit flies), which have a rapid reproductive cycle, he found that some genes could be passed from generation to generation in batches. This implied that genes might be somehow linked, and from this Morgan reached the conclusion that each species possesses its own genome, a collection of genes positioned along the chromosomes. Morgan found some two

---

*There is a story that a copy of this paper was left unread on Darwin's desk in 1866 and was then filed away, never to be studied by Darwin (Daniel C. Dennett, "Appraising Grace: What Evolutionary Good Is God?" *Sciences* [January–February 1997], p. 41).

†Humans have twenty-three pairs of chromosomes.

thousand genes in the genome of the fruit fly, and these account for all its physical characteristics.*

During this era the idea that things as tiny as genes might be isolated was wild fantasy because the microscopes of the day had insufficient magnifying power to see them. But, as technology improved scientists were able to probe the gene and to learn that it contained large numbers of proteins and huge biochemicals called nucleic acids. It then appeared certain the mechanism by which characteristics could be transferred from generation to generation lay with one of these biochemical types, proteins or nucleic acids. But which?

The answer came quite indirectly. During the 1920s, in London, a British bacteriologist named Fred Griffith had been working with the bacterium responsible for pneumonia, pneumococci. His work centered on two types of pneumococci, an infectious and a noninfectious strain. He also learned that both types were killed by heat, and of course, after the infectious bacteria were killed, they could no longer carry pneumonia. When he injected mice with either the dead infectious cells or the living noninfectious strain, the mice were unaffected. But he noticed something very strange. If he injected the mice with both the dead infectious strain *and* the living noninfectious bacteria, the mice developed pneumonia.

Griffith tried to explain this but reached the wrong conclusion. He suggested that the living noninfectious bacteria ingested a substance he called pabulum that came from the dead infectious strain, but he failed to note that later generations of the originally noninfectious bacteria continued to be infectious after their ancestor had been mixed with the infectious strain, so his theory could not be correct. It took another researcher to solve the mystery and to propel the macromolecule DNA into the biochemical spotlight. This was a Canadian bacteriologist named Oswald Avery, who worked at the Rockefeller Institute in New York, where he doggedly chased the problem for almost fifteen years.

Avery, a painfully shy and retiring son of a clergyman, had stumbled

---

*We now know there are something like one hundred thousand genes in the human genome. These are scattered among our forty-six chromosomes. The Human Genome Project has now mapped the position of each of these genes.

upon Griffith's work in a paper called "The Significance of Pneumococcal Types" published in the *Journal of Hygiene*. Although he was skeptical of Griffith's theory, Avery soon became enamored with the sheer elegance of the man's experimental findings and fascinated by the strangeness of his claims. It is said that when Griffith was killed in an air raid in 1941, Avery obtained his photograph and kept it on his desk at the Rockefeller Institute until the day he left.

As much as Avery admired Griffith, he also concluded early in his own researches that the idea of pabulum was nonsense. He replicated Griffith's experiments, and after careful analysis of the chemical processes involved, he concluded that DNA was the material responsible for converting the living noninfectious strain of pneumococci into the infectious variety by mutating the chemical nature of the DNA in the genes of the living bacteria. This mutated form was then passed on from generation to generation of bacteria and infected the mice with pneumonia.

This was the first clear evidence that DNA was the molecule responsible for transferring genetic information, but Avery had an enemy in the institute where he worked, a vigorous supporter of the idea that proteins, and not DNA, were responsible for the communication of genetic information.

By the time he clashed with Avery, Alfred Mirsky was already the author of several hundred papers and proclaimed himself the "in-house genius" at the Rockefeller Institute. He blocked Avery's theories at every turn. He gave lectures both at the institute and outside it disclaiming Avery's carefully researched findings, and he wrote paper after paper dismissing his rival's ideas and experiments.

The reason for Mirsky's vehement attacks has never been explained. It may be that his only motivation was a refusal to accept another scientist's experimentally supported theories because they clashed with his own cherished beliefs. Whatever Mirsky's agenda, Avery, who had no inclination to argue back, let alone challenge his colleague, was driven to stress-related illness and took early retirement in 1948. He spent the remaining seven years of his life with his brother in Nashville, Tennessee, and he never returned to science, or to the Rockefeller Institute.

But Mirsky had been wrong. The fundamental flaw in his thinking was his conviction that DNA was a simple and rather uninteresting chemical. He based this opinion upon experiments he had conducted during the early

1940s. Even when, at the height of his campaign against Avery, he repeated these experiments with more refined equipment and found deoxyribonucleic acid to be an incredibly complex molecule, he steadfastly refused to alter his arguments or to give Avery credit.

Oswald Avery was never to know that the Nobel committee in Stockholm had been ready to award him a Nobel Prize for his studies but the proposal had been blocked by objections from Mirsky, still then active at the Rockefeller Institute. However, Avery did live to see enormous change in his discipline. During the last years of the 1940s, it was discovered that DNA contained four bases, adenine (A), thymine (T), guanine (G), and cytosine (C), and that these appeared in unequal amounts in the molecule, a fact that made it even more complex than previously believed and adding further weight to the idea that DNA could carry genetic information.

In 1949 Erwin Chargaff at Columbia University in New York discovered that DNA contained equal amounts of A and T and equal amounts of G and C. He stated these equalities in what became known as Chargaff's second rule, and this was to play a significant part in Crick and Watson's elucidation of the structure of DNA.

Happily, Crick and Watson's triumph came before Avery died, for it was in part his triumph too. What Mirsky had to say about the elucidation of the structure of DNA was not recorded.

By 1951 THE biochemical foundations for understanding the structure of DNA had been laid, and there was some strong evidence DNA was indeed the molecule at the heart of genetics. Crick and Watson knew that DNA was a large, complex molecule, that it was probably helical, that it contained four bases—A, T, G, and C—as well as sugars, phosphates, and water. They also knew it had to be a regular structure because it could be crystallized for use in X-ray diffraction. Furthermore, they had a body of evidence from Avery's experiments to support the fact that DNA, and not proteins, was responsible for carrying the genetic code, and they had the huge advantage that Crick was on very friendly terms with Maurice Wilkins, who had access to the best X-ray diffraction pictures of DNA in the world, pictures far superior to those in Linus Pauling's possession.

Watson was confident and let everyone around him know it, declaring they could elucidate the structure with relative ease. But, as well as showing off to friends in their labs and among fellow scientists who gathered regularly in their local pub, the Eagle, he and his partner asked questions and picked brains. Crick and Watson were never shy about applying the ideas of others to blend with their own thoughts on DNA and the mechanisms of genetics. And they were never troubled by displaying their ignorance so long as it enabled them to learn something useful. At a brief, embarrassing meeting with Chargaff, they let slip that they did not know his famous rules and even muddled up the characteristics of the four bases of DNA.

However, from the foundation of biochemistry as it was in 1951, Crick and Watson soon concluded that the DNA structure must be a long chain of nucleotides, large biochemicals containing a base, a sugar, and a phosphate. In addition, they quickly learned from discussions with Wilkins that the molecule of DNA was surprisingly thick, much thicker than the single helix (or alpha-helix structure) Pauling had found for proteins. This clearly indicated that DNA would have to contain at least two chains. But how many chains were there, and how were they linked? The nucleotides contained bases, and Avery had shown there were four varieties, A, T, C, and G, but which nucleotides possessed which bases?

These were just a few of the questions requiring urgent answers. But others were even more difficult to answer. Even if they could match bases with nucleotides, how were these nucleotides bonded to each other in the chain? And were the bases pointing inward or outward?

The answers to these questions would come gradually. Some solutions were obtained through the application of logic; others by helpful nudges in the right direction from those with whom Crick and Watson talked; and as we shall see, some others came from analysis of X-ray diffraction images clandestinely obtained from King's.

In late October 1951, some two months after he and Watson began working together on the DNA problem, Crick invited Maurice Wilkins to lunch in Cambridge to learn the latest from the King's lab. Crick, Odile, Watson, and Wilkins ate crowded around the kitchen table of Crick and Odile's tiny apartment.

Naturally, Crick and Watson were keen to talk about work, and for a while Wilkins was happy to oblige, telling them that he had some improved images of DNA and that he was now convinced the molecule was helical.

But before long Wilkins became quite morose, and with Crick's prompting he began to tell them about the dreadful situation with Rosalind Franklin, who had now been at the lab for almost a year.

The relationship between Wilkins and Franklin had deteriorated so much Franklin would not even share her results with the man who was officially her boss. She had become dogged in her work and even more pertinacious. Isolating herself in her lab, she had worked throughout 1951 on the best samples then available, material Wilkins had passed on to her. She was preoccupied with what was known as the alpha form of DNA, a crystalline type containing very little water.*

Based upon these experiments, Franklin was far from convinced DNA had a helical structure because some of her pictures gave indefinite images and there were no signs of the telltale pattern that she knew demonstrated the helical configuration. But from Wilkins's quietly anguished words, Crick and Watson discerned a few useful clues. First, they were reassured that Franklin was not making huge strides forward. Furthermore, the King's team were clearly not cooperating, which meant that they were not gaining insights from any collaborative effort. But, perhaps most precious of all, they learned that Rosalind Franklin was due to deliver a seminar at King's three weeks later in which she was going to describe her findings from the previous year of research.

November 1951 was cold and London bedecked in smog. Talk was of thousands of British POWs in Korea, trouble brewing in Suez, and the astonishing return of Winston Churchill to Number 10, Downing Street. For the past three weeks, Watson had used all his spare time to teach himself the basics of X-ray diffraction technique, the very science in which he was supposed to be engaged at the Cavendish. He wanted to gain as much as he could from Franklin's talk.

Watson remembered the seminar as an incredibly dry affair. Franklin summarized her recent findings and investigations to a gathering of some fifteen colleagues, none of whom felt disposed to ask questions at the end of the talk, and after an hour the meeting disbanded and Watson left the seminar room chatting quietly with Wilkins. The two of them set off through the dense evening smog for dinner in Chinatown.

---

*This name, the alpha form, should not be confused with the alpha-helix.

Back in Cambridge the next day, Crick pumped Watson for information on what he had learned. But he was soon disappointed. He had not known that his friend never took notes. It had done him little harm in the past, he claimed, but Crick now had to trust Watson's memory for the important details of what Franklin had said. Watson recounted the numbers of types of chemical groups Franklin had found in her samples of DNA and described how her diffraction images reaffirmed the existence of the four bases, the nucleotide chain, and the phosphate and sugar groups, along with water, just as they had thought.

Watson told Crick that Franklin believed the bases to be on the *outside* of the structure. However, in actuality Franklin had suggested that DNA might be helical, that it could contain up to four chains, that in some way these bases may link the chains, and, most crucially, that the bases were on the *inside* of the structure. Inevitably, perhaps, Watson had forgotten the finer details.

Crick was beginning to regret he had not taken the trip to King's himself, but he had to take his colleague's information at face value, and the two men set about trying to build their first model of DNA based upon the new data.

For a week they slaved over a model constructed around a central framework of three helices. They had chosen a triple helix with little supportive evidence. Franklin had suggested anything from two to four helices, and they coupled this suggestion with Wilkins's experimental finding that the molecule of DNA was thicker than an alpha-helix. With typical audacity, the moment the model was finished and stood proudly on a table in their shared office, Crick was on the phone to Wilkins, inviting him to view their proposed structure. To Crick and Watson's surprise, Wilkins not only agreed to come up to Cambridge the next day but told them Franklin and her assistant, Raymond Gosling, both wanted to come with him.

Crick and Watson felt confident. They met their colleagues at the main entrance to the Cavendish and escorted them straight to the office. Crick gave a ponderous talk describing their efforts and conclusions, then stood back to let the guests view the model. Franklin, Wilkins, and Gosling walked around the four-foot-high construction and began asking questions. Then, suddenly, Franklin let out a deep sigh. Crick looked at her, puzzled.

"Where are the water molecules?" she asked, unable to betray a sarcastic edge to her voice.

Crick pointed to them.

"But that's nothing," Franklin retorted. "That's nowhere near enough. You would need at least ten times as much water to hold the structure together." She stood back with her arms folded, then shot Wilkins an angry glance, as if to say: "Why have you wasted my time bringing me here?"

Wilkins immediately flushed and suggested that if they were quick they might catch the early train back to London.

Watson had misinterpreted Franklin's data, and through either lack of understanding or fault of memory, he had passed on the wrong figures for the water content of the molecule. The X-ray diffraction pictures indicated that each molecule of DNA needed ten times as much water as he had proposed. Just as important, the bases had to be on the inside of the nucleotide backbone, not the outside.

It was an embarrassing occasion, and even Crick and Watson felt humiliated, but worse was to come. Incensed by the nerve Crick and Watson had shown in trying to build a model of DNA from her findings and annoyed that she had wasted a day traveling to Cambridge, Franklin immediately complained of the incident to Sir Lawrence Bragg.

The next day Crick and Watson were called into Bragg's office and severely reprimanded. Until then Bragg had heard only unconfirmed rumors about Crick and Watson's growing obsession with finding the structure of DNA and had managed to ignore them. Now he had been embarrassed by their encounter with the King's researchers. To understand this, we have to look at the difference in approach between British scientists of the time and their counterparts in most other parts of the world. Bragg and many of his peers still cherished the notion of gentlemanly behavior in science and did not believe it ethical that more than one team or individual should work on any specific problem at the same time, even when all parties were competing openly.

As we shall see, this attitude guided Bragg only when it really suited him. He did not like Crick, seeing him as too clever for his own good, and he had clashed with him a few months earlier when the younger man (still without his doctorate) had had the temerity to question Bragg over a technical matter in front of colleagues. The fact that Crick had been right about the issue had only made things worse. Although it would have been difficult for Bragg to find a way to terminate Crick's association with the Cavendish completely, he was not going to help him, so, when he heard Crick and Watson were treading on the toes of other scientists, particularly workers as respected as Wilkins and Franklin, he was especially irritated.

Bragg immediately forbade the pair from doing any further work on DNA and demanded they return to their official research topics, never to interfere with the work of others again. In conclusion, he insisted they hand over their models to the King's researchers, despite Crick and Watson's protests that Wilkins and Franklin were not the slightest bit interested in model building.

Frustrated and angry, Crick and Watson could do nothing but accept Bragg's demands. The Cavendish is made up of a closely knit community of scientists, each knowing what others are doing and all working in a relatively small complex of buildings. Crick and Watson could not have continued constructing their precious models without someone noticing, and they could not have used the machine shop to have parts made for them. But almost as soon as they returned, seething, to their shared office, they both realized all might not be lost. Bragg, they reasoned, could stop them from conducting experiments and building models, but he could not stop them from *thinking* about DNA.

Of course, they would have to be careful what sorts of questions they asked colleagues, even over drinks at the Eagle, and they could not let on to any of the experimenters at the Cavendish that they were still pursuing their dream, but they could hypothesize, and they could talk to each other about DNA so long as they at least gave the outward appearance of returning to their official researches.

Even so, the winter of 1951 was a bleak time, during which Crick and Watson made no progress. Crick returned to investigating the properties of hemoglobin, and Watson was obliged to turn his attention to the nature of the tobacco mosaic virus known as TMV. But the "official" work they were obliged to do was also related to their pursuit of the structure of DNA. Watson's work with Max Perutz on TMV involved elements of ideas he and Crick had been considering in their attempts to model DNA. Like all viruses, TMV contains nucleic acids, which of course also provide some of the building blocks for DNA. What Watson was learning about TMV might, he quickly realized, prove of use in his and Crick's efforts.

Meanwhile, Crick's investigations, centered on X-ray diffraction studies, were also linked to his interest in DNA. Spending his days analyzing X-ray diffraction images was making him much more aware of the subtleties of the technique, and this skill would soon prove invaluable.

Yet the pair naturally felt thwarted and frustrated. Crick and Watson were particularly aggrieved by Bragg's insistence that they give their models to the King's workers. They both knew Franklin and Wilkins were scornful of the use of models. Franklin viewed such methods as little more than children's games, and, that it was one thing for the world's most famous chemist to use models, but quite another for the likes of Crick and Watson. Beyond this, the Cambridge pair had little faith in Franklin and Wilkins elucidating the structure of DNA and concluded that the Nobel Prize would inevitably go to Pauling.

Yet in this respect they were quite wrong. Unknown to anyone but her research assistant, Franklin was the one making great strides forward. For the past year she had been conducting experiment after experiment with the dehydrated or alpha form of DNA but had been disappointed with the ambiguity of the X-ray diffraction images she had produced. Early in 1952 she turned her attention to some of the other DNA samples Wilkins had provided her, DNA in the *beta* or *hydrated* form. The beta form of DNA is colloidal, and in some respects more troublesome to work with, but overcoming practical difficulties, by the spring of 1952 Franklin had obtained the best pictures of deoxyribonucleic acid she or anyone had ever seen, clear, sharp images of the diffraction pattern.

The most significant aspect of these pictures was that the pattern showed conclusively that DNA was helical. The best image was formed in one particular frame labeled "Picture 51," which shows the telltale characteristics of the helical form, in particular a broad, dark X crossing the center of the image.

Yet Franklin told no one of her results. Only Gosling knew of these pictures. Furthermore, spitefully and calculatingly, she not only kept this crucial discovery secret from her superior, Wilkins, but made an official lab announcement that DNA was definitely *not* helical. Her dislike for Wilkins was so intense and his support for the helical structure of DNA so well known, that Franklin found particular delight in expressing precisely the opposite view, even when she knew it was untrue.

Franklin's pettiness toward Wilkins came to a peak three months later, after she had completed a series of further experiments to confirm the helical nature of DNA. In July 1952 she had a card bordered with thick black ink circulated through the biophysics department at King's. It read:

> It is with great regret that we have to announce
> the death, on Friday 18 July, 1952, DNA helix
> (crystalline). . . . It is hoped that Dr. M. H. F. Wilkins will
> speak in memory of the late helix.

Perhaps Franklin really thought she was adding a little levity to the gloomy atmosphere in the Biophysics Department, but Wilkins was not amused. By then he probably knew this "joke" was founded on a lie because since the spring he had been surreptitiously duplicating Franklin's analytic work on DNA and had some stunning images of his own to prove that DNA was a helix.

Meanwhile, spring had brought nothing new in Crick and Watson's approach. Throughout the first months of 1952 they had thought little about their colleagues at King's but had grown increasingly nervous about Pauling. They had heard nothing from California, and this alarmed them more than leaked facts might have. Pauling, they knew, tackled problems head-on and in secret before shocking everyone with a surprise paper.

And come May their worst fears were realized. Through their immediate superior at the Cavendish, Max Perutz, they heard that Pauling was planning to visit England to attend a meeting of the Royal Society. They thought it inevitable that he would try to meet with the King's workers, perhaps even visit their labs. Once there he would have access to their X-ray diffraction results and any slender advantage would be lost.

Convinced the battle would soon be over, Crick and Watson reached a new emotional low. But, on the eve of Pauling's planned visit, they heard startling news. Pauling had been stopped at Idlewild Airport outside New York and had his passport withdrawn by U.S. government officials. Considered too threatening to the capitalist cause, perhaps even a potential defector to Russia, he was forbidden to travel.

While the scientific establishment in England protested this mistreat-
ment, Crick and Watson gave private thanks to the vigor of Joe McCarthy's
imagination and celebrated at the Eagle. The next day they found their
relief made them even more determined to crack the DNA problem, almost
as though they believed Pauling's misfortune had been an omen. Bragg's
edict prevented official work on the project, but now they knew their
thought experiments and probing had to be stepped up a gear.

And, in stark contrast to their labors of the previous winter, Crick and
Watson's endeavors during the summer of 1952 quickly bore intellectual
fruit. Crick made one of the most important breakthroughs in the race.
One evening in the Eagle pub, he struck up a conversation about the structure of DNA with a friend, John Griffith, nephew of Fred Griffith. John
Griffith was a mathematics postgraduate who in his spare time had
recently begun thinking about applying mathematics to biochemical problems. One of these had been how mathematics might explain the way bases
such as those found in DNA could bond and arrange themselves within
macromolecules. He told Crick he had discovered that the only logically
consistent arrangement for bases in DNA was for G to attract C and A to
attract T.

Of course, this was precisely what Chargaff had found via a chemical
route a few years earlier, a biochemical result then familiar to most graduates. But Crick, who had only recently turned to biology, was unaware of
the importance of Chargaff's rules. When Griffith mentioned his idea, and
the fact that Chargaff had also predicted such combinations, Crick became
very excited, feeling that this would lead to deeper answers to many of the
puzzles he and Watson were struggling with.

Not only did it feel right but Crick's excitement came from a sudden,
inspired revelation. Chargaff's and Griffith's conclusions coincided with
one of the key requirements of any structure that would, in Crick's opinion,
satisfy Nature. For any structure of DNA to be the correct one, it had to
explain how the molecule carried the genetic data associated with reproduction; the chemical arrangement of the molecule had to be capable of communicating the "code of life." With C bonded to G and A bonded to T, the
two strands of a double helix could each contain a vastly complex, varying
arrangement of C, G, T, and A bonded to their counterparts on the other
strand. In a moment of dazzling insight, Crick realized that, because of the

ways the bases bonded, the two helices of DNA could "unzip," carry the code, and pass on instructions during reproduction.

In retrospect, we may view this moment as the turning point in Crick and Watson's efforts. Crick had taken a fact from Griffith (and Chargaff) and, applying great imagination, had used it to explain a crucial aspect of the role DNA should play. They were still a long way from producing a model or even creating a logical arrangement for deoxyribonucleic acid, but they had taken a giant leap forward. Then, in the autumn of 1952, Crick and Watson had another piece of good fortune. Linus Pauling's son Peter joined the Cavendish to begin his postgraduate studies, and he was assigned to their office.

Peter Pauling was a likable extrovert who stood out against the austere backdrop of Cambridge. He was bright, charming, and enthusiastic, and for Crick and Watson he was also an invaluable source of information about the latest ideas fermenting in the mind of his father, working six thousand miles away. Peter Pauling was never disloyal to his father, but he saw no reason to keep what he knew to himself. Father and son corresponded often, and if there was anything new to tell, Peter was happy to talk to about it.

But, for his first few months of his time in Cambridge, he had little to report. It seemed his father was brooding on the problem, and there was nothing but a dull silence from the great man's team. Then, one day in the middle of December 1952, Peter strolled into the office, sat at his desk, put his feet up, and waved a piece of paper at his colleagues. It was a letter from his father that outlined what Pauling thought might be a promising structure for DNA, which he hoped to publish early in the new year.

Watson later described the scene: "No details were given of what he was up to, and so each time the letter passed between Francis and me the greater was our frustration. Francis then began pacing up and down the room thinking aloud, hoping that in a great intellectual fervour he could reconstruct what Linus might have done. As long as Linus had not told us the answer, we should get equal credit if we announced it at the same time."[24]

Christmas that year was an agony for them. As hard as they tried, they could get no more news from across the Atlantic, and this frustration continued into January. But then, early in February 1953, Peter Pauling received another letter from California, in which his father mentioned the structure of DNA and news that he was about to publish, ending by saying he would send a draft copy of the paper to his son as soon as it was ready.

A few days later Peter walked into the office with a copy of his father's paper. "Peter's face betrayed something important as he entered the door and my stomach sank in apprehension at learning that all was lost," Watson recalled years later.[25] With growing anxiety the pair rushed through the main body of the text and concentrated on the summary and conclusion to see what the great man had to say.

And they were staggered. Not only had the world's most famous and accomplished chemist described what Crick and Watson knew to be an impossible structure but along the way he had made some of the most elementary errors imaginable. Pauling's most obvious mistake was to describe DNA as a *triple* helix, but there were many other faults with his model. First, his structure was too tightly packed; the chemical groups on the inside of his triple helix structure could not have been fitted into the space they were allocated. Second, he had made basic mistakes in the way he described how some of the groups bonded. He had used what are known as hydrogen bonds to link them, but to make this work he had added atoms to some of the molecules in positions in which they simply could not exist. Third, Pauling's structure said absolutely nothing. It was a lump of atoms bonded incorrectly; and, even if the laws of chemistry had allowed for these arrangements, his proposed structure could not have acted as the agent for passing on the genetic code. For both philosophical and scientific reasons, this last failing was the gravest.

To Crick and Watson, it was simply startling that Pauling had written what he had. "The appropriate emotion was pleasure that a giant had forgotten elementary college chemistry," Watson later declared.[26] After the euphoria came another strong emotion, fear. To both Crick and Watson, it was obvious that the moment Pauling realized his error (and perhaps he already had) he would stop at nothing in making up for the inevitable knock to his pride. A humiliated Pauling was, they knew, a very dangerous beast. They were convinced their rival would have the correct structure within weeks. If they were to have a chance themselves, they would have to drop everything and concentrate on DNA, no matter what Bragg said.

A few days after Pauling's paper arrived, Watson paid another visit to King's. He knew that he had to gather more data and resolve once and for all the nature of the double helical structure he and Crick were sure acted as the backbone of the DNA molecule. To do this he would have to talk Maurice Wilkins into parting with some detailed information, perhaps even some new X-ray diffraction images.

Arriving early, Watson wandered along the dark corridors of the basement lab complex and found Rosalind Franklin's room. The door was ajar, and he nudged it gently. Franklin was working intently at her bench. Watson coughed quietly, and she looked up, eyeing him with suspicion. He wandered over. Taking out Pauling's paper, he waved it in the air. "Have you seen this?" he inquired. Franklin was naturally interested and couldn't resist reading it. But then Watson made a serious mistake. He leaned over Franklin's shoulder and started harping about the stupid mistakes Pauling had made and boasting about how he and Crick knew the structure must be helical, possibly a double helix, yet Pauling, the world's greatest chemist, had gotten it all wrong.

Watson did not know quite how vehemently Franklin had been opposing the helical form, simply because of her distaste for Wilkins and his enthusiasm for this arrangement. As Watson crowed, Franklin grew increasingly angry. A moment later, she whirled on him, furious. "Suddenly," he recalled, "Rosy came from behind the lab bench that separated us and began moving towards me. Fearing that in her hot anger she might strike me, I grabbed up the Pauling manuscript and hastily retreated to the open door."[27]

Fleeing from Franklin's anger, Watson was thrown, quite literally, into the arms of Maurice Wilkins, who had just pushed open the door to the lab. The two men headed off along the corridor to the relative sanctity of Wilkins's office, and over a cup of tea the King's researcher suddenly decided to tell Watson all about Franklin's latest conclusions and the work of hers he had duplicated. Then he opened a drawer and took out an X-ray diffraction image of the beta form of DNA, a duplicate of the stunningly clear Picture 51 Franklin had obtained over six months earlier. "The instant I saw the picture my mouth fell open and my pulse began to race," Watson recalled.[28]

For Crick and Watson, this piece of data was everything they could have wished for. It told them DNA was definitely helical, but it also gave them crucial information about the form of the helix and could allow them to calculate the way the chains of the helix were arranged and how far apart the turns would be. They both realized that this was the vital clue to the structure. Combined with their own ideas, it could allow them to elucidate the structure. But time and the authorities were not on their side. If they were to succeed, they would have to get Bragg's blessing. Crick especially felt pessimistic, but when the time came for the interview, both Crick and Watson were amazed by how recent events had changed their boss's attitude. With Pauling so clearly in the race, the need for gentlemanly conduct in science appeared to vaporize.

Bragg and Pauling were old rivals. Bragg had been chasing the structure of the alpha-helix for proteins, and a few years earlier he had published a paper with Perutz that had missed the point almost as badly as had Pauling's recent description of DNA. A year later Pauling published his protein model and startled the world of biochemistry with its simple elegance. Bragg had felt cuckolded, and now, facing the specter of Pauling's gaining the greatest biochemical prize in history, his sense of ethics became remarkably flexible. Suddenly he viewed the contest as between Britain and America and not the Cavendish versus King's College. The ethical questions were, of course, the same, but they had been overpowered by patriotism and pride.

Bragg's approval was absolute. Not only did he give Crick and Watson moral support but he provided them with the entire range of facilities available at the Cavendish, including the full resources of the workshop that could fashion model parts for them and the X-ray diffraction equipment, along with technical support.

But by now Crick and Watson needed little in the way of equipment. They had the essential data from the X-ray diffraction pictures taken at King's; they knew how the bases paired as well as many of the dimensions of the double helix. What they needed was time to think and to build model after model until they could obtain a structure that was chemically accurate and explained the mechanism of reproduction.

Meanwhile, unknown to Crick and Watson, John Randall at King's had redressed the problems he had initiated between Franklin and Wilkins. Franklin had been encouraged to find a research position elsewhere. By the time Crick and Watson had received Bragg's blessing to return to work on DNA, Franklin was in her final weeks at King's. But for her this actually made little difference, because she was planning to continue her efforts to elucidate a structure for DNA until the day she had to leave, then to work on the project somewhere else. However, Franklin had no inkling that

Wilkins had passed on to Crick and Watson crucial information gathered from her experiments.

We can follow her research into the structure of DNA during those final weeks at King's because Franklin wrote detailed notes in her lab book. By early February 1953, she had decided to go as far as she could with the notion of a helical structure for DNA, and during the following few weeks she concentrated on a mathematical analysis of the X-ray diffraction images she had obtained with a view to establishing clear proof of a double helix backbone.

Meanwhile, Watson and Crick were racing ahead still faster and further. By the middle of February, they had constructed a model of the basic double spiral. They knew the angle of the turns and their separation. What they needed now was to explain the way the bases were bonded to the central framework of nucleotides. They knew that adenine (A) must link with thymine (T) and that cytosine (C) was attracted to guanine (G), but they had no idea how they bonded or if these bases were positioned on the outside or the inside of the helices.

Placing the bases on the outside of the helices seemed an easier, if ugly option, but placing them on the inside also presented problems. Because of the restricted space inside the helices, and the fact that the bases would repel each other if they were too close, finding a suitable arrangement for them required careful rejigging of the model, and they had no clear idea how this could be achieved.

Meanwhile, Pauling had learned of Crick and Watson's efforts. The Cambridge pair heard this news through Peter Pauling, who had received a letter from his father, saying, "I am checking over the nucleic acid structure again, trying to refine the parameters a bit. . . . It is evident that the structure involves a tight squeeze for nearly all the atoms. I hear a rumor that Jim Watson and Crick had formulated this structure already sometime back, but had not done anything about it. Probably the rumor is exaggerated."[29]

Two days later Watson clearly felt he and Crick were closing in on the target. To his mentor Max Delbrück, then at Cal Tech, he wrote: "Today I am very optimistic since I believe I have a very pretty model, which is so pretty I am surprised no one has thought of it before." Elsewhere in the letter, Watson displays his characteristic fiery attitude toward scientific rivalry. "I had started on DNA when I first arrived in Cambridge," he

declares, "but had stopped because the King's group did not like competition or co-operation. However, since Pauling is now working on it I believe
the field is open to anybody. I thus intend to work on it until the solution is
out."[30]

But, although they did not realize it, Watson and Crick still had one
last barrier to overcome. They had been working with structures for two of
the four bases that were, in a small but vital way, incorrect. Using these, they
could never have elucidated the structure for DNA.

They only learned of this crucial fact during a casual conversation with
Jerry Donohue, a colleague in the chemistry department who specialized in
the structure of organic bases. Over coffee in the common room of the
Cavendish, Watson had described the problems he and Crick were experi-
encing in their efforts to bond the cytosines to the guanines and the
thymines to the adenines, and as he talked he had drawn the structures of
the four bases on a scrap of paper between the coffee cups. When Donohue
saw the structures, he immediately realized Crick and Watson had been
using the wrong structural form for the bases.

The structures for the bases Crick and Watson had been using were
accepted by most chemists of the time and appeared prominently in all the
basic chemistry books the pair had on their office shelves. These structures
are called the enol form of the bases, and they contain a group in which a
hydrogen atom is bonded to an oxygen atom. But, as Donohue pointed out,
in spite of what the textbooks said, modern research had shown that this
form was not so likely to be found in Nature. The bases, he explained, could
take up another configuration by undergoing what is known as a tautomeric
shift, in which the hydrogen moves to another part of the molecule. This
alternative structure is called the keto form. For Crick and Watson, this
made a vast difference because it allowed the bases to be bonded in their
appropriate pairs using hydrogen bonds of just the right length to enable
them to pack together inside the double helical backbone of the molecule.
Now all the pieces were falling into place.

But Franklin was also heading in the right direction. She outlined her
latest ideas in her lab book entry for February 23, written just a day or two
after Watson's conversation with Donohue. In this report she described the
correct diameter for the molecule and reasoned that the bases were positioned on the inside of the double helix backbone. Not surprisingly, she had

the same separation between the turns of the helix as Crick and Watson. What she did not conclude in time was the essential nature of the molecule, the way it is composed of complementary helices. This was the great step Crick had taken to explain how such a structure could carry something as complex as a genetic code. As well as this, she had no time to study how the bases might pair and how they could pack inside the helices.

Rosalind Franklin would almost certainly have addressed these problems given time, but time was a luxury she was denied. A few days after her final notes, written carefully in her lab book on Tuesday, February 24, Crick and Watson were home free. On Saturday February 28, the final pieces of the puzzle were put in place. They had a logically consistent theoretical model in which the four bases were fitted into the interior of the helices and bonded correctly. At last they could order the parts from the workshop and begin constructing their model.

That lunchtime, with the final chemical symbol in place and the last remaining calculation done, Crick and Watson could contain their feelings of triumph no longer. Running to the Eagle pub, to everyone within hearing distance they proclaimed that they had discovered the "secret of life."

A week later, on Saturday, March 7, the model was completed and stood proudly on the central desk in their shared office. That afternoon, in his office at King's College, Maurice Wilkins, ignorant of the latest developments in Cambridge, wrote a letter to his friend Francis Crick in which he said: "Thank you for your letter on the polypeptides. I think you will be interested to know that our dark lady [Rosalind Franklin] leaves us next week and much of the 3-dimensional data is already in our hands. I am now reasonably clear of other commitments and have started up a general offensive on Nature's secret strongholds on all fronts: model, theoretical chemistry, and interpretation of data, crystalline and comparative. At last the decks are clear and we can put all hands to the pumps! It won't be long now, Regards to all, Yours ever, M."[31]

Crick received the letter on Monday, but instead of phoning Wilkins he decided to pass on his own momentous news through official channels and asked his senior at the lab, John Kendrew, to call, while he and Watson started writing their paper.

The King's team and Rosalind Franklin took the news with magnanimity, and Wilkins wrote Crick another letter in which he good-naturedly called the Cambridge pair "a couple of old rogues." The only request Wilkins and Franklin made was that they might each publish a paper alongside Crick and Watson's planned announcement, adding what they had discovered and contributed to the elucidation of the structure. This Crick and Watson agreed to immediately, and what was arguably the most important biological breakthrough of the century was announced with not one paper but three, all published in the April 25, 1953, issue of *Nature* (issue 171).

Wilkins's paper, "Molecular Structure of Nucleic Acids: Molecular Structure of Deoxyripentose Nucleic Acids," coauthored with two other King's workers, A. R. Stokes and H. R. Wilson, described the experimental data gathered over three years and verified that closely related molecules to DNA (deoxypentose nucleic acids) also had a helical structure. Franklin's paper, written with her assistant Raymond Gosling and titled "Molecular Structure of Nucleic Acids. Molecular Configuration in Sodium Thymonucleate," presented her best X-ray diffraction images along with data gathered during her time at King's, all of which fully vindicated Crick and Watson's model.

Linus Pauling, who had dived in early with a totally wrong structure, had nothing to contribute.

Crick and Watson's paper, titled "Molecular Structure of Nucleic Acids: A Structure for Deoxyribose Nucleic Acid," was a mere nine hundred words long and contained just one diagram. The paper had been typed by Watson's sister Elizabeth, who was visiting her brother in Cambridge, and the drawing illustrating the proposed structure for the double helix was by Crick's wife, Odile.

The paper began modestly: "We wish to suggest a structure for the salt of deoxyribose nucleic acid (DNA). This structure has novel features which are of considerable biological interest." The authors went on to describe their proposed structure for DNA and concluded with an oblique reference to the huge biochemical importance of the molecule. "It has not escaped our notice," they wrote, "that the specific pairing we have postulated immediately suggests a possible copying mechanism for the genetic material."[32]

This may seem coy, but in private Watson at least knew the value of what he and Crick had achieved. To persuade his sister to spend a Saturday typing the article because the lab secretary was away, he told her with absolute conviction that she was "participating in perhaps the most famous event in biology since Darwin's book."[33]

He was almost certainly right.

*W*HAT THEN MAY WE conclude about the method of Crick and Watson's discovery? Of course, the elucidation of the structure of DNA has led to the rapid advancement in genetics we are witnessing today. Deoxyribonucleic acid is the raw material of genetics, and understanding how it functions was the first step into the modern era of genetic science. Based upon the breakthroughs of the 1950s and 1960s, geneticists are now altering the world in much the way physicists revolutionized their science with the introduction of quantum theory during the early part of the twentieth century. These changes have had an impact on many of us already and will continue to play an increasingly important role in all our lives.

Such are the most significant outcomes of Crick and Watson's efforts. But what should we conclude about their place in the pantheon of science? What is to be made of their methods? Without them, would the structure have remained a mystery for long? And did the way in which the discovery was made and the process by which the world learned of it have a great effect upon the development of science in general and of genetics in particular? Finally, given slightly different circumstances, could Rosalind Franklin or Linus Pauling have beaten Crick and Watson?

First, we should consider, Crick and Watson themselves. They are clearly very clever men, but their methods were unlike those of any other theoretical scientists mentioned in this book. They were different in two crucial ways. First, Crick and Watson did almost no experimental work of their own, and second, they were not in any sense experts in the area of research they had chosen to investigate. Of course, they had reached doctorate level (in all but name in Crick's case) and so were able to focus their intellects and to distill information in a highly sophisticated way, but when they began their project in 1951, neither Crick nor Watson could have been considered a professional biochemist.

As we have seen, many scientists in history have been motivated by drives other than the desire to learn. Indeed, this idea lies at the heart of scientific rivalry, but Crick and Watson seemed free from any scruples or ethical barriers when it came to being first to a discovery. They didn't actually steal ideas or information, but they were remarkably adept at being in the right place at the right time in order to maximize their opportunities. In this way, their approach may be compared with that of Thomas Edison, although he worked entirely as a practical scientist.

It should be emphasized that Crick and Watson, as maverick as their methods may have been, are certainly not intellectual lightweights. Indeed, it might be argued that their achievement was all the more stunning because of their backgrounds and the fact that they did no original experimental work to elucidate the structure of DNA. Furthermore, each of them went on to do important work after their revolutionary discovery. Each became a cornerstone of the global scientific establishment. For several years Watson was head of the U.S. Human Genome Project, probably the most prestigious job in genetics. In 1977 Crick established an important lab at the Salk Institute in San Diego. Here he has become preoccupied with the study of human consciousness. Together with their huge contributions to science, including between them a vast collection of scientific papers and academic books, each has written a successful and controversial memoir.

In some ways, Crick and Watson had the common touch. They could view a problem from the perspective of outsiders and were not confined by a certain rigidity of technique displayed by both Wilkins and Franklin. There can be no clearer indication of this than by comparing the ways the three parties wrote their papers for *Nature*. Wilkins's and Franklin's accounts were packed with jargon, whereas Crick and Watson's was in plain English and contained as few technical expressions as possible. Yet Crick and Watson delivered an equally thorough description of their findings.

Perhaps because of their shunning of orthodoxy, some of Crick and Watson's fellow scientists despised the pair from the beginning and never stopped resenting them after they became scientific icons. Watson himself is fond of relating a story about how, two years after the discovery, he was climbing in the foothills of the Alps and encountered a chemist named Willy Seeds, who had worked with Wilkins at King's. As they approached each other along a path, Watson made to stop for a chat, but Seeds simply said: "How's Honest Jim?" and walked on.[34]

Erwin Chargaff, whose rules Crick and Watson applied in their development of the structure of DNA, never accepted the pair as bona fide scientists and was unimpressed with what they did. "You want to know who really deserves the credit for the DNA discovery?" he once asked a researcher. "It's Friedrich Miescher, and he described his discovery in a paper written in 1871." When the researcher admitted he had never heard of Miescher, Chargaff retorted angrily. "No, I was sure of that, but you've heard of James Watson and Francis Crick, Nobel laureates, our mass-media substitutes for saints."[35]

But Chargaff was quite wrong. In 1869 Miescher stumbled upon a substance of high molecular weight, which indicated a complex structure, that was found in cells. He went on to make the remarkably prescient hypothesis that this material (DNA) could carry the genetic code, but he did not elucidate the structure of DNA or come remotely close to doing so. One can only assume that Chargaff was demonstrating professional jealousy in this statement, or that at some point in the distant past Watson or Crick had in some way offended him. He had met the pair as they were struggling to model DNA, and they had let slip their ignorance of his rules. Perhaps he never forgave them for that.

Others have attacked Crick and Watson since they became globally famous. The biochemist Colin MacLeod has commented: "Some day perhaps you will enlighten me about the earthshaking significance of the double helix, etc. If it hadn't been worked out on a Tuesday, it would have happened in some other laboratory on Wednesday or Thursday."[36] This comment too should be taken with a large dose of salt. MacLeod was a colleague of Oswald Avery, and he said this after Watson's *Double Helix* had reached the best-seller list.

Some fellow scientists never dropped their conviction that the Cambridge pair had treated the King's team badly, that they had used Franklin's and Wilkins's results without letting them know and had somehow misled Wilkins into divulging information he had spent years gathering. One crusading commentator, the writer Candace B. Pert, has gone so far as to say: "But the truth of the matter is, the two men visited Franklin's when she was out of town and persuaded her boss to let them take a peek at her data. In what must have been a moment of incredible rationalizing they stole Franklin's findings and got away clean, tossing her a bone of acknowledgement in their seminal paper."[37]

This statement is plagued with such stunning inaccuracy it loses all credibility: Only Watson visited Franklin's lab on the occasion in question, the "bone" was a shared announcement of the discovery in *Nature*, the use of the word *stole* is thoroughly misleading, the sarcastic reference to "rationalizing," which implies the data were all that mattered in elucidating the structure, is also prejudicial; and so it goes. Yet if we ignore the hysterical tone and the misplaced vitriol contained in this comment from a book written in 1997, almost half a century after the events, it clearly illustrates how the circumstances of the discovery remain emotive.

And Watson is especially sensitive about this. In his famous book, he delivered a rather flippant, if immensely entertaining account, but recently he has become somewhat prickly over the issue and angry about the attacks he and Crick have suffered. In the process of defending their actions (something he does not really need to do), he has acquired a reputation for shock comments and public rebuttals. In a speech at the Harvard Center for Genomics Research in 1999, he announced:

> There is a myth which is that, you know, that Francis and I basically stole the structure from the people at King's. I was shown Rosalind Franklin's X-ray photograph and "Whoooa! That was a helix!" and a month later, we had the structure; and [that] Wilkins should never have shown me the thing. I didn't go into the drawer and steal it. It was shown to me, and I was told the dimensions, a repeat of thirty-four angstroms, so you know, I knew roughly what it meant and it was that the Franklin photograph was the key event . . . psychologically, it mobilized us back into action. The truth is that we should have got the structure in the fall of '51. There was enough data. We wouldn't have been able to say with finality that it was right because, uh, that came with Rosalind's X-ray work, that was the proof it was right.[38]

What is almost certain is that Wilkins must have known what Crick and Watson were doing. He and Franklin had visited the Cambridge lab and seen their early failed attempt to construct a model of DNA. Wilkins and Crick had been good friends for many years; Wilkins must have been aware how determined Crick could be. It is possible Maurice Wilkins was blinded by his ongoing battles with Franklin, and this might have caused him to miss the threat presented by Crick and Watson, but this was not the fault of the Cambridge team.

Most strikingly, those who should have been most upset about Crick and Watson's tactics were the most supportive of them after their announcement of the discovery while those most critical had always stood on the periphery of the race. Furthermore, no one could say Rosalind Franklin was unable to defend herself.

But Colin MacLeod's comments also lead to my second and subsequent questions about the discovery of the structure for DNA: Would the structure have remained a mystery for long? And did the way in which it was discovered and the process by which the world came to know of it greatly influence the later development of science in general and genetics in particular?

There are no definite answers to these questions. Some commentators, like MacLeod, claim that the structure of DNA was ripe for elucidation, that all the pieces of the jigsaw were available to many scientists at the time and all that was required was for someone, or some group, to put the pieces together. Others suggest the puzzle may not have been fully solved for several years.

What is perhaps more interesting is how the way DNA was discovered affected the course of scientific progress. Crick has commented: "It is the molecule that has style, quite as much as the scientists."[39] What he meant is the idea that because the discovery had such an impact both in science and in the public domain, it spurred on other scientists to take the ideas he and Watson had illuminated and develop them in many rewarding ways. If the news of the structure of DNA had trickled out piecemeal in papers from many scientists, the discovery might not have made anything like the impact it did.

To fully understand this, we must try to appreciate just how revolutionary and revelatory Crick and Watson's paper was. The announcement made the newsstands just a few weeks before the coronation of Queen Elizabeth II and the conquest of Everest by Edmund Hillary and the Sherpa Tenzing Norkay, and it created equally bold headlines around the world. In that more innocent age, the modeling of DNA, the "molecule of life," soon gained an almost mythical status. It is easy for us to forget that through global media and communications we have become accustomed to revolutions in science and technology and jaded by what now seems a bombardment of discovery and innovation.

With the singular exception of Charles Darwin's evolution theory, Crick and Watson's discovery was the first example of purely theoretical science making an impact in the media and stimulating the interest of the lay public. Until the twentieth century, news of scientific advances rarely filtered out to the public, and when it did it was appreciated by very few people. The use of the first atomic bombs had an enormous impact upon society and changed the attitudes of many toward scientific discovery, but this influence took effect only when nuclear testing had become well established. In 1945 the bombing of Hiroshima and Nagasaki using a new "super-weapon" was not viewed by the public as "science" in any real sense. The bomb was linked with "war engineering," "weapons development," a product of aggression, work driven by politics and the need to survive. Even if most people had no understanding of the scientific details, the public knew that Crick and Watson had found the structure of the "molecule of life." Just as Darwin's ideas, first stated publicly a century earlier, were immediately significant to every educated person at the time, the work of Crick and Watson was perceived as science for you and me, and it promised to allow the unwrapping of one of the great unresolved mysteries of human existence. Also, like Darwin's theory, it dared to intrude into the nebulous borderland between science and religion. It was dangerous, bold, and seductive, what today we would call sexy.

But if not Crick and Watson, then who? Clearly Rosalind Franklin was close, but she had missed several key elements of the solution. She suffered the great handicap of working alone; with a collaborator she trusted, she may well have beaten Crick and Watson to the prize. Wilkins would have made a perfect intellectual partner for her, but by ill fortune they were entirely mismatched personally, which made it impossible for them to work together.

In some ways Linus Pauling was at a similar disadvantage. He had many competent colleagues with whom he could share ideas, but none of these people could offer him the close relationship Crick and Watson shared. He was also in a sense stifled by his own elevated position in the world of science, in that many of his subordinates would have been reluctant to question his authority or challenge his ideas. How else could such a horrendous paper as the one he wrote early in 1952 have slipped through the net? It is also questionable how close Pauling really was to determining the structure. He was a little startled by Crick and Watson's final and correct model and appears to have been thinking along quite different lines. Early in 1953 he had been preparing to modify his theory based upon a triple helix.

For our heroes the elucidation of the structure of DNA was a life-altering experience. In 1962 Crick, Watson, and Wilkins shared the Nobel Prize for their efforts to describe the molecule of deoxyribonucleic acid. Sadly, Rosalind Franklin died of cancer in 1958, aged just thirty-seven. According to the rules of the Nobel Prize committee, the prize may go only to living scientists and cannot be split more than three ways. Many have wondered what the judges in Stockholm would have decided if Franklin had lived. The debate, although perhaps a pointless one, continues to this day.

A friend of Franklin, Sir Aaron Klug, who was awarded the Nobel Prize in 1982 for his work in genetics, used his acceptance lecture to honor her memory. "Had her life not been cut tragically short, she might well have stood in this place on an earlier occasion," he said.[40] Watson did not hesitate when asked recently who he thought would have been given the prize if cancer had not claimed Franklin's life. "Crick, myself, and Rosalind Franklin," he replied.[41]

In reality, such questions are academic. By virtue of their intelligence, daring, and intuitive grasp of science, their aggression, egotism, and a fair share of good fortune, it is the names Crick and Watson that will forever be cited as the unravelers of one of Nature's greatest and most potent mysteries.

# SEVEN

## Reaching for the Moon

Secrecy in science robs it of the main element
that keeps it healthy: scientific public opinion.
—*Peter Kapitza*

*175 miles above America, October 1957*

*I*T WAS JUST AN IRRITATING BLEEP, bleep, bleep, but it meant
the world had changed irrevocably.

On October 4, 1957, the flight of *Sputnik 1*, or *Prostveishiy Sputnik* (Simple
Satellite PS-1), shattered the sense of superiority held by most Americans.
It was the opening shot in what was to turn into a battle within the Cold
War, a battle that was also the most audacious, expensive, and inspirational
scientific endeavor in history, a battle that, in fewer than a dozen years, led

from a satellite weighing 184 pounds to a man walking on the surface of the Moon.

The space race was begun by the Russians, and they won victory after victory, first after first, until at least the early 1960s, but gradually their lead became eroded by inefficiency and infighting, wasted on bureaucracy, and superseded by superior engineering and vast injections of cash from the coffers of their American rivals.

In the beginning the Russians grabbed the limelight and stole the headlines. The morning after the launch of *Sputnik 1*, the front page of *The New York Times* carried a three-line, banner headline covering the entire page that read:

> SOVIETS FIRE EARTH SATELLITE INTO SPACE;
> IT IS CIRCLING THE GLOBE AT 18,000 M.P.H.;
> SPHERE TRACKED IN FOUR CROSSINGS OVER U.S.[1]

*Pravda* rather prematurely proclaimed: RUSSIANS WON THE COMPETITION.[2] In France the headline of *Le Figaro* declared: MYTH HAS BECOME REALITY: EARTH'S GRAVITY CONQUERED. Beneath this they gloated: "Americans had little experience with humiliation in the technical domain" but were now "disillusioned."[3]

It was not only the media who vaulted the achievement. "The success of the Russian Sputnik," wrote Martin Summerfield, a distinguished academic at Princeton University, "was convincing and dramatic proof to people around the world of the real prospects of space travel in the not-too-distant future. The fact that twenty-three-inch sphere weighing 184 pounds has been placed in an almost precise orbit indicates that a number of important technological problems have been solved."[4]

But the pain and humiliation was felt most profoundly by the leading space engineers in the United States. They were more frustrated by the news than the average American citizen because they knew the headlines could have been very different. They knew more than anyone that it was the American program that could have begun the space race with just such a famous victory.

The American administration, whose leaders were ultimately responsible for funding the ideas put to them by the country's most eminent scientists, had plenty of warning of Russian advances in space engineering. As early as 1946 the U.S. Army had paid a group of major American engineering corporations to submit preliminary reports on the feasibility of what they called earth-orbiting satellites. One from the Rand Corporation of Santa Monica, California, titled "Preliminary Design of an Experimental World-Circling Spaceship" suggested that "the achievement of a satellite craft by the United States would inflame the imagination of mankind, and would probably produce repercussions in the world comparable to the explosion of the atomic bomb."[5] But the army's money was wasted, the reports ignored.

During the following decade numerous learned and popular papers on earth-orbiting satellites appeared in publications ranging from scientific journals to tabloid newspapers. One of the most respected scientists in the United States, and the man who went on to mastermind the space program in America, Wernher von Braun, wrote in 1952: "The United States must build a manned satellite to curb Russia's military ambitions."[6]

In July 1955, at the Sixth Congress of the International Astronautical Federation in Copenhagen, a Russian delegation there only as observers announced their plans to launch a satellite during the International Geophysical Year (IGY), set to begin on July 1, 1957. "From a technical point of view," declared Leonid Sedov, the head of the delegation, "it is possible to create a satellite of larger dimensions than that reported in the newspapers which we had the opportunity of scanning today. The realization of the Soviet project can be expected in the comparatively near future. I won't take it upon myself to name the date more precisely."[7]

*The New York Times* picked up on this story and ran a rather obscure four-hundred-word piece on the subject with the headline "SOVIET PLANNING EARLY SATELLITE. Russian Expert in Denmark Says Success in Two Years Is Quite Possible."[8]

But the warnings continued to be ignored. President Eisenhower had no interest in space travel and failed utterly to see its potential propaganda value. Keith Glennan, a senior NASA administrator who met Eisenhower in 1960, was quoted as saying after a frustrating talk with the president: "I told him something of the costs that appear to be involved in Project Apollo. He expressed himself once more as having little interest in the manned aspects of space research."[9]

One of Ike's most important political guides, Nelson Rockefeller, realized the importance of talk about spaceships and pushed for them, making it clear to the president that letting the Russians beat the United States into space would have "costly consequences." "This is a race that we cannot afford to lose," he prophesied.[10] Yet even Rockefeller's influence had little effect because many of Eisenhower's key advisers believed precisely the opposite and supported the president's Luddite attitude. In 1959, just two years before Yuri Gagarin entered the history books as the first man in space, Eisenhower's chief science adviser, George Kistiakowsky, said of NASA's attempts to develop the Mercury Project: "It will be the most expensive funeral a man has ever had."[11]

Meanwhile, respected journalists were starting to stir up exasperation. *Newsweek*'s Edwin Diamond wrote: "No amount of soft soap can gloss over the dismal fact: The U.S. is losing the race into space, and thus its predominance in the world."[12] But even this appears to have had absolutely no effect on U.S. policy. Eisenhower's deputy secretary of defense, Don Quarles, succeeded in convincing the president that it would be seen as a hostile act if the United States were to launch the first satellite, an idea that showed a lamentable misreading of the signs even within the supercharged environment generated by Cold War paranoia.[13] Then, to compound his error, after the announcement of Sputnik's success, Quarles not only claimed the Russians had done the United States a veiled favor by opening up the peaceful use of space but admitted that American scientists could actually have beaten the Russians all along if they had decided to.[14]

And Quarles should have known. In September 1956, over a year before the launch of *Sputnik 1*, Wernher von Braun was working on the latest rocket designs for the U.S. Army at Huntsville, Alabama. Von Braun and his team had been developing the Redstone missile and others into rockets that could carry payloads beyond the atmosphere. A few months earlier, von Braun had conducted some calculations that told him his newly designed four-stage Jupiter-C rocket would have sufficient thrust to generate the escape velocity of seventeen thousand miles per hour, the speed needed to carry a payload into Earth orbit.

Somehow news of these calculations leaked out and was passed onto the Pentagon. By 1956 von Braun had been a U.S. citizen for a year and had worked for the U.S. Army for a decade, but he and his German colleagues who had built the Redstone rocket were still not entirely trusted by many within the United States government. Major General John Medaris, then commander of the Redstone Arsenal working at the Pentagon, was alarmed by the idea that von Braun might deliberately alter the flight plans of his next launch to send a missile into orbit and claim later it was a mistake. The U.S. administration, aware that the Soviets were tracking every test launch, did not want the world's first satellite to be an empty nose cone of a missile developed by the military. If the nose cone contained no scientific equipment, its launch might be viewed as a threatening gesture. To prevent this, Medaris arranged for a specially trained team to secretly fill the fourth stage of the Jupiter-C with sand so it could not achieve escape velocity.

On September 20, 1956, von Braun's rocket rose into the sky, traveled more than 3,000 miles out over the Atlantic Ocean, and reached an altitude of 682 miles. But its velocity fell short of the escape velocity by less than a thousand miles per hour, so it failed to reach Earth orbit before reentering the atmosphere and burning up. As one space historian has commented: "America owned the future, but America's bureaucrats were blind."[15]

As a consequence, America hardly heard the starting gun that began the race into space and lost the first rounds of the contest. But a few years after the propaganda failures of the late 1950s, the old men of the Eisenhower administration were superseded by a more energetic and visionary government. And, as time would prove, the first off the block is not always the one who wins the race.

THE PRINCIPLE OF JET PROPULSION was first documented during the first century A.D. Hero of Alexandria demonstrated its power by rotating a metal wheel using the force of a jet of steam fired from a series of outlets on its perimeter. But little was done with the idea until the Chinese of the thirteenth century started to use the same principle to produce "fire arrows" or small rockets attached to arrows, which on the battlefield they fired at advancing soldiers.

By the mid-eighteenth century, the military use of rockets had grown considerably more sophisticated. The armies of the Indian leader Hyder Ali sent panic through the ranks of British soldiers on the battlefield by firing eight-foot-long rockets made from bamboo and iron tubes. The British learned the lesson and made rocket weapons an important part of their own armory. In 1806 they used rockets to destroy a French invasion fleet at anchor in Copenhagen harbor and in the process leveled half the city. Six years later the Royal Navy turned this firepower on the Americans during the short-lived War of 1812. The spectacle of rocket explosions on this occasion even made its way into the lyric of "The Star-Spangled Banner" with the words "rockets' red glare."

But by the twentieth century, military rockets had become unfashionable and were largely replaced by powerful artillery pieces. By World War I, rockets were almost redundant and huge field guns like the colossal Big Bertha (or Paris gun), used by the Germans to terrorize the French capital, were far more important and impressive. Big Bertha could fire a shell weighing 1,764 pounds a distance of six miles.

Yet as the military uses of rockets faded temporarily, their potential peacetime uses were beginning to be taken seriously. Perhaps the most important innovator in early rocketry was the Russian mathematician and engineer Konstantin Eduardovich Tsiolkovsky, who in 1911 wrote: "Mankind will not remain on the earth forever, but in the pursuit of light and space will at first timidly penetrate beyond the limits of the atmosphere, and then will conquer all the space around the Sun."[16]

By this time Tsiolkovsky was already considered a visionary thinker by other rocket enthusiasts; in a book published in 1903 and quaintly titled *Exploration of Space by Rocket Devices*, he had outlined the basic ideas behind the use of rockets for space exploration. With this book Tsiolkovsky established the theory of rocketry and presented the mathematics engineers would employ in designing successful rockets, including equations governing the ratio between fuel consumption and mass of the rocket and the relationship between speed of exhaust gases and thrust. But although, more than anyone, he deserves the epithet "father of the rocket," Tsiolkovsky was a consummate theoretician who worked solely with equations and words; he did not build a single rocket.

An American engineer named Robert Goddard was the first to fly a practical rocket not designed for military use, and he began testing devices soon after Tsiolkovsky published his theories. By 1919 Goddard had written a definitive work on the practicalities of rocket design titled "A Method of Reaching Extreme Altitudes," which was submitted to the Smithsonian

Institution. For the most part this was a serious and carefully researched piece of work, but toward the end of the book Goddard suggested outlandish ideas such as the possibility of sending manned rockets to the Moon. When these ideas attracted ridicule from his peers, he was widely discredited within the scientific establishment. Goddard, who responded by declaring that all he was trying to do was "to get this thing off the ground," carried on, and by 1926 he had launched the first liquid-propellant rocket.[17]

This was an enormously important breakthrough because Tsiolkovsky had shown that solid fuels such as those used in ancient military rockets (and today's fireworks) could not be used to carry a rocket into space. Using solid fuel, the rocket would always be too heavy for the thrust produced, but liquid propellant provides a usable power-to-mass ratio.

Even before this launch, others around the world were starting to take the idea of rockets into space seriously. In 1923 a German enthusiast, Hermann Oberth, appropriated some of the ideas from Goddard's Smithsonian paper and turned them into a book he called *The Rocket into Interplanetary Space*, which investigated different means of propulsion and even considered designs for space suits and many of the serious practical problems associated with humans living and working in space.

Goddard was angered by Oberth's book and convinced the author had stolen many of his ideas and trivialized them. Oberth had done precisely that, but *The Rocket into Interplanetary Space* became a best-seller and an extremely influential book. From the 1920s the idea that rockets might one day allow humans to leave the earth finally captured the public's imagination. Unfortunately, at that time dreams far outstripped technology, and it would be another three decades before real science caught up.

TWO MEN STOOD at the center of the space race. One of these was a Ukrainian, a man obscured from the public gaze by layers of government obfuscation. He was known only as the chief designer, his birth name, known to few until after the fall of Communism, was Sergei Korolev. The other, in almost every sense his opposite, lived in the United States; he was a naturalized American, German by birth, an extrovert, an evangelical crusader for space technology, outspoken, lauded, a media-friendly scientist,

Wernher von Braun. In many ways these two men may now be seen to epitomize the cultures they represented.

Korolev once confided to a colleague that he believed he and von Braun "should be friends," but the two men never met, nor did they write or speak to each other.[18] It was they, more than anyone alive, who knew the potential offered by space travel and they were both daring and visionary enough to sell the concept to their respective governments. Each respected and in a sense feared the other. According to one of his senior engineers, Korolev said of his rivals: "The Americans are at our heels, and the Americans are serious people. He wouldn't use the word *Amerikantsi* but *Amerikan-ye* as if these weren't just American residents but the entire American culture we were competing with. He didn't mean this as an insult but as a show of respect for the competition."[19]

During the late 1950s, technological progress was in a steep ascendant. Transportation and improved communications were starting to open up the globe (although the real revolution in global communications was to come as a result of the space race), and television and pop music were becoming increasingly influential in cultural change. Politically, the world was a more dangerous place than it ever had been. Only a small group of nations possessed hydrogen bombs and delivery systems, but this technology was beginning to proliferate.

And because of improved communications and the growing power of the media, the escalation of nuclear arms and the potential for global annihilation became a very public issue—at least in the West. Science fiction also played a role in equipping people mentally, encouraging a language of science, and strengthening the communication of technology at the frontier (including the means for mass destruction).

In the same way, since the end of World War II, Westerners had become excited about the explosion of innovation. *Consumerism* became a byword inextricably linked with scientific and technological progress. The ultimate symbol of this trend, amplified by science fiction in books, magazines, and movies, was the idea that humans would soon travel into space. On both sides of the Iron Curtain, space travel was perceived as the new frontier, the ultimate challenge.

One of the great proselytizers of space travel, Wernher von Braun, was born in 1912 in the East Prussian town of Wirsitz (now within the borders of Poland). His father, a baron, was an ambitious politician, and when he was appointed minister of agriculture and education in 1922, the family moved to Berlin.

Between the crushing defeat of World War I and the beginning of the depression of the late 1920s, Berlin basked in a golden age for culture and counterculture and was a preeminent center for artistic innovation. Berliners danced to the new sounds of jazz, the galleries were filled with the innovative work produced by young artists drawn to the city, and the cafés buzzed with the conversation of writers, playwrights, actors, filmmakers, and designers. Berlin had, Brecht, Schönberg, Nabokov, Berg, Adler, Klee, and Kandinsky. All of them were observed and recorded by Christopher Isherwood and W. H. Auden, writers witnessing a society at a precipice.

Berlin also promoted the work of scientists and engineers. During the 1920s Albert Einstein lived and worked in the city along with Werner Heisenberg, Erwin Schrödinger, and Max Planck. Germans were keen engineers and fond of innovation, and Berlin was the intellectual epicenter for societies and groups who wanted to create a sparkling high-tech future. Spurred on by the Futurist movement and the Bauhaus, young men were drawn to what had up to then been viewed as pure science fiction, ideas such as space travel, rocketry, and missile design.

When he was twelve von Braun was inspired by Hermann Oberth's book, *The Rocket into Interplanetary Space*, published the previous year. To most people this would have seemed a work closer to Jules Verne's novels than to real science, and it did contain much that was far-fetched, but it was mostly based upon known facts and Oberth was a competent scientist and engineer.

Reading *The Rocket into Interplanetary Space* changed von Braun's life. Turning its pages, he discovered his vocation, and, according to legend at least, he knew immediately that he wanted to build a rocket that would one day take men to the Moon. Amazingly, unlike almost everyone else who has such wild fantasies, von Braun made his dreams reality.

By the late 1920, von Braun had made all the right contacts for the development of his career. He was a cultured, handsome, and wealthy aristocrat, and he was immensely ambitious. "He had blue eyes and light blond hair and one of many female relatives compared him to the famous photograph of Lord Alfred Douglas of Oscar Wilde fame," one associate wrote years later. "His manners were as perfect as rigid upbringing could make them."[20]

In 1929, at the age of seventeen, von Braun joined a group calling itself VfR (*Verein für Raumschiffahrt*) or Society for Space Travel, headed by Hermann Oberth, and here he was introduced to well-educated and ambitious young men who also fantasized about sending rockets to the Moon. They met to discuss designs and wrote papers on the future of space travel. Inevitably, some of their theoretical ideas were misguided, but a few plans may now be seen as the earliest steps toward practical rocketry in Europe. During the late 1920s and early 1930s, members of the VfR were discussing and writing about such things as space shuttles, space stations, even designs for Mars landers and such exotica as interstellar craft.

Von Braun became a prominent member of the Society for Space Travel, and his designs for rockets and launch systems became increasingly sophisticated through the early years of the 1930s. But the VfR remained a group of enthusiastic amateurs until, in 1932, the society drew the attention of the Nazi Party, then on the threshold of genuine political power in Germany. The Nazis appointed a senior officer named Walter Dornberger to see if their ideas might offer serious application in weapons development. Dornberger, who had a master's degree in engineering, was not impressed with the lack of professionalism he found, but he did notice von Braun. "I had been struck during my casual visits to Reinickendorf [the Berlin suburb where the society met to launch its rockets] by the energy and shrewdness with which this tall, fair student with the broad massive chin went to work and with his astonishing theoretical knowledge," Dornberger wrote. "When General Becker later decided to approve our army establishment for liquid-propellant rockets, I put Wernher von Braun first on my list of proposed technical assistants."[21]

Within a year von Braun was employed by the German Army as a civilian rocket engineer, and in 1936 he began working at a rocket development site and launch area in Peenemünde on the northeast coast of Germany. Here he was to stay for the duration of the war.

By the outbreak of World War II, von Braun's belief in the vast potential of rocketry was shared by many high-ranking officials in the Nazi Party. Hitler himself distrusted such modern weapons (he considered tanks "cowardly"), but the technophilic Albert Speer was most enthusiastic, and he persuaded the Führer to pour money into rocket development.

Many have criticized von Braun for his collaboration with the Nazis, for developing the "revenge weapons" the V1 and the V2, which together killed some three thousand Londoners toward the end of the war. But such criticism is misplaced: Von Braun was a scientist first and a weapons designer second. He knew that if he hoped to gain support to develop his scientific ideas, he would have to build missiles that found important application to the war effort. The record is clear; like his countryman Werner Heisenberg, von Braun was never a Nazi sympathizer and he did not join the party. Von Braun was no different from his counterparts in Allied countries; he merely played his part in helping his nation attempt to win the war.

Ironically, von Braun was distrusted by some in the Nazi Party. Heinrich Himmler had him arrested in 1944, and he was imprisoned for three weeks before Dornberger and Speer secured his release. The reason for his arrest was Himmler's misguided conviction that von Braun was preparing to escape to Britain with the secrets of the V1 and the V2.

In May 1945 American troops captured von Braun after he and his team had fled the site at Peenemünde. The group were allowed to travel to Britain to discuss their designs with their former enemies, and the British government tried to persuade them to stay to work on rocket development there, but von Braun knew the financing necessary for such research would probably not come from a country that had been made almost bankrupt by the demands of war. Instead, they returned to mainland Europe and the custody of the U.S. Army, and in the autumn of 1945 von Braun and many of his team chose to move to the United States, where they were eventually granted U.S. citizenship. The scientists were stationed at the army's largest rocket research center, Fort Bliss in Texas, where they were provided with the best facilities available and a commission to continue the research they had begun during the war.

In 1947 von Braun returned to Germany to marry his eighteen-year-old cousin, Maria Louise von Quistorp, whom he had known since she was a child. He returned to the United States with his young bride, and a year later the couple had the first of three children (Iris Careen was born in 1948; another girl, Margit Cecille, arrived in 1952; and a boy, Peter Constantin, in 1960).

At Fort Bliss, von Braun and his colleagues immediately began to make great strides forward with their designs, and with U.S. Army funding they were able to build increasingly powerful rockets. The most important of these was the A-4 missile, which had been in development in Germany in 1945. It has been claimed that the early version of the A-4 being modeled in Peenemünde had been only two years away from its maiden flight and that before the war ended the Nazis had been drawing up plans to use this device to deliver a warhead to New York. Working for the Americans, von Braun and his team could turn their expertise away from threatening the United States and apply it to the development of the earliest intercontinental ballistic missiles (ICBMs), which were soon to become such a prominent feature of the Cold War.

However, while they were spending most of their time and resources on designing and building missiles for the army, von Braun and his team were also given research funds for the development of rockets for purely scientific work. Much of this work ran in parallel, so that the rockets designed to carry nuclear warheads to Moscow were the same as those used to study high-altitude flight, and later models propelled the first American missile. By 1949 the A-4 had been built and launched successfully, and this was followed by the more sophisticated A-9 and A-10. By 1950 von Braun had developed a still more advanced rocket named Hermes, and this led to a powerful multistage rocket, the Bumper, which was in theory capable of reaching the upper atmosphere. Five further years of development produced the Jupiter rockets, devices von Braun knew could carry payloads into orbit.

In 1950 von Braun and his family moved to an army research facility at Huntsville, Alabama, and began developing further programs for both civil and military use. From here he produced a series of designs for bigger and better rockets that led to the multistage launchers capable of lifting satellites into orbit. By 1954 he had risen to the position of chief aeronautical engineer at the Huntsville site and had become a direct rival to his counterpart who lived and worked some five thousand miles away in Moscow, Chief Designer Sergei Korolev.

BY THE TIME von Braun was born, in 1912, Sergei Korolev was six years old and living with his mother and grandparents in the small Ukrainian city of Nezhin. His father had deserted them two years earlier. Von Braun had been inspired to pursue an interest in rocketry after reading a book; Korolev was awakened to the majesty of flight by one of the first men to make a living by exhibiting aircraft, a traveling aviator who turned up in Nezhin with a circus in 1913, when Korolev was just seven.

Korolev was a lonely child who had few friends, and the idea of flying immediately captured his imagination. "He didn't have any childhood acquaintances of his own age," his first teacher recalled, "and never knew child's games with little friends. He was often completely alone at home. They [his mother and grandparents] didn't allow him to run in the street. The gate was always bolted. He would sit a long while on the upper cellar door and watch what was happening in the street."[22] Korolev himself was more direct about his early life. "I did not have any childhood," he once told a relative.[23] Contemporaries also describe his intellectual precociousness and how he was a dedicated student who showed a particular early aptitude for mathematics.

His mother, Maria, remarried in 1916, and Sergei's stepfather, Grigory Mikhailovich Balanin, encouraged him in his studies and introduced him to a wider world of literature and learning. Balanin was himself a successful engineer and an intellectual who was appointed to the prestigious position of head of the power station in Odessa a few months after marrying Maria Korolev.

The family moved to Odessa late in 1916, just as the Bolshevik Revolution was at its height. Because of Balanin's status, they were relatively isolated from much of the turmoil created by the popular uprising against the tsar, but Korolev certainly witnessed the instability of the time: the constant passage of troops of the revolution and the counterrevolution, British warships followed by the French navy in Odessa's harbor. One of Korolev's school friends recalled how the early 1920s, the aftermath of the revolution, was a period of poverty and deprivation that touched even the relatively privileged. "This was a menacing time," he wrote. "The hunger of 1921 had not yet been forgotten. Apartments were heated with cast-iron stoves. Often they were fueled with household furniture. Electrical lighting was still greatly limited. Kerosene lamps seemed a luxury, and we had to make do with homemade carbide or oil lamps. The water lines were always breaking. On the streets you could hear the click of the 'woodies' of the Odessa residents—sandals with soles made out of boards held together with strips of leather."[24]

It was a childhood very different from that of von Braun. But Korolev was also learning about rockets and the potential of space travel. He did not know of Oberth's book, but his stepfather introduced him to the works of Jules Verne and H. G. Wells. By the age of twelve he too had become obsessed with the idea of building aircraft and rockets, and, like von Braun, he had begun dreaming of interplanetary travel. A friend from this time said: "Korolev was the most mature, . . . energetic, and independent among us. He was guided by a clear goal in life."[25]

And Korolev pursued his goal with great energy. In 1923, when he was seventeen, he joined a society dedicated to exploring the potential of rocketry and aeronautics. The group had the unwieldy name the Society of Aviation and Aerial Navigation of the Ukraine and the Crimea (OAVUK). It was very different from the VfR von Braun was to join a few years later. The OAVUK had a million members and was really a vast club for technically minded young men and women that provided the authorities with another opportunity to indoctrinate its young. In contrast to the VfR, it was strict, authoritarian, and fueled by political activism.

For the bright and ambitious, the OAVUK could provide an entrée into the prestigious Zhukovsky Academy in Moscow, the college at which the great Andrei Tupolev lectured part-time while designing military aircraft. But Korolev did not qualify because he had not trained as a pilot. Instead, in 1924 he enrolled at a college generally considered second best, the Polytechnical Institute in Kiev.

Once there, he worked his way through college earning a few rubles as a roofer and builder using skills he had acquired between math and physics classes at the vocational school in Odessa. A contemporary at the Tech was Igor Sikorsky, who a decade later emigrated to the United States, where he became the creator of the first commercially successful helicopter and founded an immensely successful avionics corporation.

After two years in Kiev, Korolev was able to join a course at the Moscow Higher Technical School, the MVTU. And although he had not made it to the Zhukovsky Academy, the MVTU was almost as famous and had by then also secured Tupolev as a teacher. There Korolev studied flight mechanics, aerodynamics, materials, and engine design. But the college also prided itself on the fact that it encouraged students to experiment, to build real aircraft, and even to fly them themselves. However, Korolev was most attracted by the presence of Tupolev, whose work he had admired for years. During his time at the MVTU, Korolev established a close bond with Tupolev, and the renowned designer was to have a profound influence upon the course of the younger man's career.

But although he was obliged to spend his time thinking about the design of conventional aircraft, Korolev had not forgotten about rockets and space vehicles. In 1927, while studying at the MVTU, he attended the First World Exhibition of Interplanetary Apparatus and Devices in Moscow. The organizers had on display models of rockets from around the world, designs from the desks of Goddard in America, Oberth in Germany, and the Russian innovator of rocketry, Tsiolkovsky. As a center piece, a huge moonscape had been constructed with a star-studded backdrop and an image of the earth rising above the cratered lunar surface. This moonscape was strikingly similar to the now famous photograph of earthrise over the lunar horizon, an image captured by the astronaut James Lovell aboard *Apollo 8* as it emerged from the dark side of the Moon some forty years after this exhibition.

But for a while at least, Korolev's dreams of space travel had to be put on the back burner. In order to qualify as an engineer he had to apply himself to the rigidly authoritarian academic and professional system. Two years after the First World Exhibition of Interplanetary Apparatus and Devices, he found himself at Factory 22 helping to construct a plane named TOM-1 for the Soviet air force. But his ambitions were unchanged. Two years after entering Factory 22, he met and became friends with a scientist named Fridrikh Tsander, who was also obsessed with the idea of using rockets for space exploration: As team leader at the factory, Tsander was able to guide Korolev into serious rocket research.

Tsander was in his mid-forties when he met Korolev in 1931. By then he was already an accomplished rocket designer and had written a prophetic book titled *Flights to Other Planets*, which had been published to great acclaim in 1924. In 1931 he was head of a collective, the Group for Studying Reaction Propulsion. Officially, this was dedicated to the evolution of aircraft propulsion systems, but in their spare time many of the group's members worked at the cutting edge of rocket research and design.

A contemporary said of Tsander: "He was the first man in the USSR to take the first steps to turn astronautics into an applied science."[26] In other words, Tsander was the first in the Soviet Union to reverse the bias that rocket research was little more than science fiction, the first Soviet engineer to bring rocketry under the mantle of "real" science. His intellect matched his imagination, so he was no starry-eyed fantasist, and he knew exactly how space could one day be explained and exploited given appropriate time and finance.

Korolev began to work with Tsander after hours on a series of rocket designs. The most important of these was a liquid-propellant rocket they named GIRD-09, but, tragically, in 1933 Tsander died after a protracted illness just as the GIRD-09 was to have its final laboratory test. Undaunted, Korolev took over the team and worked tirelessly to perfect the rocket for its launch. In August 1933, Korolev realized Tsander's goal: reaching an altitude of over 400 yards, the GIRD-09 set a new Soviet record.

Of course, this effort was not in the same league as some of Goddard's achievements of the previous year, nor did it come close to the sophistication of von Braun's devices built only a few years later, but considering Korolev and his colleagues designed and constructed the GIRD rockets entirely in their spare time from odd scraps, it was a remarkable achievement.

In another parallel with his later rival, at almost the same time von Braun was attracting the interest of the Nazis, Korolev was approached by officials from the Soviet government. Dornberger's Soviet counterpart was Marshal Mikhail Tukhachevsky, who had been charged with finding fresh talent among the enthusiasts of rocket science. He had heard of the development of the GIRD rockets and was excited by the success of the GIRD-09.

Tukhachevsky was quick to realize the potential of rockets, once declaring prophetically: "The significant and important results [of Korolev's Group for Studying Reaction Propulsion] create unlimited opportunities for firing shells of any size and range." These, he believed would "eventually lead to the solution of the problems of high-velocity stratospheric flight."[27] And within months of the successful launch of GIRD-09, Marshal Tukhachevsky had secured funding for Korolev's group to develop rockets officially alongside their more conventional work.

Korolev had married his childhood sweetheart, Lyalya Vincentini, in 1931, and in 1935 they had a daughter named Natasha, but he spared almost no time for family life. To fulfill his enormous ambition, he worked incredibly hard and expected the same of his subordinates. "I was one of four engineers who worked every day with Korolev beginning from 1937," one of his colleagues recalled many years later. "Sometimes we worked twenty-four hours straight, but usually we would start at eight in the morning and keep going until 9:00 or 10:00 P.M. We couldn't go home as long as he was there. It was exciting, like a sport, the technical challenge was stimulating."[28] Despite the workload and his official responsibilities, like von Braun in Germany, Korolev could still find time to think beyond missiles and weapons delivery systems and began to consider the practical aspects of designing rockets for space travel.

But then, disaster. During the early hours of November 2, 1937, Korolev was arrested by the KGB. After a brief hearing he was charged with collaborating with Wernher von Braun and his team in Germany and sentenced to eight years in a labor camp. The charge was pure fabrication; Korolev had simply become another victim of Stalinist paranoia. Yet he was lucky to have escaped with his life. During the Great Purge between 1936 and 1938, Stalin sent an estimated 3 million men and women to their deaths and more millions to labor camps for contrived crimes against the state.

Korolev's destination was the most infamous camp, Kolyma, in Siberia, later immortalized in Alexander Solzhenitsyn's *Gulag Archipelago*. For more than two years at Kolyma, Korolev worked outdoors, eighteen hours a day in subzero temperatures, on a near-starvation diet. He was saved from joining the statistics of camp deaths by the intervention of Tupolev, who had been asked to draw up a list of the most gifted engineers in the Soviet Union to work in his Moscow *sharashka* (a guarded complex where engineers were forced to work for the military). Tupolev had placed Korolev close to the top of the list.

The two men worked together from 1940 until the end of the war. Korolev was forbidden to leave the compound but was allowed rare visits from his wife and the daughter he had not seen since his arrest. He was also engaged in work he enjoyed. Under Tupolev's leadership he was developing bombers and fighters, but when time allowed he was once again drawing up plans for the rockets he dreamed would one day send Russians into space ahead of any other nation.

With the end of the war, von Braun again vicariously influenced Korolev's life. Soon after the German researchers had been captured by American troops, the Russian army found the deserted missile testing site at Peenemünde. In August 1945 everything changed again for Korolev. Without warning, he was released from the *sharashka*, made a colonel in the Red Army, and posted to Peenemünde. There he became part of a team assigned to the task of working out exactly what the Germans had achieved and investigating how their work might be replicated and developed.

Korolev arrived in Germany in September 1945, missing von Braun by just four months. The Germans had abandoned the site in a hurry and left behind a vast collection of papers, equipment, and parts, enough to keep Korolev and his colleagues busy deciphering the material for seven years.

This find altered the political map of the world. The Americans were unable to retrieve the material at Peenemünde because the area had fallen under Russian control, so the Russian scientists had sole access to von Braun's work, giving them all the information they needed to develop their own missile capability. This culminated in the production of a vast arsenal of ICBMs, with which the Soviets were able to challenge the military might of the United States. The Los Alamos scientist Klaus Fuchs gave the Soviets nuclear secrets; von Braun's work provided a delivery system.

The Russians at Peenemünde simply could not believe what they had found. Von Braun and his colleagues had been decades in advance of Soviet plans and far ahead of their counterparts in the United States or Britain. They had designed missiles that could travel thousands of miles and multistage rockets that were the precursors to launch vehicles capable of carrying a payload into orbit. Along with these were plans for rockets with wings similar to the cruise missile designs of the 1970s and even antecedents of shuttle vehicles that could be launched from pads but were able to land like aircraft.

Von Braun was now working in America, but, his ideas were also in Korolev's hands, and Korolev was every bit as talented and ingenious as von Braun. When he returned to Russia in 1952, Korolev's career blossomed. He became a central figure in the development of Soviet ICBMs and by 1954 had risen to the position of chief designer. Beyond a circle of friends and family, this epithet would serve as his name for the rest of his life. From the 1950s, as his power grew, Sergei Korolev retreated behind a politically engineered mask.

THE LAUNCH OF SPUTNIK sent a wave of anxiety across America and received startled responses from around the world. NEXT STOP MARS?

was the rather optimistic headline in the *Manchester Guardian* in Britain, and the newspaper's editor went on to write: "The achievement is immense. It demands a psychological adjustment on our part towards Soviet society, Soviet military capabilities, and—perhaps most of all—to the relationship of the world with what is beyond." And pinpointing one of the underlying fears felt but at first unspoken in America, the piece went on: "The Russians can now build ballistic missiles capable of hitting any chosen target anywhere in the world."[29]

But as well as alarming the public and some of the more aware sections of government departments, the launch also galvanized the media and the scientific community to try to solicit government action. However, just as Roosevelt had prevaricated about the establishment of the Manhattan Project, most government officials were caught unprepared by *Sputnik*. As we have seen, they were oblivious to both the propaganda coup the Russians had created and the potential military and technological advantages of space exploration.

In his memoirs, George Kennan, the U.S. ambassador to the Soviet Union during the late 1950s wrote: "It [*Sputnik*] caused Western alarmists . . . to demand the immediate subordination of all other national interests to the launching of immensely expensive crash programs to outdo the Russians in this competition. It gave effective arguments to the various enthusiasts for nuclear armament in the American military-industrial complex. That the dangerousness and expensiveness of this competition should be raised to a new and higher order just at the time when the prospects for negotiation in this field were being worsened by the introduction of nuclear weapons into the armed forces of the Continental NATO powers was a development that brought alarm and dismay to many people besides myself."[30] He wrote this in 1972, three years after the first Moon landing.

Even when the Russians repeated the trick a month later with the launch of *Sputnik 2*, weighing half a ton, there was still no clear directive or guidance from the top in Washington. This prompted a frustrated von Braun to tell a journalist: "We should declare a moratorium on obstructive inspections and monitoring by scientific committees and let the people building the Atlas, Titan, Thor, and Polaris work in peace. I just wish someone had the authority to tell me. 'All right, we'll leave you alone for two years, but if you fail we're going to hang you.'"[31]

Von Braun knew how difficult it would be to master space travel, but he also saw it as absolutely imperative that America succeed in the eyes of the world. As early as 1952 he had written in one of his many popular articles: "The Soviets claim a head start but the advantage in the competition to conquer space probably rests with us—if we move quickly."[32]

And the political opponents of the government, already gearing up for the 1960 election and sensing public anger, were quick to take an anti-Republican stance. In a speech on January 7, 1958, before America had answered the Russian challenge, Lyndon Johnson declared: "Control of space means control of the world.... From space the masters of infinity would have the power to control the earth's weather, to cause drought and flood, to change the tides and raise the levels of the sea, to divert the gulf stream and change temperate climates to frigid. There is something more than the ultimate weapon. That is the ultimate position—the position of total control over earth that lies somewhere in outer space . . . and if there is an ultimate position, then our national goal and the goal of all free men *must* be to win and hold that position."[33] Although most of this is nonsense, composed by a speechwriter who had perhaps been reading too much science fiction, it does indicate how space had so quickly become immensely important in the political world.

Even so, others in important scientific and political positions were still blind to what was happening. Eisenhower's scientific advisers were most to blame because they should have known better. Tired conservative thinkers to a man, not only did they miss the scientific value of space exploration but when they did have the reality of its importance thrust upon them, they handed out very poor advice.

Their worst mistake was not trusting von Braun. He was the most famous space scientist alive, a regular on the lecture circuit who had written extensively about space travel in popular journals and learned magazines for many years. As the pioneer of modern rocketry, he was also immensely respected within the scientific community. Von Braun and his team were also more experienced than any other rocket designers. Since 1945 they had developed further the technology that had given the Nazis the V1 and V2, and if it had not been for the misguided interference of the Pentagon, they could have beaten the Soviets to the prize of having the first satellite in space. But even as late as 1957, after they had been working in the United

States for a dozen years, Our Germans', as they were dubbed, were not entirely trusted by the government.*

"There was always a lingering resentment at the Washington end toward von Braun and his team," one historian has written. "There were always rumors that von Braun would someday be head of NASA. But there is great sensitivity in Washington about racial and ethnic interests. . . . Von Braun would never have been given a political position."[34] It was to be many years before von Braun's opponents could accept that, without him, America might have taken a decade or more to match *Sputnik*, and that this would almost certainly have led to the Soviets landing a man on the Moon first.

But in 1958 both Eisenhower's scientific and political advisers chose to sideline von Braun and instead to invest interest in the scientists working with the U.S. Navy, whose rocket, *Vanguard*, was markedly inferior to von Braun's new Redstone rocket, then under development at Huntsville.

The navy satellite mission took to the launch pad at Cape Canaveral under the full glare of the media's cameras on December 6, 1957, just over two months after the launch of *Sputnik*. As the ignition sequence started, tens of millions of viewers sat before their black-and-white TVs watching expectantly. At T minus 1 second, the rockets fired up, *Vanguard* lifted off the ground, and then it crumpled in a fiery heap, sending a shudder through the space center and an incandescent ball of flame spreading from the launch pad. As *Vanguard* smashed to the ground, someone shouted out:"Look out! Oh, God, no!"[35]

The media were merciless and the American public bemused. The *London Daily Herald* carried the headline OH, WHAT A FLOPNIK![36] At the United Nations in New York, Russian delegates proposed to their American counterparts that they might like to consider a Soviet program that offered technical assistance to backward nations.

Von Braun had predicted the navy project would fail. Before the launch, in a face-to-face meeting with the then secretary of defense, Neil McElroy, he declared: "*Vanguard* will never make it. We have the hardware

---

*Some of the original team at Peenemünde had escaped capture by the U.S. Army and ended up working with the Soviets when the rocket site was occupied. Many of these scientists became East German citizens and worked on the Soviet space program during the 1950s and 1960s. In the West they were referred to as Their Germans.

on the shelf. For God's sake, turn us loose and let us do something. We can put up a satellite in sixty days, Mr. McElroy. Just give us a green light and sixty days!"[37]

On that occasion McElroy had shown him the door. Only after the fiasco of *Vanguard* did Eisenhower's advisers finally switch support to the army, and in response, von Braun had his rocket on the launch pad by the end of January 1958. This time the launch, on January 31, 1958, was kept secret. No journalists were invited; the TV cameras were kept away. But they need not have worried. Von Braun's Redstone lifted off perfectly, and a few minutes later, America's first satellite, *Explorer*, reached Earth orbit. Only then were the press informed. Suddenly "Our German," Wernher von Braun, was a national hero. Though he was still resented and distrusted by many political and military figures, no one could ignore his achievement.

But even this success had a bitter edge. America had, after all, struggled to match what the Soviets had accomplished almost half a year earlier. Furthermore, Russian efforts had not slowed; in fact, their scientists were forging ahead still further and faster. It was now abundantly clear that if the United States was to compete, and it had no choice but to compete, the entire idea of space research from theory to government checks had to be reevaluated and a coherent strategy established.

Eisenhower was lamentably ill-equipped for the task of stemming public anxiety or facilitating a kick start to American efforts to compete with the Russians. Aging and technophobic, he simply did not understand the issues. A sure indication of this fact comes from the recent revelation that Eisenhower's government investigated the feasibility of exploding a nuclear warhead on the surface of the Moon in order to demonstrate to the world America's military and technological prowess.[38] However, the effort, begun in May 1958 and based at the Armour Research Foundation in Chicago, was abandoned after a year of research because it was realized that the risks involved in attempting to transport a nuclear warhead to the Moon were so large they outweighed the possible publicity benefits. If the rocket exploded on the pad (a very real possibility given the track record of the time), the effects would have been catastrophic.

But aside from the practical implications of this crazy idea, it was a proposal that was both cynical and deeply flawed politically because it carried none of the glamour or sense of daring associated with the Soviet efforts up to that time. It is a precise indication of how the thinking of the U.S. administration was out of step with the mood of the time.

Fortunately for the future of space research, Eisenhower's vice president, Richard Nixon, thought in a very different way. Aware of the potential of space travel, he was soon taking an avid interest in developing a coherent space program. He, more than any administrator in Washington, realized the propaganda value not only of space exploration but of staging a space race—a competition in high-powered technology coupled with a irresistible show of danger and daring. He was also confident that although the Americans had been second off the launch pad, they, and only they, could afford the long-distance race that would define the space effort.

Nixon realized from the beginning that if America was to show the world it could beat the Russians at their own game, it would cost enormous amounts of money and vast human resources. This would be no friendly tussle but a war, the most public, media-orientated manifestation of the Cold War, every bit as important as important as the stockpiling of missiles and nuclear warheads.

Vice President Nixon pushed hard to negate Eisenhower's floundering, and NASA was created because he undertook the task of organizing the scientists and politicians and forcing these strange bedfellows to work together. Nixon and his advisers understood that a successful space agency must be a civilian organization, independent of a purely military agenda.* The result, the National Aeronautics and Space Administration, was created on October 1, 1958, three days short of the first anniversary of the launch of *Sputnik*. A space engineer named Robert Gilruth became its first head. America was still far behind, but at least the first constructive move had been made since giving the technical reins to von Braun, and Nixon's plan would eventually prove to be one of the most important factors in turning America's humiliation to eventual triumph.

During those early days of the space race, one of the most serious problems facing American scientists and politicians was that the United States approached the contest with complete openness. Of course America was not inferior to the USSR technologically, the world simply perceived it to

---

*NASA has always provided launch facilities for military missions, but the Defense Department does not control the agency.

be. Furthermore, because of the different ways in which the political structures of the two countries operated, this perception gap widened rapidly.
What lay at the heart of the propaganda war surrounding the space race was
the way the Soviets told the world only about their successes and kept very
quiet about their failures. With the exception of the successful launch of
*Explorer* in January 1958, all American civilian launches were open to public
scrutiny. The U.S. media knew as much about the terrible failures as about
the glowing successes.

Combined with this, the Russians preferred deliberately dramatic
adventures in space and turned partial successes into publicity stunts. They
launched the first dog into space, they sent ever bigger satellites aloft and
made rather desperate attempts to launch probes to crash-land on the lunar
surface, all in a struggle for one-upmanship rather than for scientific
advance.

Mastering the sound bite decades before the term was coined, after the
launch of the gigantic *Sputnik* 3, Nikita Khrushchev declared that "America
sleeps under a Soviet moon." Yet no mention was made of the four failures
for every successful Soviet launch during the first full year of the space race,
1958. In America public accountability meant that U.S. citizens and the rest
of the world could keep track of the space program. Eighteen successful
launches, nineteen failures during the first three years of the race.

The space race was a numbers game almost from the beginning. Initially this was based solely upon how many satellites each side could launch
successfully, but gradually the nature of the launch and the type of probe
became almost as important : How large was the object? What did it do?
What was its destination? But this score sheet too was soon outmoded.
After the first round of launches on each side, ambitions grew rapidly. By
the end of the 1950s, only twenty-seven months into the Space Age, one
objective began to dominate and would soon become the only goal that mattered: Who would be first to send a man into space and return him safely?

In America they were called astronauts. In Russia, cosmonauts. But
there were greater differences between these two groups than mere names.
Most cosmonauts and astronauts were on friendly terms. During the 1960s
Russian spacemen traveled to the United States for publicity tours, and
they sometimes met on neutral ground in Europe and the Far East. Defiance of death was their stock-in-trade, and even if they spoke different lan-

guages and had been brought up in very different societies, they had an immediate mutual understanding. But true to the character of both the nations behind these men, cosmonauts and astronauts held very different roles. They underwent similar physical training, and both were initially selected from military backgrounds, but until at least the 1980s, cosmonauts were really little more than passengers aboard their craft. Operators at Mission Control and automated systems aboard the spacecraft were the real pilots. "We believed that we could do everything automatically, even with the cosmonaut present," one of Korolev's chief engineers, Boris Rauschenbakh, has recalled. "The United States put men in control first."[39] By comparison with their Soviet counterparts, astronauts were true pilots of their vehicles, and from the earliest flights they had significant command over their craft.

In America, the first seven astronauts—Alan Shepard, Gus Grissom, John Glenn, Scott Carpenter, Walter Schirra, Gordon Cooper, and Deke Slayton—quickly became household names. Their faces appeared on the covers of magazines, and they rivaled the top pop stars of the day in popularity. They were known to womanize, to party hard, and to work harder.

In stark contrast, the first group of cosmonauts were anonymous; only two or three of them became famous after their groundbreaking missions had been successfully completed. Like the chief designer himself, these men were often kept in shadow, and the earliest cosmonauts were essentially little more than human guinea pigs. Only Yuri Gagarin was recognizable globally and vaunted as a national hero within the Soviet Union. Russian capsules were even equipped with a guidance system that required a code from Mission Control designed to prevent any wayward cosmonaut attempting to defect with his precious spacecraft.

Both cosmonauts and astronauts were prepared for the risks involved in space exploration. But the American space administration was incredibly cautious and, some believed, overprotective of its crews. The medical consultants employed by NASA insisted upon every conceivable check on the health of the astronauts, and before approving a manned mission they pushed for more and more test flights using monkeys to help eliminate as much risk as possible.

This concern was laudable but became so exaggerated it eventually slowed NASA's efforts to compete. Medical advisers seemed to lose sight of the fact that by its very nature space travel was an incredibly dangerous undertaking and that a certain level of risk could never be totally eliminated. Furthermore, they appeared to forget that all of the seven original astronauts had been test pilots long used to risking their lives flying jets every bit as experimental and dangerous as the vehicles designed to take them into orbit. The difference, and a crucial factor in the decision-making process at NASA, was that space capsules flew under media scrutiny. If the first American to lift off into orbit was killed, it might have spelled the end of the race.

The Russians showed far less caution. They tested their vehicles with dummies and dogs, they had teams of medics supervising each step of the training and preparation of their cosmonauts, but at every level of the administration of the Soviet space effort there was far greater concern for secrecy than for safety. Quite simply, the Russian administration knew that if a cosmonaut died during a mission no one need ever find out. Because the American program was public, any risk an astronaut might die on the pad or during the mission had to be avoided at all cost.

By the end of 1960, NASA was actually in a better position than the Russians to launch a manned mission. It had been claimed in 1958 that the U.S. space program was four years behind the Russians, but this gap had closed dramatically, so that by December 1960 there was little between them.[40] But the significant difference in approach toward safety was one of the most important factors in determining the nationality of the first man in space.

On April 12, 1961, Yuri Gagarin was immortalized by a mission lasting just 108 minutes in which he orbited the earth and returned safely. He had had almost no control of his craft; in fact, his flight did not even comply with the true definition of a takeoff, flight, and landing because during the final stages of reentry a malfunction forced Gagarin to eject from his vehicle, and he parachuted to the ground. None of these details was known outside the USSR for many years, however, and in the eyes of the world they would not have mattered anyway; Russia had won another round in the fight.*

But it could have been a very different story. Korolev had supervised a

---

*During a private moment before the launch, Korolev gave Gagarin the secret code that allowed him to take over the craft in an emergency, overriding the antidefection device.

series of missions to test the *Vostok* spacecraft in which Gagarin was to make his triumphant flight, but these had suffered a terrible failure rate. In July 1960 two dogs had been killed when a prototype of the rocket to launch the *Vostok* exploded on the launch pad, and in December two more dogs died when their *Vostok* capsule burned up during reentry.

However, the greatest disaster in the history of space exploration took place in October 1960, just six months before Yuri Gagarin's flight. At the Baikonur launch site, a rocket exploded on the pad, killing 165 people, including a government minister, Marshal Mitrofan Nedelin. Only a month earlier Korolev had applied for a manned launch and succeeded in obtaining official approval; ironically, one of the signatories had been Marshal Nedelin. If it had not been for these setbacks, Gagarin would probably have flown before the end of 1960, but if he had, he might have died in his capsule.

Of course, for decades the world knew nothing of the Baikonur tragedy. To an American public still not fully recovered from their second-place performances of the 1950s, the Russians had simply pulled off another amazing coup. All the public really cared about were the images of a human in space for the first time and the fact that his space suit bore a red star. As one of the more outspoken astronauts, Alan Shepard, put it: "We had 'em by the short hairs, and we gave it away."[41]

Once again the press pulled no punches. *The New York Times* declared: "It will be hailed as one of the great advances in the history of man's age-old quest to tame the forces of Nature."[42] An ecstatic headline in *Pravda* went further: GREAT EVENT IN THE HISTORY OF MANKIND, it announced.[43]

This latest Russian adventure could not have come at a worse time for the United States. Nixon had lost the 1960 election in part because the public believed Eisenhower had failed to even the score in the space race. At the beginning of 1961, the forty-three-year-old John F. Kennedy had become the thirty-fifth president of the United States, and his first months in office had been fraught with one political crisis after another. The humiliation of Gagarin's triumphant flight came a mere five days before Kennedy made one of the greatest political errors of his life when he endorsed U.S. support for a group of exiles' attempt to invade Cuba. The mission was a total failure, and Castro's Communist forces prevailed.

Of course, Kennedy was magnanimous in public. At a press conference the day after the flight, he said: "We, all of us, as members of the human race, have the greatest admiration for the Russian who participated in this extraordinary feat." He then went on with uncharacteristic caution: "I indicated that the task force which we set up on space, way back last January indicated that because of the Soviet progress in the field of boosters, where they have been ahead of us, that we expected that they would be the first in space, in orbiting a man in space. . . . We are carrying out our program, and we expect to hope to make progress in this area this year ourselves."[44]

But behind the scenes the president was furious. Clearly NASA had to perform, and fast. Kennedy wasted no time before stirring up a reaction. Two days after Gagarin's flight, the president called a meeting with NASA chiefs and insisted they find a way to catch up and quickly overtake the Russians, because, he said, "the United States intended to maintain its position of world leadership, its position of eminence in commerce, in science, in foreign policy, and in whatever else might develop from space exploration."[45]

Five days later he called upon Vice President Johnson, chairman of the newly formed National Space Council, to file a set of recommendations for action as soon as possible. The following day, April 20, only hours after hearing of the disaster in Cuba, Kennedy wrote to Johnson, placing him "in charge of an overall survey of where we stand in space. . . . Do we have a chance of beating the Soviets by putting a laboratory in space, or by a trip around the moon, or by a rocket to land on the moon, or by a rocket to go to the moon and back with a man? Is there any other space program which promises results in which we could win? . . . How much additional would it cost?"[46]

Within days Johnson had canvassed both von Braun and the chief of NASA, Bob Gilruth. Von Braun had been working for almost three years on the U.S. manned space program; now, with political pressure growing and an angry and confused public becoming increasingly bitter, he knew his moment had come. Without hesitation he told the vice president that the United States had an excellent chance of beating the Russians to the Moon.[47] Unlike Korolev, he knew he was not punching above his weight; NASA was ready, the technology was in place, all von Braun needed was the funding to make it happen.

The first tentative step was to create what became Project Mercury, a series of launches that would send a tiny, frail-looking capsule carrying one man into orbit atop another Redstone rocket developed from a customized ICBM (the Jupiter-C) similar to the one that had launched the first American satellite into space three years earlier. Since 1958, over one million man-hours of thought and effort had been applied to bring this rocket to a level of sophistication that would allow it to launch a man into orbit.

By May 5, 1961 all was ready. Before an estimated TV audience of 45 million, Alan Shepard took the elevator to the top of the rocket and walked across the narrow connecting bridge to the entrance of the Mercury capsule he had named *Freedom 7*. "In Cocoa Beach [a few miles from the launch pad] people left their homes to stand outside and look toward the Cape," an eye-witness recounted. "They went to balconies and front lawns and back lawns. They stood atop cars and trucks and rooftops. They left their morning coffee and bacon and eggs in restaurants to walk outside on the street or on the sands of the beach. They left beauty parlors and barbershops with sheets around their bodies. Policemen stopped their cars and stood outside, the better to see and hear. Along the water, the surfers ceased their pursuit of the waves and stood, transfixed, swept up in the fleeting moments."[48]

The rocket rose majestically into the air, and a few minutes later *Freedom 7* jettisoned its boosters and followed a ballistic trajectory sending it into space for a few moments but without attaining orbit. But it was enough, it was a "space flight," and America was back in the race.

After Shepard's flight, and with a little time bought, the politicians and the scientists at NASA and von Braun's domain in Huntsville were finally able to see a path to the Moon. NASA had been working hard to close the Russian lead, but the Soviets were still winning the propaganda war. They had become adept at producing one great publicity stunt after another as well as skillfully concealing their expensive errors. To push America into the lead in the space race, NASA knew it would have to do something truly dramatic, but, most crucially, it had to be something not merely spectacular but attainable: If promises were to be made to the American public, they would have to be kept.

Since he was a young boy, von Braun had been dreaming of building rockets that would take men to the Moon, and during the 1950s he had spent increasing amounts of his time designing vehicles and refining his ideas. During those days he had been years ahead of his time, and the most he could hope for was to proselytize his ideas in the pages of popular magazines and technical journals. By 1961 politics and technology had finally caught up with his imagination, and he knew it. With public opinion pushing their government and a president and vice president aware of the propaganda value of a space race, von Braun was the man for the time.

In discussion with Gilruth and von Braun, Kennedy was made to realize there could be only one target for the United States, a project that would be awesome in concept and dramatic in its execution but at the same time feasible, one they could complete before the Russians, even though they had been beaten at every hurdle so far. Gilruth told the president: "It's got to be something that requires a great big rocket, like going to the moon. Going to the moon will take a new rocket and a new technology. If you want to do that, I think our country could probably win because we'd both have to start from scratch."[49]

Kennedy was young, and after the political disaster of his first few months in office, he was becoming desperate for good publicity; such an audacious plan held a deep attraction. It was certainly fantastic but not so fantastic his enemies and critics could accuse him of mere political posturing. His announcement of intent came toward the end of his next speech to Congress, on May 25, 1961: "Recognizing the head start obtained by the Soviets with their large rocket engines, which gives them months of lead time, and recognizing the likelihood that they will exploit this lead for some time to come in still more impressive successes, we nevertheless are required to make new efforts on our own. . . . I believe this Nation should commit itself to achieving the goal, before this decade is out, of landing a man on the moon and returning him safely to earth."[50]

Gilruth had told the president that the Russians had "a big rocket and that was about all," but the United States had been building a powerful space science infrastructure since the creation of NASA. The space agency had in place a carefully linked network of companies, each able to take responsibility for different aspects of the rocket, and it had designs for the vehicles that could take men to the Moon and back. But most significant, America had the cash for the long haul. As time would tell, once the juggernaut had picked up enough pace, the corporate might of the United States would be unstoppable. To NASA scientists, it was one thing flinging a man in a slingshot orbit around the earth and back again but quite another bringing together a vast array of technological, engineering, and pure science elements with the finesse to land a team of men on the surface of the Moon and bring them back safely, then to repeat this process over and over again.

The Russians were not slow to realize they had tugged the tail of a sleeping monster. They immediately began pouring money and resources into their own project to land a man on the Moon. But Korolev and his workers faced three serious problems not shared by Gilruth or von Braun.

First, the Russian government promised Korolev financial support, but the resources they provided never came close to matching requirements. "As always, military problems stood first," Korolev's close colleague Sergei Kryukov wrote many years later. "Necessary money could only be obtained for strengthening national security. . . . Therefore the lunar programme had to be realised 'by the way.' . . . This decision could hardly be judged reasonable. The United States of America, seeking to take from us the role of space leader, set for itself a special goal: to be first to land on the Moon. The Soviet Union should have either answered this challenge or decisively turned away from it and gone its own way. We did neither one nor the other, and thus did not concentrate our efforts on the solution of the 'lunar' problem."[51]

Working on the principle that they could later acquire more funding for a project that had already cost hundreds of millions, Korolev's team initially proposed a budget of 457 million rubles to finance ten moon rockets. But it soon became clear this was such a drastic underestimate that even the plan of gradually persuading the government to provide more funding would prove inadequate. To succeed ahead of the United States, Korolev would have needed at least twenty times his original figure, and although he performed the near miracle of gradually acquiring 4 billion rubles from the Kremlin (approximately $10 billion in 1970, or $40 billion today), this was still insufficient. By comparison, Bob Gilruth had budgeted with amazing precision. In 1961 he proposed a budget of $24 billion ($100 billion today) for the Apollo Project, and this was almost exactly the final bill for the Moon landings.

The second problem facing the Russian engineers was their constant need to fight the cumbersome Soviet bureaucratic system. In the United States, NASA enjoyed an amazing degree of autonomy and relied upon the government only for financial support. In the USSR, every decision Korolev made had to be approved by government officials, even though most of them had little understanding of the technology or the science involved. Korolev was an expert manipulator of the Russian system and enjoyed a warm personal relationship with Premier Khrushchev (he had a direct telephone link to the Kremlin), but even so he was faced with constant frustration caused by underfunding and red tape.

However, perhaps the most damaging obstacle to its success was rivalry within the Soviet space program. Korolev was the most favored engineer, but he faced competition from others of equal brilliance who were every bit as passionate about space travel. Until the chief designer had notched up a series of spectacular successes, he managed to retain his title only through a combination of aggression, foresight, versatility, and luck.

At least four other brilliant engineers were simultaneously fighting for recognition, and, most damaging, they were acquiring government funding. The four were Vladimir Chelomei, Valentin Glushko, Vassily Mishin, and Mikhail Yangel. Each headed his own division, and each was working on projects that bore no relation to Korolev's or the others.'

Of these four the most successful was Chelomei, who was not only a very talented scientist and engineer but an astute political operator. In 1959 he had persuaded Premier Khrushchev's son, Sergei, to join his lab. Sergei Khrushchev was a fine engineer in his own right, but Chelomei wasted no time in cultivating their relationship and thereby succeeded in forging a link with the Kremlin to rival Korolev's. According to one observer, by the early 1960s, if Chelomei had wanted the Bolshoi for his project he would have gotten it.[52]

Chelomei had designed a three-stage rocket and capsule under the code name UR-500K (later known as the Proton) that was capable of sending a single cosmonaut to orbit the Moon and return to Earth without landing on the lunar surface. To Korolev, this was a second-rate idea; he wanted the Moon itself, not merely a circumlunar flight. But through Sergei Khrushchev, Chelomei acquired funding and support for the project, hundreds of millions of rubles that could have been injected into Korolev's program. From the early 1960s the Proton project was up and running and competing with Korolev's own manned program.

Some have claimed that Korolev has been given too much credit for the Russian space program. One Soviet journalist, Mikhail Rudenko, has gone so far as to say: "He belongs in a large circle which includes also Vladimir Glushko, Mikhail Yangel, Valentin Chelomei, others. It is a stereotype that it was mostly Korolev who built the Soviet space program."[53]

But others feel very differently. A close associate of both Korolev and Chelomei, the space engineer Vsevolod Avduyevsky has claimed: "It is absolutely correct that Korolev was the founder of cosmonautics in the Soviet Union."[54] And most commentators agree with Avduyevsky. They point out that, for the sheer breadth of his work within the space industry, and the fact that he designed and built rockets as well as capsules, Korolev earned not just the title of chief designer but credit for being the master-mind behind the Soviet space program.

Like von Braun, during the 1950s Korolev had been too far ahead of his time, and although he had friends in high places, many of his early ideas had been dismissed by influential enemies within the Kremlin. In September 1956, before the launch of the world's first satellite, Korolev had drafted proposals for a manned mission to the Moon detailing the technical and financial requirements. This had been summarily rejected not only by government officials but by the scientific advisers to the Kremlin. "The Ministry of Defence," complained one of Korolev's colleagues on the Russian lunar program, Gyorgi Vetrov, "was always cool on space exploration, seeing it as a direct threat to the defensive capability of the country."[55]

But after the success of *Sputnik* and the triumph of Gagarin's first manned flight, the Soviet government realized it had been right to allow the chief designer to push through his more manageable ideas—satellites and simple capsules in low-Earth orbit. And when the United States declared its intention to reach the Moon before its great rival, the Kremlin again allowed Korolev to attempt to realize his dreams. His triumphs were still fresh memories, and to the Soviet government it seemed obvious they should now go for the greatest prize of all.

And, like his counterpart in Huntsville, Alabama, chief designer Sergei Korolev had been ready for years. Now, he too was a man for the moment. Gagarin's flight had generated a huge escalation in the race, eclipsing completely the spontaneous reaction of the United States to the launch of *Sputnik 1*. Beginning in 1961 the gloves were off, and both rivals established

programs for achieving the incredible goal of being first to land one of its citizens on the surface of the Moon.

$\mathcal{F}$ROM THE START, the deep motivations of the two foes had been
quite different. It appeared to be a race between equals, militarily powerful
nations with totally opposite ideologies, clashing political systems. To the
worried American public, the Russians appeared to be running away with
the trophies, and the Russians made as much of this misconception as they
could.

A closer analysis shows that the Russians were never as advanced as
they appeared, but, further, the Soviet government always perceived space
exploration purely as a weapon. Korolev knew this and exploited it to allow
him to further his research, once writing: "The realisation of performing a
manned expedition to the surface of the Moon must be considered to be the
main goal of a programme to explore and master the Moon."[56]

The Russian government had almost no interest in the scientific, cultural, or future economic value of space travel. The space program created
jobs, but then, under the Soviet system, technicians employed building a
Vostok capsule or a launch vehicle could just as easily have been put to
work on collective farms or in factories building tractors or cars. Furthermore, from its outset the Russian space effort was a smoke screen to conceal a huge military inferiority, a bluff to make them seem stronger than
they were. The Russians were behaving like an isolated rebel encampment
that put dummy soldiers on the turrets with guns pointing at the approaching enemy, in late 1961 they even had their turrets and fortress, the Berlin
Wall.

This ethos became clear only many years later, when it was revealed
that between 1961 and 1963 President Kennedy twice put it to Premier,
Khrushchev that it would be better for both nations if they cooperated in a
joint attempt to reach the Moon. Twice his suggestion was rejected.*

With hindsight, the reason for such reluctance seems obvious. The

---

*Kennedy's plan filled the NASA scientists and administrators with horror. They realized
(as the politicians almost certainly did not) that the equipment used by the two rivals was
totally incompatible. Furthermore, NASA scientists had little faith in Soviet technology.

Russians were bluffing about their space and military capabilities, and if they had agreed to a joint space program, the Kremlin would have been forced to reveal the number of missiles they had. "What do they have to hide?" Khrushchev's son, Sergei, once asked his father. "If we cooperate with the Americans," he replied, "it will open up our rocket program to them. We have only two hundred missiles, but they think we have many more."[57]

The United States was always the more advanced technologically, and although both space-faring nations were heavily in the debt of German scientists, the Russians had deciphered the papers left at Peenemünde while the Americans were working with the authors of those papers. America also possessed the infrastructure and what proved to be the best political and socioeconomic system in which to nurture a space program. However, to establish a race, the United States had needed the impetus of rivalry from an ideological (and potential military) foe. Consequently, the Moon became the "safe" playing field against a backdrop of genuine nuclear threat.

*A*MERICA'S MERCURY PROJECT RAN from May 1961 to May 1963, and it was an unqualified success. Six astronauts took their tiny capsules (only 74 inches wide) into "space." The first two launches, Alan Shepard's *Freedom 7* and Gus Grissom's *Liberty Bell 7*, made suborbital flights, but in February 1962 an estimated TV audience of 100 million watched John Glenn become the first American to reach orbit.

These missions were pure experimental science. They tested the capabilities of the rocket designed by von Braun and built by the Chrysler Corporation as well as assessing the efficiency of the capsule constructed by McDonnell. But the flights also determined how the human body withstood the g-forces of takeoff and the effects of zero gravity, and how astronauts dealt with the psychological and emotional pressures placed upon them.

Even then the Russians were performing spectacular feats in space. As the American astronauts chalked up a few more orbits with each flight (three for Glenn, six for the third American in orbit, Walter Schirra), the Russians launched two craft, *Vostok III* and *Vostok IV*, on consecutive days, and the capsules passed within three miles of each other before completing sixty-four orbits. And, as the two superpowers battled it out in space, the headlines in the world's press were filled with news of the Cuban missile crisis, tensions over the Berlin Wall, and nuclear tests. To many at the time, interest in space travel was a little misplaced, an academic exercise; the human race, they thought, might not survive to see the exploitation of this amazing new technology.

Even though the United States was now matching the USSR, and in real terms overtaking them, many still saw the United States as shadows of the Soviet scientists. The British radio astronomer Sir Bernard Lovell, head of the Jodrell Bank radio telescope, where many satellites were tracked, commented: "I think the Russians are so far ahead in the technique of rocketry that the possibility of America catching up in the next decade is remote."[58] And in the United States, the mastermind of the hydrogen bomb, Edward Teller, was reported to comment: "There is no doubt that the best scientists as of this moment are not in the United States, but in Moscow."[59]

But the critics were entirely wrong. By the end of 1963 and the closing of Project Mercury, the race was effectively over. The United States had organized a program for achieving Kennedy's goal of reaching the Moon by the end of the decade. Project Gemini was gearing up to further the work developed during Project Mercury, and the infrastructure for a Moon landing was in place. A collection of corporations was responsible for the elements that produced both the launch vehicle and the capsule. The plans of von Braun were developed by teams of engineers and technicians and the nuts and bolts put together by tens of thousands of workers in each state of the union.* By the time of Project Apollo, four hundred thousand men and women were working on the American space program, while in the USSR an estimated five hundred thousand civilians were involved in their rival projects.

On November 22, 1963, John F. Kennedy was due to give a speech to the Dallas Citizens Council in which he planned to highlight the advances of the American space effort. Had he lived to see the day's end, he would have told the council: "We have regained the initiative in the exploration of outer space—making an annual effort greater than the combined total of all space activities undertaken during the fifties—launching more than 130 vehicles

---

*The first NASA administrator, James Webb, insisted that from the outset the organization would have a contracting policy whereby work was to given to every state.

into earth orbit—putting into operation valuable weather and communications satellites—and making clear to all that the United States of America has no intention of finishing in second place."[60]

Aside from the main target of reaching the Moon, the U.S. space agencies had already begun to influence the future. During the summer of 1962, NASA had launched the world's first communications satellite, *Telstar*, and the first transatlantic pictures had been transmitted. Ironically, one of the most famous early uses for this new technology was to broadcast around the world images of the cavalcade of the American president as the shots rang out.

In the USSR, Korolev and his workers were floundering. The Soviet hierarchy had created a system that was inefficient, expensive, and clumsy. The space program had been damaged irreparably by the establishment of parallel but unconnected projects that drained precious funding, and it had become demoralizing for those who worked in it. Any sense of purpose or solidarity was lost, and the Soviet effort had become a dark reflection of NASA. The Russians had entered the space race with a distorted attitude. The ability to develop and sustain a program that would take men to the Moon lay beyond their resources: Continually underfunded, they were surviving upon image alone. Inevitably, cracks began to appear.

As Project Gemini reinforced NASA's image and the American space engineers practiced techniques that would be employed during the complex lunar missions—how to build safer craft, how to dock in space, and how to maneuver in orbit—the Russians found themselves beset with problems. In October 1964 Khrushchev was deposed and replaced by Leonid Brezhnev. Khrushchev had understood almost nothing about science and the technicalities of what Korolev and his team were doing, but at least he had liked and encouraged his chief designer. Brezhnev was far less receptive, and Korolev never established an easy working relationship with him.

And as Korolev's political support grew weaker, the American system became stronger each month. It was now becoming clear that the United States had called the bluff of the Russians and was trumping them. From the beginning of the race in 1957 until the end of Project Mercury, the Russians had made a show of their successes and been far from reticent about their ideas and plans. Suddenly, Moscow fell silent and, with the single exception of being the first to conduct a successful space walk in March

1965, for almost two years the Soviet manned space program disappeared from the headlines of the world's newspapers.

In the United States authorities had taken great care to ensure they would only make a show of catching the Russians when they were sure they had. Some had argued that NASA's medical advisers had cost the agency early victories, but others were beginning to believe such cautiousness was paying off. All the Mercury and Gemini missions were unconditional successes, and they pushed the U.S. manned space program far into the lead; this then allowed politicians and spokespeople for NASA to express confidence about the plans for Apollo and a Moon landing before the end of the decade. Meanwhile, Korolev had nowhere left to go.

With the fall of Communism in the late 1980s, Korolev's plans and designs for the Russian mission to the Moon became available to researchers, and the most striking first impression they give is how similar they were to NASA mission plans and how they evolved in an almost identical fashion.

Perhaps this should not be surprising; as young men both the chief designer and the creator of the V2 had been weaned on the same ideas about traveling to the Moon, the same science fiction writers, and books written by the same pioneers of rocketry. Furthermore, the program of both nations had been built on the foundations laid at Peenemünde. Both von Braun and Korolev originally conceived of a mission in which a large rocket would be constructed in Earth orbit. Once completed, this vehicle would travel to the Moon and take up lunar orbit, where a smaller craft would be deployed to make the trip to the lunar surface. This was called the earth orbit rendezvous (EOR) method.

However, at NASA this scheme received almost immediate opposition. It was too expensive, too cumbersome, and created a vast array of technical problems, many of which were considered insoluble in the time available. Fortunately, a young engineer named John Houbolt, then working under von Braun, had a radically different plan he called the lunar orbit rendezvous (LOR) method.

In this scenario a single, very large rocket would launch all the parts required for the mission. These included two compartments for the astronauts, a command module, and a lunar module, which were docked together at the top of the rocket. Once in orbit these capsules would detach themselves from the main thrusters and fly to the Moon with the three astronauts in the command module. Taking up lunar orbit, the lunar module, with two of the crew aboard, would descend to the surface of the Moon. At the end of the mission, the upper part of the lunar module would take off and rejoin the command module, which had been piloted by the third crew member. Once docked, the two men from the surface would join the pilot of the command module, the lunar module would be jettisoned, and the command module thrusters fired to return to Earth.

This was first proposed by Houbolt in late 1961, and although few were happy with von Braun's EOR method, almost everyone thought the LOR approach crazy. To make things worse, von Braun thoroughly opposed it. But Houbolt was sure his method was right. Realizing the only way he could make the heads of NASA see it his way would be to tirelessly promote the scheme through a series of well-placed presentations, he took his proposal to every manager, team leader, and senior designer in the organization. By the summer of 1962, with the Mercury missions in full swing, Houbolt had managed to convince almost everyone, except von Braun and his closest colleagues.

In June that year a meeting of NASA's most senior executives and designers was called at the Marshall Space Flight Center in Huntsville. Intense discussions about the best method of traveling to the Moon lasted most of the day, and for much of the time von Braun sat quietly following the flow of argument, EOR versus LOR. Houbolt was given a platform to present his LOR plan, and those behind the EOR method, von Braun's people, argued their case. Toward the end of the day, having weighed up the arguments against EOR, von Braun took the floor. Everyone around him believed he would present a cogent argument to oppose Houbolt's scheme and then describe again his own technique. But instead he said: "It is the position of the Marshall Space Flight Center that we support the lunar orbit rendezvous plan."[61] Von Braun had finally been persuaded by Houbolt's well-presented and ingenious scheme and, without even consulting with his staff, many of whom had been fighting the EOR cause all day, he simply switched sides.

Korolev's work did not proceed so smoothly. There was no bright young spark on his team who could see a better path than the earth-orbit-rendezvous method, and the Chief Designer spent far too long following his

problematic scheme. By the time he realized the EOR approach was
unworkable, he had lost two years of development. By 1964, when he made
the decision to adopt a scheme similar to Houbolt's, it was already much
too late. Switching to the lunar-orbit-rendezvous method, Korolev was
immediately faced with further setbacks. The most serious new problem
was that the N-1 rocket (the most powerful Russian rocket ever built and
leading candidate for a launch vehicle using the EOR approach) was found
to be inadequate for the task of lifting the machinery needed to reach the
Moon using the LOR method.

In the United States, von Braun had been working on the development
of the Saturn V rocket. This employed a cluster of five F-1 rockets similar
to those that had individually powered the early manned flights and satellite launches. Together, these five rockets produced some 7.5 million
pounds of thrust. Korolev's N-1 could not match this, and to heap on still
more problems, the Russian lunar and command modules were significantly heavier than those designed by NASA. So even the 7.5 million
pounds offered by the enormous Saturn V would have been insufficient for
Korolev.

It was this fault in the scheme, coupled with the mire of Soviet red tape,
muddled thinking at the top, and confusion over the real purpose and direction of the their space program, that was to end Russian dreams of landing
on the Moon before the Americans. To develop the rocket power, the delicacy of design, and the elaborate communications and guidance systems
that allowed Neil Armstrong and Buzz Aldrin to step onto the Moon, the
U.S. government was spending what seemed to the Russians almost unlimited amounts of money. And because the administration in the USSR used
a scattergun approach to program development, only a fraction of the available funds were directed specifically at the Russian equivalent to Apollo—
Korolev's N-1/L-3 program. The alternative was Chelomei's circumlunar
project, which was progressing well but could be no match for a complete
lunar mission including a landing.

Korolev never recorded his personal feelings about the Soviet failure,
but a close colleague, Boris Chertok, recalls that "He realized though, that
it would be difficult to beat the Americans, who were pouring huge
amounts of money into the Apollo project. . . . Some of the Soviet Mar-

shals were opposed to the race. . . . Marshal Nedelin's death had been a big blow to the program because he had supported the Moon project."[62] But another Korolev ally, the physicist Andrei Sakharov, remembers: "I had just learned from a foreign broadcast that the Americans had used a gigantic Saturn rocket to boost a nineteen-ton space station into orbit, a long stride on their way to the moon. I couldn't resist asking Korolev if he had heard the news (knowing of course, that we had nothing to match the American rocket). Korolev smiled, put his arm around my shoulder and, using the familiar form of address, said to me, 'Don't worry, we'll have our day yet.' "[63]

But Korolev's day had passed. His triumphs had filled the front pages of the world's newspapers—the first satellite in space, the first man in space, the first woman in space, the first space walk. Now, overworked, underresourced, and frustrated, Korolev was unable to solve the design problems he was facing, and by 1965 the Chief Designer was a sick man. In January 1966 he was admitted to a Moscow hospital for an operation to remove bleeding polyps from his intestine. He died on the operating table.

Korolev's death, it has been claimed, was the result of the very bureaucratic nightmare that had plagued him throughout his career. "Korolev was the victim of a botched operation," a historian of the Soviet space program, James Oberg, has written, "of malpractice brought on by the defects of Soviet science bureaucracy."[64] Such bureaucratic incompetence, overcomplex political hierarchies, and unreliable technology had, in part, stopped him from fulfilling his ultimate dream.

It has been claimed that "after his death in 1966 there was no one equivalent to Korolev."[65] But that did not stop every senior engineer and designer in Russia trying. And so an unseemly scramble immediately ensued for the top job in the program. Korolev was barely cold before each of his former rivals had tried to force through his own program, but the surprise successor to Korolev was the relatively mild-mannered Vassily Mishin, who made the totally unexpected decision to continue with both Korolev's plans for the N-1/L-3 program and Chelomei's Proton project.

Although there were several senior scientists in the Soviet space program who desperately wanted Mishin's new appointment and resented him enormously, it was actually a thankless job. Mishin had been chosen primarily for diplomatic reasons, to try to find a way for the rival groups of designers in Moscow to cooperate, but his appointment was another misjudgment. Within months of Korolev's death, there were no fewer than twenty-six divisions in the Soviet space administration, most of whom would not only dismiss the thought of cooperation but preferred not to communicate with one another at all. However, by this time the Soviet space program, with its string of famous early successes, had absorbed so much money it was a locomotive that could not be stopped. With the loss of Korolev, it no longer had a driver capable of controlling its course.

As the chief designer's state funeral was being arranged, an astronaut named Neil Armstrong was completing a stunningly successful mission with David Scott aboard *Gemini 8*, and this merely reminded the Russian engineers how far behind they had slipped. From June to September of 1966 other U.S. successes followed. *Gemini 9, 10,* and *11* all laid the groundwork for the procedures and techniques crucial to the success of the Moon missions. But then, before final triumph, tragedy struck both competitors in the space race.

The year 1967 will be remembered for its space disasters. It began terribly for America with the launch pad deaths of Gus Grissom, Ed White, and Roger Chaffee in the capsule of *Apollo 1*. Then, in April, cosmonaut Vladimir Komarov, piloting a new Soviet craft, *Soyuz 1*, was killed when he crash-landed after the parachute of his module failed to open. The Americans did not attempt another launch for nine months, but the Russians grounded Soyuz for a full eighteen months, a delay that dragged them still further behind in the race.

By the time another Russian manned mission could be mounted, American astronauts were only a year away from the Sea of Tranquillity. And as the Apollo crews edged ever nearer to their goal, the Russians were faced with mounting catastrophe. Unmanned test capsules plummeted to Earth or went off course in orbit, and successive N-1 launches were either aborted or malfunctioned. On July 3, 1969, a loose bolt fell into an oxidizer pump aboard an N-1, causing an explosion in one of its thirty first-stage engines. Fire broke out immediately and spread with horrifying speed. The cables holding the rocket to the gantry snapped, and the enormous rocket collapsed, destroying the entire launch pad. No Western researcher has yet managed to establish the death toll that day.

Seventeen days later, America had won the race.

THE SOVIET SPACE PROGRAM continued to be plagued with troubles after the early N-1 disasters. Funding was slashed, and morale evaporated; yet it was to take the Russians a full five years to lower their sights from the Moon. In 1971 three cosmonauts were killed when their capsule burned up during reentry, and a year later an unmanned N-1 rocket exploded a minute and a half after liftoff. Then, a few days before the Russians were to attempt a manned lunar mission in August 1974, the entire N-1 program was scrapped and Vassily Mishin fired.

Mishin's successor as head of the Russian space program was Valentin Glushko, formerly a bitter opponent of Korolev's ideas and one of his most energetic rivals. As soon as he took over the program, he set about expunging all trace of Korolev's most ambitious effort, the N-1 rocket. "He incinerated every notion of the N-1 with a hot iron," wrote one of Korolev's colleagues, Sergei Kryukov.[66] As far as the government was concerned, the entire program to reach the Moon was never begun. Only in 1989 was its existence officially admitted by the Gorbachev administration.

To understand why Russia failed and America succeeded in reaching the Moon, we should turn to Korolev's immediate successor, Vassily Mishin, who, after Korolev himself, was perhaps closer than anyone to the epicenter of the science and politics that determined the Russian side of the race. "The [U.S.] programme was not cloaked in secrecy, which facilitated the free exchange of data between all the interested organisations and the flow of information was organised not only vertically (from the higher organisations to the lower, and vice versa) but also horizontally, i.e, between contractors," he wrote. "The free flow of information made it possible to track and monitor the course of the work."[67] He was later even more direct, declaring: "Our country could not afford to spend the kind of money that was spent by the United States for the Apollo/Saturn programme."[68]

But these were just some of the reasons. The Russians merged civilian scientific space exploration with their military efforts and were therefore forced to shroud their world in mystery. Those who worked within the manned space program enjoyed the luxury of zero public accountability and were answerable only to their military and political leaders. This offered initial inspiration and led to global media triumph, but it gradually became a destructive force. The Russian space program was entangled in a web of political and military considerations, which led to sloppy thinking, unnecessary risk taking, and a plan of development that was totally unfocused. From this foundation came loss of reputation and the eventual implosion of the Soviet lunar effort. During a twelve-year race the Russians went from having a three- or four-year lead to running a program that was five years behind the United States.

Yet perhaps the most important reason for Soviet failure was the wasteful duplication of effort by rival research groups, each trying to capture the limited imagination of the premier and his marshals. This cost billions of rubles and millions of precious man-hours.

This outcome is another sad reminder of the negative effects of certain rivalries. On the one hand, anxiety that any failure might allow an opening for Chelomei, Glushko, or any of the others must have fueled Korolev's many late nights at the lab. On the other, it is clear that by diverting funds from a single agreed upon course, the Russian space engineers, running a race against a common enemy, were destroying their own chances. Such a waste of resources far outweighed the advantages of internal competition.

Many people still claim the U.S. and Soviet space program represented and continue to represent a waste of money and human resources. This is quite wrong. Indeed, the race to the Moon directly influences the technology of today as much as Tesla's great battle with Edison, the building of the atomic bomb, or Lavoisier's radical chemistry.

The Russian effort to send cosmonauts to the Moon was a complete failure, but the work that went into producing the hardware designed to get them there was not wasted. These designs were developed and refined for use in other areas of space exploration. Russians may be seen as the also-ran in the race for the lunar surface, but they have a superb record in other aspects of space exploration. They now lead the United States in the technology required to sustain long trips into space, and their cosmonauts have notched up many more hours in space than their American counterparts. The

United States is providing the majority of the funding for the International Space Station, but much of the conceptualization of the project derives from the vast experience the Russians have accrued developing and living on their space station, *Mir*, the first module of which was launched in 1986.

But, to justify the importance of the space race, we need to look at the ways it has influenced industry that is only indirectly related to space exploration, and these are manifold. First, both programs pushed computing forward by at least a decade. Second, both programs drove the miniaturization of electronic components (which has become synonymous with the incredibly rapid evolution of computer and communications hardware). Third, the space race has led to rapid advancement in materials technology—the development of new fabrics, plastics, and alloys. Fourth, significant medical advances have been stimulated by what has been learned from manned space flights—better pacemakers and defibrillators, improved physiotherapy techniques, and the development of drugs that could be manufactured only under zero gravity provide just a few examples. Many of these developments have generated entire industries unimagined before the 1980s. Among them, just these advances have created millions of jobs and generated hundreds of billions of dollars in global revenue.

Because of the space race, we have weather satellites that have saved perhaps millions of lives, while global communications have radically altered business, culture, and entertainment. Other satellites are used to help locate natural resources, to monitor wildlife, to assist navigation, and, of course, to explore the universe. Even if we considered the global earnings from just one of these applications, say the communications industry, in one year they would exceed the entire budget for the Apollo program.

A further result of the excitement and inspiration produced by the Moon landings, and one that is often ignored, is the way they influenced an enormous number of young people to become interested in science. This in turn stimulated America to pay greater attention to the teaching of science. Throughout the 1960s and 1970s, the U.S. government became increasingly aware that American scientists had played a key role in the Cold War and that, for the future welfare of the country, more bright and interested students should be encouraged to don lab coats. Aside from this, the sheer glamour of space exploration captured the imaginations of perhaps millions of young people throughout the world, who found a new excitement in science and decided to pursue their interest academically. This was science fiction made reality and, for a time at least, the rapid advances in space technology became a great source of inspiration and gave hope for the future of all humanity.

The role of rivalry in this kaleidoscope of scientific and technological advancement was absolute. If there had been no rivalry, there would have been no serious effort by either the United States or the USSR to reach the Moon within a clearly defined time frame such as that imposed by President Kennedy in May 1961.

It is perhaps ironic that, in the aftermath of World War II, and in part thanks to the work of the Los Alamos scientists who developed the atomic bomb, the world entered a political framework in which two superpowers struggled to imprint their ideological systems just as the technology became available to express that struggle as a race into space. In past centuries humans dominated the globe through the exploration and exploitation of distant lands, but by the end of the 1950s, the new frontier lay beyond the earth itself. And within the minds of those holding the purse strings and the reins of power, this new frontier was also a place where a propaganda war could be fought.

But space exploration turned out to be a far more productive venture than anyone could have imagined in 1957. In those days the only people with real vision were the scientists and engineers who made the whole thing work; the politicians were, as always, guided by short-term aims and showmanship. The dreamers believed space exploration would open up the universe for human beings, and perhaps one day it will. Admittedly, for those who imagined walking on Mars or their grandchildren exploring the nearest star systems, the space effort has proved a disappointment. We have not exploited even the one place beyond the earth where humans have stood, and the plans to travel to Mars suffer endless delays and revisions. But beyond these dreams the space race has had a massive impact upon everyone alive today.

Now the political ambitions that drove the space race are gone, and we should be thankful. The rivalry that fueled the American and Soviet efforts was built upon shaky and short-lived foundations. Although the leaders who walked the halls of the Kremlin as *Sputnik* flew overhead might never have imagined it, the Soviet system collapsed a mere thirty-two years later.

But this rivalry built upon human fear and paranoia and merely supported by the right thinking of men like Korolev and von Braun could never have allowed us to reach beyond the preliminaries. Science blended with politics could propel men into orbit; it could even land a dozen men on the surface of the Moon at a cost of $100 billion in today's terms, but no more. The space race provided the platform NASA used to change all our lives; yet, to go further, to reach the planets and the stars, to colonize and to exploit the galaxy and beyond, to make our home among the stellar fields, a new impetus will be needed, a fresh motivation found.

Space exploration will survive and evolve only through self-financing. NASA and the many emerging private space agencies will support themselves through the commercialization of space travel and the exploitation of resources beyond Earth. The money earned from space already far exceeds that spent on it, but the rewards are spread among the industries of the world. Future generations may see space exploited thanks to a new rivalry that will develop among pan-global industrial giants. Space travel will be financed by drug companies, computer giants, and communications industries; stocks in planetary commodities will make the FT Index and the Fortune 500.

Neil Armstrong's first step onto the Moon was a great leap for mankind, but one day soon interplanetary travel will become commonplace. Humans will live and work wherever life may be sustained. And, as our descendants walk through the ravines of Mars and kick up moon dust in the Sea of Tranquillity, they may choose to forget the grand rivalry that took them there.

# EIGHT

## The Battle of the Cyber-Kings

It is pretty to see what money will do.
—*Samuel Pepys*, Diary, 1667

THE CYBER-KINGS, the men who run the computer industry, were once thought of as geeks, and geeks used to be impoverished geniuses understood by few and used by the powerful. But the commercial world and the world of technology have been revolutionized by the computer. It has made a few rare individuals some of the richest people who have ever lived, comparable to Renaissance pontiffs and the demigods of the first media revolution, the Hearsts and Rockefellers. These men, among whom Bill Gates, Larry Ellison,

Marc Andreessen, and Scott McNealy form the highest echelon, wield almost unimaginable power, greater by far than any other corporate leaders, greater indeed than many heads of state. They are driven, obsessive individuals with what appear to be superhuman reserves of energy, determination, and self-belief. They are also, predictably perhaps, egotistical and often vicious.

Bill Gates has written two best-selling books about computers and the computing industry, *The Road Ahead* (1995) and *Business @ the Speed of Thought* (1999). But, in a move reminiscent of Newton obliterating all reference to Leibniz or Hooke from his *Principia*, Gates does not once mention the names Ellison, McNealy, Andreessen, Oracle, or Sun Microsystems, all very big contributors to the computing world, but also his greatest rivals.

Larry Ellison is fond of quoting Genghis Khan: "It is not enough to succeed; everyone else must fail."[1] He and his peers talk of "cutting off the oxygen supply of rival companies," and they see themselves as generals in an endless bloody battle. Using a *Star Wars* analogy, Scott McNealy of Sun Microsystems once declared: "There are two camps, those in Redmond [the headquarters of Microsoft], who live on the Death Star, and the rest of us, the rebel forces."[2]

These few men are changing the world we live in, but their power and perhaps their public images have earned them few friends. Log on and type in "Microsoft," "Sun," "Oracle," or "Gates" and you will find thousands of Web sites dedicated to opposing what these corporations and individuals do. Such feelings are not limited to obsessively jealous individuals with too much time on their hands; the cyber-kings attract particular opprobrium and vitriol from some of the people they work with, their peers in the industry. One anonymous IBM executive has said: "I'd like to put an ice pick in Gates's head."[3]

*M*ICROSOFT HEADQUARTERS, known as the "campus," based in Redmond, about ten miles outside Seattle, Washington, is as clinical and clean as the inside of a PC rolling off the production line. Here the nerd is king and the "king nerd," the CEO of Microsoft, Bill Gates, is in complete control.

It is a few days before Thanksgiving 1995, and as I arrive on the campus in the cab paid for by Microsoft and booked by a Microsoft "fluffer," I think through the story of how this steel and Perspex complex arose from a confluence of zeros and ones, digital spaghetti.

The Greeks had an obsession for numbers that has survived all subsequent change; indeed, civilization might have foundered without it. The idea of the computer was certainly familiar to Greek mathematicians, the abacus a cherished tool. Millennia later, others, most prominently Charles Babbage, designed machines that could accelerate computation by mechanical means. But remarkably, true computing, the science of digital manipulation, can trace an ancestry only back to World War II.

Now, even though most of us can recall a time when *digital, modem,* and *hard drive* were not words we thought about or used often, it is difficult to imagine a world without computers. If all the computers in the world shut down tomorrow, we would not simply regress to a time before they were commonplace—there would be no return to a 1950s lifestyle—such an event would mark the end of civilization itself. For today we are totally dependent upon computers. The way electricity is delivered to our homes and our factories, the way essential commodities such as gasoline and food reach us is controlled entirely by automated computerized networks. If the world's computers were to shut down, our gas, electricity, and water supplies would soon stop; sewage systems would fail; the financial markets would collapse in disarray; communications would break down; there would be no TV, no radio; military and law-enforcement establishments would be thrown into confusion; hospitals could function only on a basic level. And from all this, anarchy would soon follow.

When I leave the cab and ten minutes later sit facing Bill Gates, it is easy to forget that this man has done much to create such a world.

As we speak, Bill Gates rocks in his seat. At first, this is off-putting, but soon I stop noticing. He is intense and focused; each word he chooses carefully. Initially, it is hard to imagine this man at the apex of a global corporation employing seventeen thousand people, possessing a personal fortune in excess of $15 billion, but gradually you become aware of Bill Gates's power, his inner steeliness beneath a veneer of blandness.*

He claims he has not yet even begun to fulfill his ambitions, sees himself as a "young kid trying to find out about the world," and still finds immense pleasure in the day-to-day running of the multinational giant he controls. He offers an image of modesty untainted by power, but the mask keeps slipping,

---

*This was in 1995. Latest estimates from the start of the new millennium place Gates's fortune at around $50 billion and growing almost exponentially.

and beneath it occasionally a monumental ego slides into view, an ego dazzling in its intensity. An example is the moment he talks of his recent purchase of Leonardo's *Codex Hammer* for over $30 million. "It was on show in Venice immediately after I bought it," he says nonchalantly, "and I attended the grand opening . . . actually the exhibition had been opened a few days before, but I was there, so it was the grand opening," he adds without a trace of irony.[4]

We talk about his house then under construction on the shores of Lake Washington. We discuss his book *The Road Ahead*, and whether or not Microsoft has a "far-future" research team busy developing futuristic ideas.* I ask him about his marriage to his former employee Melinda French and his hopes for the future of Microsoft. He tells me about his love of three-hundred-dollar-a-round poker, his favorite material possession—his Porsche 959—and he becomes animated when asked about his recent vacation in China, during which he apparently did not touch a computer for three weeks.

This is an exciting time for Microsoft, and for Bill Gates. The company has just launched Windows 95, and a few weeks earlier it announced that the product was selling globally at a rate of one million copies a week. Even though Microsoft faced an army of critics, many of whom were screaming that Gates was making the biggest personal fortune in history by reselling to people something they already had, it is easy to see how the head of Microsoft could feel contented. "I've had above average luck," he admits and, leaning forward, adds quietly: "This is the most exciting time in history for doing what I do. People say I always want to live in the future, but that's not strictly true. I'm thrilled by what's happening now."[5]

Yet, at that very moment, Bill Gates's world was not the exclusively wonderful place he would have had us believe. Microsoft was indeed riding high on the FT Index of companies (in the preceding quarter it had reported revenues of over $2 billion, a rise of 62 percent over the previous year), but, at the heart of an industry in which Gates and his company were undisputed masters, powerful and exciting developments beyond these spotless Redmond walls threatened to tip Microsoft into obsolescence. The world was changing so fast even one of the great manipulators of the zeitgeist was in danger of being outpaced. The digital monster was threatening to turn on its Dr. Frankenstein, and one of Gates's bitterest rivals was making ready to employ the creature for his own ends.

---

*Suprisingly, perhaps, they don't; or if they do, it is secret.

. . .

$\mathcal{M}$ANY OF THE GREAT names in the history of science made attempts to produce computers. At the age of nineteen, Blaise Pascal constructed a machine that could add and subtract numbers containing up to eight digits, and thirty years later, in 1673, Gottfried Leibniz offered a calculating machine to the Royal Society. The problem with all such machines built before the twentieth century was that they worked by the operation of cogs and wheels that mechanically crunched numbers. A machine that possessed worthwhile calculating power required many of these cogs and wheels, so construction proved immensely difficult; every element of the machine had to be handmade and integrated to a phenomenal level of accuracy.

By the early twentieth century it was possible to build electronic devices that handled numbers and could calculate, but although these could be made smaller than their mechanical antecedents, they were not really any more efficient because they operated on a decimal system—they were *analog* machines. To conduct calculations, analog machines use the numbers 0 through 9, which requires ten electrical values, each assigned to one of the digits. Although this seems simple, it is impossible to construct electrical circuits able to handle ten different impulses, making electronic analog computers a practical impossibility.

The solution to this barrier came from a book published in 1854 in which the author introduced a radical way to manipulate numbers. The book was called *An Investigation of the Laws of Thought*, written by the son of a Lincolnshire cobbler, George Boole. This text explained how all numbers could be defined in terms of zeros and ones, or *binary*.*

---

*In binary, base 2 is used (as opposed to base 10 in the decimal system). This means that each position in a numeral represents a particular power of 2 instead of powers of 10 as used in the decimal system. In base 2, or binary, all positive integers are represented only by a string of 0s and 1s, for example, the binary number 1110001. In binary, as in all number systems (including the decimal system), each digit in the number represents a place value. So, reading from right to left, the first digit represents the number of units. In binary, there is only either a 0 or a 1 in any of the place values, unlike in the decimal system, where there are the digits 0 through 9. The second digit from the right represents the number of 2s (or 10s in the decimal system), the third, the number of 4s (i.e., $2 \times 2$); this would be equivalent to the 100s in decimal notation. The fourth number from the right represents the number of 8s ($2 \times 2 \times 2$), which is comparable to the number of 1,000s in the decimal system, and so on.

This method found a resonance in practical electronics. A switch is either on or off, either current flows or it doesn't. Any quantity may be expressed as a number, but this property in itself is unusable, so this number must first be converted into a binary number. This may then be translated into an electrical impulse and used by a computer: Each on or off or each 1 or 0 is then referred to as a bit.

The first modern computer, the first to use binary, was a machine called Colossus built at Bletchley Park, England, in December 1943. Colossus was designed by the great Cambridge theoretician Alan Turing and employed to crack the ENIGMA code used by the Germans to encrypt instructions for their U-boat fleet then harassing shipping in the Atlantic.*

After the war, the potential of computing began to excite some of the great minds of the age. John von Neumann, who had been instrumental at Los Alamos, turned his full attention to the theoretical development of computing. He advanced the theory of automata and showed how to construct efficient computer memory, developing a machine called MADAM (Manchester Automatic Digital Machine), which, in 1948, became the first computer to have a memory stored electronically. Soon after this, Turing began to explore the theoretical basis for such exotica as artificial intelligence.†

And as Turing and other theorists worked to conceptualize the fundamental elements of computers, the evolution of electronics was developing in parallel so that the theories could be rationalized and made practical. The thermionic valve was the essential component of all electrical devices before the 1960s, and it evolved into the transistor, the chip, and eventually the microchip.‡

---

So, in binary, the number 2 is represented by the digits 10. The decimal number 3 is equivalent to 1 times 2 plus a 1, or 11. The decimal number 4 is equal to two 2s, or 100.

*The Royal Navy captured a German U-boat and an ENIGMA machine in May 1941. After the code was cracked at Bletchley Park, the Allies were able to communicate false instructions that were interpreted as genuine by the German naval command. This action helped change the course of World War II.

†Sadly, these two leading figures in the early development of the digital computer both died young, before the full impact of their creations could be realized. Turing, who is only now beginning to be seen as one of the great geniuses of the twentieth century, committed suicide at the age of forty-two in June 1954 after he was prosecuted for homosexuality. The equally brilliant and visionary Neumann died aged fifty-four in 1957.

‡The thermionic valve had been developed from Edison's discovery (during the 1870s) that electricity flows from a heated filament to an electrode in one direction only.

As computers gained greater memory capacity, they could be used with increasing sophistication. The first computers to enter commercial use emerged in the 1950s and 1960s, devices that employed a means of input (in those days a vast collection of punch cards), a processor (containing the memory), and an output (a printer and, later, a visual display unit). Two new words quickly entered the language: *hardware*, to describe the computing machines themselves, and *software*, the ancillary components that could be installed to make the machine operate.

As the earliest computers left the laboratory and found application where they could be afforded—government offices and the "computing departments" of large corporations—few could have visualized the multitude of uses they would be put to. For at least the first decade in which they were used outside the laboratory, computers remained simply number crunchers, devices used to save worker-hours by carrying out large and complex calculations.

International Business Machines (IBM) was founded by the American inventor and entrepreneur Herman Hollerith after he had devised a punch-card tabulator and sorter during the late 1880s. His system was chosen for the 1890 census, and he soon established his company as the market leader in commercial analysis. The computer was ideally suited for the kind of work for which IBM was becoming famous, and soon the company was not only using the computers designed and constructed in university labs but developing its own machines and operating systems, a move that (in terms of revenue) propelled it into the top echelon of the world's corporations.

But computers could do more than grind numbers. During the 1960s those at the cutting edge of research in computer technology realized there would one day be a clear distinction between two major types of computer use. On the one hand, the large organizations, the military, the accounting departments of multinational corporations, the health authorities would all continue to need large machines that could manipulate huge quantities of information, the sort of material that once filled acres of filing cabinets and occupied thousands of clerks.\* Quite another type of machine would be employed by individuals, who would one day use computers tailored to suit their personal needs (including the ability to word-process and to manipu-

---

\*Today such functions are taken care of using networks rather than single huge machines.

late small, particular sets of data). Perhaps, in their more imaginative moments, computer designers dreamed of people playing games on desktop machines, using them to communicate across the globe, even talking to them and listening to their digital replies.

Government and big business made large demands, and one of the key functions of the computer became the manipulation of what is now called a database. By the late 1970s, as the PC was being postulated, IBM (Big Blue, as it was nicknamed), which by then had a payroll of four hundred thousand and an annual turnover of tens of billions of dollars, began to spend millions of dollars developing the first practical database capable of storing and manipulating large retrievable amounts of data—facts, figures, statistics. To do this, a means to interconnect blocks of information based on specific criteria was needed. In other words, computer programs were required that could *manage* data, not just conduct calculations, however complex.

A database is really a large, complex list of facts and information. Examples include telephone directories, bibliographic references, and flight guides; they are different from simple lists because databases can be used to extract a specific group of disparate facts from the full collection of data in the memory.

The man responsible for working out the theory of the database was an IBM researcher named Ted Codd, who in 1970 wrote a paper titled "A Rational Model of Data for Large Shared Data Banks." It was a highly academic work, but it laid the foundations for what would become perhaps the most heavily utilized aspect of computer technology. By 1977 a group of researchers in IBM calling itself System R had developed Codd's theory and molded it into experimental software. But, unfortunately for IBM, the inventiveness of the researchers was not matched by those responsible for developing experimental ideas into marketable products. It was not until 1982 that IBM had a commercially viable database. By that time, as we shall see, Larry Ellison had captured the huge global market with his own software and established his computing empire.

And as IBM allowed others to capitalize on one of its most important creations, another IBM idea, the personal computer, was making its way through the company's research and development department. While Ellison found his way into the IBM empire and extricated an experimental application with which he could eventually build a business to rival Big Blue itself, Bill Gates concluded that the bigger picture lay with the PC. Working in a style quite different from Ellison's but with equal chutzpah and ingenuity, he too found a way of empowering his own business interests with the groundwork offered by IBM. Gates and his programmers began to devise the software to make the PC a machine the world would adopt and grow to depend upon.

*W*ILLIAM HENRY GATES III was born into a prosperous Seattle family on October 28, 1955. His father was a successful lawyer and president of the state bar, his mother, a socialite and director of a bank. Gates's great-grandfather had been a civic leader in the early days of the city and established the National City Bank in Seattle. By the time of Bill's birth, the Gates family had links with commerce, banking, education, and law, and could number as friends many of the wealthiest people in the country, including the then richest American, Warren Buffett, who remains one of Gates's closest friends and confidants.

Bill, an only son, was nicknamed Trey by his maternal grandmother, Adelle, after the cardplayer's term for third. Grandma Adelle was also a great influence on the boy's preschool development. She introduced him to the world of books and constantly encouraged him to "think smart, think smart!" She also taught him to play cards, and he became a skilled bridge player before he had entered the fourth grade. According to his family, Bill was competitive even at this age, and he hated to lose. In college, cards and gambling would become something of an obsession for him, an interest that would remain long after fraternity capers and other aspects of the undergraduate lifestyle had faded into a half-remembered past.

Like many of those who gained recognition as scientists or technologists, Gates became interested in science through reading science fiction.* In 1962, when he was seven, he was hugely influenced by visiting the World's

---

*This is by no means uncommon. Stephen Hawking has claimed that he was first stung by the fascination with science because of science fiction stories he read as a child, and he remains a huge fan of *Star Trek*. Isaac Asimov was motivated to follow a career in science thanks to his reading of early science fiction classics. After years as a biochemist he went on to nurture the dreams of later generations of bright young people with his famous SF writing, and many of these readers too took up careers as scientists.

Fair arrayed around the Space Needle in Seattle. The theme was Century 21, and at the various exhibits he saw demonstrations of what their creators dubbed "gadgets of the future." He watched in awe as the monorail that now operates routinely through the center of Seattle sped past the fair, and he stood so close to Alan Shepard's space capsule *Freedom 7*, which had traveled into space just a year earlier, he could almost touch it. Later, at the IBM exhibit The World of Tomorrow, young Bill sat transfixed by demonstrations of how one day computers would speak, how they would control the flight of planes and rockets, the traffic of cities, become the sole repository of records, and even allow people to communicate around the world.

He was stung by the science bug and, as one prep-school teacher, has said: "He was a nerd even before the term was invented."[6] But even so, Gates's intelligence had no satisfactory outlet. He told teachers he wanted to be "a scientist" and declared that math was his favorite subject, but he was in danger of becoming lost because of lack of direction. Through his family, though, Gates came early to the idea of making money and the merits of commerce. At school he became interested in economics and gained an A on his report card for a project titled "Invest in Gatesway Incorporated," in which he envisioned himself the head of the corporation.

Although he was academically minded, Bill did not mix well at school and for a while became quite disruptive. On three occasions his parents were called in to consult with the headmaster because their son had consistently flouted class rules. Gates's teachers realize in retrospect that the boy was simply bored.

By the time he turned twelve his parents had become worried by his seemingly permanent isolation and only rare ability to engage with other children. He appears to have been lost in a world of his own, and when his mother asked him what he was doing staring into space for hours at a time, he would simple say he was "thinking." Eventually, he was sent to a psychiatrist, who concluded that there was nothing wrong with him and that he might gradually come out of himself.

Gates entered the elite Lakeside prep school in 1967, and a year later, with funds raised by the parents, the school invested in an extremely cumbersome, refrigerator-size "mini computer" bought from the Digital Equipment Corporation. Almost from the day it arrived, Gates spent every spare moment at the terminal with his friend Paul Allen.

Gates and Allen shared an interest in science and computers, and each was outstanding at mathematics. They had a great deal in common, but Gates's competitiveness was always an element of their relationship. When they met Allen was more knowledgeable than Gates, in part because he was two years older, but he had also read more widely and benefited from the fact that his father, Kenneth Allen, was associate director of libraries for the University of Washington. Gates learned much from Allen and respected his relative urbanity, but he always knew he could work closely with someone or even be a friend only if he was in control of the relationship. Soon after they met, he told Allen: "I'm easy to get along with...if I'm in charge."[7]

Unlike Gates, Allen was well liked at school. He was seen as cool and clever without being the model nerd. Gates was probably aware of this but unconcerned by it. He has always been supremely self-contained, and although he enjoys respect, he has never placed great importance upon being liked.

The computer at Lakeside school was an ASR-33 Teletype. It had the memory capacity of a calculator you might find today given away in a cereal box or on a novelty key ring. It had no screen, and punch cards were used to input data, but it was the best computer available on a budget smaller than the millions required for a mainframe such as those used by the military and civic authorities. It gave Gates and Allen an opportunity to learn programming techniques few American children could have imagined in the 1960s, and by the eighth grade Gates was known around the school as "the computer guy." By 1972 he and Allen had started their first business, designing and building a computerized "car-counting" machine they called Traf-O-Data, which they tried, rather unsuccessfully, to sell to municipalities along the West Coast. One classmate recalls: "He was obnoxious, he was sure of himself, he was aggressively, intimidatingly smart. When people thought of Bill they thought, Well this guy is going to win a Nobel Prize. But he didn't have any social graces."[8]

Such a mind-set has stayed with Bill Gates and has even prompted uncorroborated suggestions that he may suffer from a mild form of Asperger's syndrome, a type of autism whose symptoms include uncontrollable rocking, sudden mood swings, and a marked lack of interest in normal human social activity. Sufferers usually possess a prodigious memory and a high IQ, but an intelligence that is almost completely single-focused. Whether or not Gates suffers from this illness, the skills that have turned him into the richest man in the world and one of the creators of the computer age were apparent in boyhood. In addition to being extremely bright, he possessed the irrepressible self-confidence and aggressiveness needed to push through his ideas and to make real the then nebulous concept of software.

Gates and Allen were not merely school friends; they shared the same dreams and were driven by the same aspirations. Each was fanatical about computers, each wanted to go into business, each knew his vision of the future was sure to become reality, and each wanted to be at the epicenter of the new world computing would create. After graduation Allen attended Washington State University, and two years later, in 1973, Gates flew to Boston to enter Harvard.

He worked hard and was excited by his courses. Initially shocked to discover there were others every bit as intelligent as he and some who beat him on tests, Gates dedicated himself to yet more intense study. He enrolled in graduate-level math and physics courses, which by his second year took up about one-third of his time. But he was unconventional in his approach to work. He slept without discernible pattern, often working for thirty-six hours without a break, then sleeping for twelve hours before consuming a king-size pizza and returning to work for another long stretch.

To relax, he played high-stakes poker, a game played only by the blue bloods who could afford to lose two thousand dollars a game. But although Gates had all the makings of a successful academic, he was constantly drawn to the business world and to money and power. Rather than its being overshadowed by university life or newly discovered areas of interest, his fascination with computers had grown, and even in those days technological developments were beginning to gain pace.

While Gates had been finishing school and enrolling at Harvard, Allen had continued at Washington State University and been trying in vain to generate more business for the Traf-O-Data project. He and Gates had kept in touch, and during the summer vacation of 1974, after Gates's first college year, Allen had traveled to Boston to work with Gates at Honeywell, a relatively small computer business based near the city. As well as a vacation job that paid well, working in the computer industry provided the pair with information about selling computers and operating systems and further encouraged their belief that this young industry would soon reach critical mass—a point where the computer would evolve into an integral element of society. Gates and Allen wanted to become intimately involved with the revolution they could visualize, but at this point they still had no entrée into the commercial world. At the end of the summer, Allen returned to Seattle and Gates began his sophomore year.

One cold December morning of that year, Allen was on his way to visit Gates when he stopped at a newsstand and bought a copy of a hobbyist magazine called *Popular Electronics*. He had been buying the magazine for years, but that day he had been attracted by the cover, which carried a picture of a rectangular metal box covered with lights and switches, a device called the Altair 8080; the banner headline read: WORLD'S FIRST MICRO-COMPUTER KIT TO RIVAL COMMERCIAL MODELS.

In itself, the Altair 8080 was little more than a box of components, and it could do nothing without a simple operating system. At the time the only affordable system for the hobbyist was built around a computer language called BASIC, a skeletal code that could instruct computers to follow extremely rudimentary processes. But Allen and Gates realized that this box of tricks was a herald of the age to come. The arrival of this cheap, easy-to-assemble computer gave them the platform upon which they could build their skills. Their moment had arrived; they knew that they could produce software using an adapted version of BASIC and that this software would enable the successors of the Altair 8080 to do things previously seen only in science fiction films.

Within months Gates and Allen had formed Micro Soft, as the company was originally named, and Gates began to juggle the demands of college and his new business venture. He managed this, but only for another eighteen months. As Micro Soft became Microsoft, and he and Allen started to secure deals for their primitive operating system (software cannibalized from the Traf-O-Data project), it became clear to Gates he would have to choose between his academic career and his business aspirations.

Long discussions with his parents about his future became more intense whenever he returned home for a vacation, and Gates agonized over what to do through most of his second college year. Allen was all for launching headlong into creating computer software and breaking early into the new world of opportunity the business could offer them, but he knew that, in order to succeed, he needed Gates with him full time. For his part, Gates was still torn between the opportunity of an academic career, the aspirations of his family, and the lure of the cyber world, with its associated excitement and potential for making vast fortunes. Ironically perhaps, one of the factors in Gates's decision to leave Harvard was that during his first term there he had learned he was not the "smartest kid in the world." This had dampened any half-expressed ideas he had had of becoming a math professor. Today he claims with a knowing smile that he only took a break from Harvard and he just hasn't yet returned, but it is easy to believe that even before he left Seattle he suspected he might never complete his degree. In 1973, on the eve of his trip east, an eighteen-year-old Gates had declared, "I'm going to make my first million by the time I'm twenty-five."[9]

Bill Gates is fond of claiming that he is only in the computer business because he wants to produce "great software," that he wants to create the best product he can; but from an early age a much stronger drive was his desire to make money. Gates was a whiz-kid programmer, but he was never an inspired programmer; indeed he was never an innovator. Paul Allen is as bright as Bill Gates and was in many ways a more creative force, but Gates is a supreme salesman and marketeer. As early as 1977 he understood the best way to infiltrate the marketplace with his product, his "great software."

The question of whether Bill Gates is also a scientific force worthy of comparison with other figures drawn from the history of science will be considered at length later. What is clear is that he eats, sleeps, and breathes computers and is at the same time a driven, hard-nosed businessman.

By the age of twenty-one, Gates looked fourteen. A rather squeaky voice and a lazy attitude toward personal hygiene sustained the image of a bedraggled teenager. More than once he was mistaken for a delivery boy at his own company offices. Soon after Gates and Allen formed the company, Allen employed a secretary while Gates was away. One morning a few days later, the secretary ran into Allen's office in a panic. Some scruffy kid had paced through reception and walked into the boss's office—what was she to do? Of course, it was Gates.

Most observers of the evolution of computers since the 1960s point out that Gates and Microsoft did not and still do not innovate; they absorb the ideas of others, modify them to suit the marketplace, and make billions from the imaginations of the truly creative. To a degree this is true, but the ability to understand the ideas of others, to modify and, in many cases, greatly improve upon these ideas, to package them for a global market is a unique skill. Gates is overendowed with this skill and was fortunate enough to realize it and to apply it from an early age. However, many of those who hate Bill Gates declare that Microsoft began as a "poacher" and has never changed. Indeed, they claim Microsoft's first coup should have set alarm bells ringing throughout the industry.

The real innovator who created the first marketable software for ordinary people was a programmer named Gary Kildall, who owned a small company called Digital Research. In 1980 he created what became known as an operating system that could be used on microcomputers. This he called CP/M.

An operating system may be thought of as the fuel that keeps a computer running. Another way of visualizing it is to compare it to the electrical system in a building. Without the electrical system, the elevators would not operate, the coffee machine would be useless. The operating system provides the link between the user and the various appliances within the computer—the word processor, the spreadsheet, the games.

Within months of Kildall's innovation, a company called Seattle Computer came up with a product it christened Q-DOS (quick and dirty operating system), based upon his model. Hearing of Q-DOS, and having already met Kildall and encountered his innovative product, Bill Gates quickly realized that he could make Seattle Computer's quick and dirty operating system an indispensable addition to the industry. He also knew instinctively that the only way to succeed with a new computer system was to work with the solitary giant of the industry, IBM.

In 1980 IBM produced 90 percent of the world's computers, and its rivals were known collectively by the disparaging sobriquet the Seven Dwarfs. Operating systems (or indeed any software) need a machine to run on. It was therefore clear to Gates that if he could form an affiliation with the industry's only giant, he could be carried along on its coattails to access IBM's massive global market. Big Blue could supply the machines, he could provide the operating system.

Gates knew how to play the corporate game perfectly. Kildall did not, and this difference had a major part in making Gates a multibillionaire and Kildall a footnote in the history of computer development.

Big Blue was renowned for its formality and old-fashioned business style. Employees were required to wear "sensible suits" that constituted an unofficial uniform; they sang company songs at conferences and were even expected to drive only "approved" makes of cars. Gates had admired IBM since childhood. To him it *was* the computer industry, so he was happy to behave in any way IBM expected in order to forge a deal between his company and the undisputed master of the industry.

Kildall had no such dreams. He wanted to place his operating system with as many customers as possible, but by 1980 commercial success was no longer his prime motivator. He had already earned several million dollars from his creation, he lived comfortably with his wife (and business partner) in Monterey, and he was not the sort of character who would sell his soul to form a liaison with a corporate giant like IBM. To Kildall, IBM was the bloated dark side of commerce, and he admired Arthur C. Clarke, who had spoofed the computer company by calling the errant computer in his epic *2001: A Space Odyssey* HAL (IBM with each letter moved back one place in the alphabet).

Unsurprisingly, IBM did not approve of Kildall's approach to business. Suspicious of his working methods, the computer giant ignored the fact that his product was the best available and was blind to his brilliance as a computer scientist. Most damagingly, IBM failed to realize that Kildall's software was perfect for its new product, the PC, which in 1980 was still some way from launch. When Gates learned that Digital Research and IBM had failed to come to a meeting of minds, he did not waste a second. Microsoft bought the Q-DOS operating system from Seattle Computer for twenty-five thousand dollars, dropped the *Q*, and, for a flat fee of two hundred thousand dollars, sold the rights for its use in every PC IBM built.

This aspect of the deal was only a minor contributor to Microsoft's huge promotion through the industry ranks during the following decade. Admittedly, it placed the Microsoft name on every PC sold and was a leader for the company's other products, but, crucially, Gates pulled a masterstroke in the negotiations over the licensing of DOS. He realized that when IBM's PC hit the market, what became known as clones would appear almost immediately. Big Blue failed to realized this, so it allowed Microsoft to license the operating system to any competitor selling IBM-compatible machines. These licenses were not sold for flat fees but charged for each machine sold.

During the following decade Gates's negotiating skills over DOS made Microsoft billions and turned Gates and Allen into very rich young men. By contrast, IBM lost an estimated $25 billion because it was unable to devise its own operating system and failed to realize that its groundbreaking PC would be copied by dozens of quick-footed rivals who could move into the market the giant had opened up. With typical extravagance Larry Ellison, the man who almost two decades later was to become Gates's most outspoken rival, described IBM's failure to develop its own operating system as "the single worst mistake in the history of enterprise on earth."[10]

With the huge revenues provided by DOS, Microsoft was propelled into a rarefied level of the computer industry. But this was merely the start of the company's success. In a way reminiscent of Thomas Edison (but with his own sustainable finance), Gates repeated over and over the trick of taking the ideas of original thinkers, melding them, and adapting them for a mass market. And as Microsoft grew the media began to take notice of the company's young CEO.

Early profiles referred to him as a "programmer," but the last line of code he ever wrote was in 1982 for a text editor to be used in a Radio Shack portable computer. Gates's overarching talent is that he understands every aspect of computing. This awareness of the industry is combined with a refined business acumen. Being in the right place at the right time, possessing ruthlessness and determination as well as great technical know-how, made his and Microsoft's rise through the young industry unstoppable.

In 1983 *People* magazine put Gates on its 25 Most Intriguing list, accompanied incongruously by Ronald Reagan and Mr. T. The piece contained very little that was accurate but raved: "Gates is to software what Edison was to the light bulb."[11] Eschewing such comparisons, Gates likes to see himself as a latter-day Henry Ford, and he told *People*: "We want to be to software what IBM is to hardware."[12]

Microsoft went public in 1986, and almost overnight Gates and Allen became billionaires. Since then their wealth has grown almost exponentially, but, revealingly, Gates's fortune far outstrips that of the other founder of Microsoft. Although the two were originally thought to have begun Microsoft as equal partners, Gates has always been far richer than Allen. In 2000 Gates's estate was valued at $50 billion while Allen's was $30 billion.

The disparity began before Gates left Harvard. Allen was then in a salaried position as a programmer, which Gates believed less demanding than study, so he insisted upon a 60–40 split in his favor. The mild-mannered Allen agreed. But even then Gates was not satisfied. When he decided to drop out of Harvard and work full time with Allen developing their embryonic corporation, he renegotiated a 64–36 split. He justified this on the basis that he had walked away from what would have been a promising academic career. Again, Allen demurred without mentioning that to cofound Microsoft he too had left behind a university career.*

$\mathcal{B}$ILL GATES IS A COMPULSIVE WORRIER. He remains convinced Microsoft could collapse tomorrow and everything would be lost, and he smothers this anxiety with excessive hard work and the constant, ruthless rush toward greater wealth. Throughout the 1980s and early 1990s, as Microsoft grew to overtake the market value of IBM and before he found an anchor in marriage and fatherhood, Gates was a junk food addict. In one twelve-month period during the mid-1990s, he gained thirty pounds. Even after his marriage, times of stress meant reverting to old bad habits. When Microsoft stock took a rare and temporary nosedive in late 1995, Gates apparently brainstormed throughout the night with his most trusted col-

---

*Allen was diagnosed with Hodgkin's disease in 1982 and retired as executive vice president of Microsoft in 1983. He remains a member of the board, and has a 4 percent stake in the company. Allen lives still in Seattle, and when he is not playing in his band, Grown Men, or helping to run scores of companies, he is involved with great philanthropic works. His most important contribution to date has been the Experience Music Project (EMP)—a rock museum in the heart of Seattle costing $250 million. Dedicated to his beloved Jimi Hendrix, EMP opened in May 2000. Allen and Gates have remained close friends.

leagues. But in an interview in *Time*, he later admitted that he had spent hour upon hour screaming until he was hoarse and that it had been an "eight-cheeseburger night."

In 1994 Bill Gates paid $30 million for Leonardo's notebook the *Codex Hammer*. A year after the purchase I asked him why he had bought that particular artifact. "Because I wanted to," he replied with a brief smile, but he went on to say it was also a calculated business move because he planned to use the images in the codex for Microsoft advertising and to launch an Internet gallery. The codex has now been placed in a specially designed, climate-controlled case that sits as a centerpiece to the library in Gates's lavish Lake Washington home. Is it merely a trophy? Or did Gates want this document because he wished to be associated with the most famous intellect in history? By owning a scientific notebook of Leonardo's, does Bill Gates feel closer to some form of intellectual nexus? Is it another sign that his billions of dollars are still not enough to make him feel as though he has fulfilled his intellectual potential? Does owning the *Codex Hammer* make him feel like a member of the right club?

As I talked to Gates about the codex, I could not help wondering these things, and my eyes were drawn to a portrait displayed prominently on the wall directly above Gates's desk, a portrait of Leonardo da Vinci. Gates clearly feels an affinity with the Renaissance genius and is happy to admit that when he was in his early twenties and first read about Leonardo's life and work, he was "blown away" by it. So his purchase of a piece of work created by Leonardo himself might be explained simply by the fact that he *can* do such a thing. Many of us would like to own something produced by our heroes, but in most cases we could never contemplate having the resources to do so. Yet, as I sat face to face with this, one of the richest men who has ever lived, in what is actually a rather modest office, a man dressed casually, a man doing a convincing job of projecting the image of a thoughtful, cerebral individual who just happens to be CEO of a vast corporation, I could not help but sense that money in the bank, vast power, and the adulation of the clever young clones who fill Microsoft's Redmond offices and want to *be* Bill Gates were not enough.

Leonardo had been recognized during his lifetime as a great artist and an imaginative engineer, but he had accrued little wealth, and, although he was certainly not penniless when he died, he had remained dependent upon the support of rich patrons his entire life. Perhaps something in Gates, something almost imperceptible, admires the purity of a life such as Leonardo's, an existence in which the expression of genius was enough. If this is true, then Leonardo's life and contributions might represent for Gates an image of unadulterated creativity, a white canvas unsullied by the grime of commerce, a symbol from another world, a place he may go only when he sits in private silence and peers through the Perspex case enclosing his priceless *Codex Hammer*.

$\mathcal{B}$Y THE BEGINNING OF THE 1990S Microsoft had evolved into one of the most powerful forces in the commercial world and was perceived as the leading player in the global software market. But, in achieving this, the company had parted acrimoniously with the onetime master of the industry and Gates's childhood idol, IBM.

At the core of their dispute was the application of competing operating systems. Problems were seeded in 1986, soon after Microsoft went public. Big Blue wanted a new operating system to supersede the aging DOS and had already laid much of the groundwork for a system it called OS/2. IBM had plenty of extremely capable computer scientists and designers on the staff, the men who had developed OS/2 thus far, but for reasons never explained, senior IBM management was convinced it needed Microsoft to perfect the software. However, Microsoft was already developing its own successor to DOS, a system with the development name Windows.

Gates could not contemplate turning down IBM's approaches. He was still motivated by his obsession with the corporation and was also far from convinced that Windows was a worthy successor to DOS: Early reviews of the prototype had not been encouraging, and the product was already behind schedule. So he hedged his bets, accepted IBM's overtures, *and* continued to develop Windows.

Microsoft worked in parallel on the two projects and continually reassured IBM that OS/2 was its first priority and its own operating system was a very poor second. In 1987 OS/2 was launched with great fanfare, at least from IBM. But although it received critical acclaim within the software industry as well as the admiration of reviewers, it completely failed to inspire the public. Undaunted, IBM threw tens of millions of dollars at the product, but still it did not impress.

A few months after the launch of OS/2, Gates had convinced IBM that Microsoft should release Windows. He performed this trick by making the highest levels of the IBM hierarchy believe that OS/2 would be able to dominate the higher end of the PC market, especially those machines whose owners had installed upwards of a thousand dollars' worth of memory upgrades. Here the more sophisticated aspects of OS/2 would truly come into their own. Meanwhile, he claimed, Windows should be viewed as an ideal operating system for less powerful machines. Because of an endemic elitism in the higher echelons of IBM management, senior administrators failed to recognize that the PC would make its greatest impact in just that sector of the market Gates claimed was most suited to Windows.

But the desire to succeed with Windows was not Gates's only motivation. By this time he had grown to despise IBM. When IBM launched the first PC soon after establishing its relationship with the fledgling Microsoft, Gates had asked to be included in formal festivities. He was refused and instead received a letter that read: "Dear Vendor, Thank you for your valuable contribution."

Even after Microsoft was invited to develop OS/2, IBM's senior management acted as though Microsoft was a bit player. In 1986 a dozen Microsoft executives flew to Miami for a conference to discuss the new operating system, only to discover that Big Blue had made space for just two representatives from the company. The others had to return to the airport. When Gates came to speak, he was allocated only six minutes onstage, the time IBM had granted to very junior representatives.

Still, Gates believed Microsoft and IBM should forge yet stronger links. In 1987 he offered IBM a 30 percent stake in his company. Astonishingly, Big Blue turned it down. This was the worst decision IBM (or perhaps any corporation in history) has even made. Not only because today those shares would be worth in the vicinity of $60 billion but because the company's arrogance was perceived by Gates as the final insult. It is possible that from that moment on he began to see a time not far ahead where he would grind IBM into the dirt; any fond memories of the World's Fair of 1962 had long since dissolved.

The early form of Windows, Windows 2.0, had been justifiably mauled by the critics, and it was viewed as incredibly unreliable, but the computer-buying public embraced it almost immediately. Windows was rightly perceived as a revolutionary new approach to operating software; it was relatively easy to use and far more powerful than DOS had ever been. By 1992 it was outselling OS/2 by nine to one. It is estimated that IBM lost more tens of billions of dollars because it could not fully develop its second-generation operating system. It could only watch amazed as OS/2 was superseded by Microsoft's solo effort. And Big Blue suffered the consequences.

In 1991 IBM announced its first money-losing year since the company was established. The following year it published losses of $5 billion, the largest in American corporate history. In 1993 it beat its own record with losses of $8 billion, by which time it had shed almost half of its staff. The stock value of IBM was passed by that of the company it had paid to help create OS/2.

As Microsoft grew steadily more powerful, it acquired a growing roster of enemies; throughout the industry anti-Microsoft sentiments became more intense. Loose affiliations of some of the biggest computer companies in the world formed and dissolved in a fluid opposition to Microsoft, but their common belief was that the growing power of Gates's company would eventually strangle the industry. Some companies, often strange bedfellows indeed, would work together against Microsoft in a more thorough way, speaking out against new Microsoft products they thought inferior to less successful software designed and produced by small start-up companies. Microsoft's most vigorous opponents were eventually given the acronym NOIS—referring to Netscape, Oracle, IBM, and Sun. Gradually NOIS became NOISE—Netscape, Oracle, IBM, Sun, . . . and Everyone Else.

As each year of the 1990s passed and Microsoft seemed to acquire a shield of invincibility, its opponents became increasingly vicious, their attacks more personal. One of the unusual things about the computer industry is that many of the leading companies have CEOs with high media profiles; the interested public directly link the CEOs (men who often founded the companies) with the corporations, so that Bill Gates *is* Microsoft. The same applies to the heads of corporations opposed to much of what Microsoft does. Like Gates, these men are the mouthpieces of their companies, they are usually the major shareholders, most are multibillion-aires and men with very strong opinions.*

Larry Ellison, the man at the heart of the rivalry discussed in this chapter, was never restrained in his feelings toward Microsoft and Bill Gates. He despised the fact that Microsoft managed to infiltrate global markets with products that were, he claimed, inferior to others less commercially successful. He describes its method as following four phases. In Phase One, Microsoft dismisses as impractical an idea created by someone else. In Phase Two, it begins to admit that there may be a chance the idea could work. Phase Three starts when Microsoft releases a competing product, and with Phase Four, Microsoft claims it was their idea in the first place. "The thing I find most contemptible is Bill's lying, this thing about innovating," Ellison declared in a speech delivered at Harvard. "It makes me want to puke."[13]

Others have been equally scathing. Scott McNealy, CEO of Sun Microsystems, stated in a speech: "To kill Microsoft—that's the top priority for us."[14] The computer journalists Charles Ferguson and Charles Morris called Gates "the most hated man in the computer industry."[15] Strong in his belief that Gates and Microsoft befriend and then jettison potential rivals or anyone who threatens them, Pete Peterson of WordPerfect described Gates as "the fox that takes you across the river and then eats you."[16] Philippe Kahn, who was the CEO of Borland International, once the third largest software company in the world, was far more blunt. He claimed: "When you deal with Gates, you feel raped."[17]

But whatever the critics have said, and in spite of the aggressive attacks of Microsoft's rivals, Gates and his company seemed impervious. From the early 1990s it was almost impossible to pick up a business or computer magazine without encountering the name Microsoft or Bill Gates, and in many profiles the emphasis was not on how Microsoft and its competitors were changing the world, but on how rich Gates and his colleagues were becoming.

Admittedly, the statistics make stunning reading. In 1999 alone Bill Gates earned an estimated $20 billion, or almost $6 million an hour. Observers are fond of calculating what this means and coming up with illus-

---

*Bill Gates stepped down as CEO of Microsoft in 2000.

trations, such as the fact that if Bill Gates saw a $500 bill flutter to the ground in front of him as he stepped out of his car, it would not be worth his time to pick it up. Others have noted that, if he chose to, Gates could buy every major league team in the four key U.S. sports—football, basketball, baseball, and hockey—and he would still hold the number-one position on the *Forbes* list of richest Americans. It has been said that if Gates continues to grow richer at his present rate, he will be the world's first trillionaire by the age of forty-eight, in 2003. And, by this reckoning, in the year 2020 Gates will own everything in the world.

When Bill Gates married the former Microsoft employee Melinda French in a private ceremony on the Hawaiian island of Lanai on New Year's Day 1994, the story made the front page of *The Wall Street Journal*. At this time there was also press hysteria over Gates's house then under construction on the shores of Lake Washington. Again the interest revolved around its astronomical price tag, placed by some at close to $100 million after Melinda's last-minute modifications.

By the time I first met Gates in 1995, he was certainly the most famous man in Seattle and one of the most important figures in the United States. In his hometown everyone seemed to have a Bill Gates story; shop assistants claimed their female friends had once been on dates with him. Others would tell how Melinda Gates was fond of instructing the builders at the house to move trees lining the drive a few inches to the left only to order them moved a few inches to the right the following week. The local newspapers often carried photographs they claimed to be of a drunken Gates in one nightclub or another. In fact, it was a Bill Gates look-alike who delighted in trying to embarrass the original.

Yet Bill Gates has made mistakes and came very close to losing his remarkable touch when he almost missed the importance of the Internet. In the hardback edition of his book *The Road Ahead*, published in 1995, Gates makes no direct reference to the Internet. In fact, he refers to the *Information Superhighway*, a term that faded almost as quickly as it was coined and now carries an amusingly anachronistic ring.

The reviews of *The Road Ahead* were predictably merciless, and every one of them highlighted the fact that Gates had made no serious mention of the Internet. But the damage was far more extensive than unkind comments in the literary review sections of newspapers and magazines. Respected reporters began to criticize Gates and his company for being slow to realize the importance of what others even then knew would be a revolutionary change in computing and communication. The crisis point came when one of the chief analysts for Goldman, Sachs made the extraordinary move of taking Microsoft off his list of recommended shares. The share value of Microsoft plummeted, and Gates's personal fortune dropped by $2 billion in one day.

However, perhaps even more astonishing than this, massive error on Gates's part is the way he bounced back. Within little more than a year, he had turned Microsoft around, embracing the Internet by buying up the companies and the know-how at the forefront of the new wave. Microsoft spent hundreds of millions of dollars, but these acquisitions led to profits of tens of billions and saved the company. More than anything else Gates has achieved, this reversal of potential disaster shows the man's ability and energy.

A year after our first meeting, I met Bill Gates again when I was invited to three days of press conferences and talks in Seattle organized by Microsoft. Each day was filled with lectures on every aspect of the Internet delivered by phenomenally young computer experts all trying to look like Gates, adopting his mannerisms and referring to new products as "super good," "super cool," or "awesome," just like the boss. In private conversation these executives would describe "enemy" software (that produced by rivals) as pieces of "brain-dead shit." On the last evening of the conference, Gates himself gave a talk and peppered it with so many references to the Internet, servers, browsers, downloads, and uploads people could hardly believe that only a year earlier he had been slow to appreciate the power of the Net.

Gates's near disaster had been precipitated by two errors. First, he had been looking away. He had allowed himself to become preoccupied with the traditional flow of software advances Microsoft had initiated back in the late 1970s, and this had distracted him from the Internet revolution coming up on the inside track. The second reason was more serious, and rooted in the central ethos of the Internet.

The British physicist Tim Berners-Lee is the man seen as the originator of the World Wide Web. Scientists and academics had used electronic mail via the Internet since the early 1980s, but in 1989 Berners-Lee created the idea of a global network through which users could access documents and an interlinking web of information. By 1993 the first user-friendly browser (named Mosaic) had been produced. This allowed users to navigate the Web with ease. Then in 1995 Netscape, founded by one of the co-creators of Mosaic, Marc Andreessen, began selling the successor to Mosaic, a browser they called Navigator.

Gates's failure to realize the potential of the Net came because, in its easiest incarnation, it was seen as a free medium, a way people around the world could communicate and share information totally free of charge. Gates simply could not understand the concept of anything being free, and if the Net was truly free, he could see no reason for himself or Microsoft to become involved with it. He woke up to the power of the Internet only when the press began to proclaim it to be the next great computing and communications revolution.

Bill Gates survived these errors of judgment in a thoroughly unforgiving business environment because he employed his considerable battery of assets and skills with lightning speed. By 1995 Microsoft had vast capital at its disposal and was able to buy the very best brains in the industry. However, for one rival in particular, this moment was clearly the right one to strike with another threatening innovation. While Microsoft had been occupied trying to find a path through the greatest challenge it had ever faced, Larry Ellison, CEO of Oracle, and a man who resented Bill Gates more than anyone in the industry, had stumbled upon an inspired idea.

LARRY ELLISON HAS BEEN referred to as "Software's *Other* Billionaire," but in many ways he is a far more interesting character than Bill Gates.[18] One former employee described him as "very self-centered."[19] Another onetime colleague, Stuart Feigin, has said of him: "If he hadn't made me rich, I'd probably hate him, because he's obnoxious."[20] No one then could call Ellison "a charisma black hole," as one journalist has Bill Gates.[21]

The one thing that makes Bill Gates seem larger than life is his mind-boggling wealth, but Ellison lives the dream. Gates displays some of the trappings of immense success—his enormous house and expensive Porsche—but much of this has been affectation encouraged by his image makers; until the mid-1990s Gates largely shunned ostentation.

In contrast, Ellison was keen on extravagance even before he made his first million. Whereas Gates is most comfortable in plain slacks and a sweater, Ellison favors Savile Row suits and handmade brogues. Gates is fond of his Porsche, but Ellison maintains a fleet of some of the world's most exclusive cars. Ellison has a home to rival Gates's Lake Washington mansion and several other vast houses dotted around the globe. Gates is living with his first wife and daughter, Ellison has gone through four marriages, and when he appeared on *Oprah* in 1996, he claimed he was looking for wife number five, even though he had a steady girlfriend at the time. Gates is almost entirely monochrome, Ellison lives a Technicolor existence.

The early life of Lawrence Ellison has been obscured in part by Ellison himself. Remarkably, for a man who could manifest almost any fantasy he wishes, Ellison is so extreme he is not content to make dreams reality, he needs to turn reality into dreams; and he does this most conspicuously with his own past. According to Ellison, he grew up in one of the worst areas of the South Side of Chicago, where street violence was common. Deserted by his mother, he was raised by relatives and given a ramshackle education. However, almost all of this is exaggeration.

Born in August 1944, Ellison was indeed adopted by his mother's aunt and uncle after his father walked out on the family, but they were a relatively wealthy and stable couple who lived in the affluent North Side of Chicago. His adoptive father, Great-uncle Louis, had been a Russian immigrant who changed his name to Ellison after Ellis Island.

According to Larry, Louis was a millionaire property developer who lost his fortune in the Great Depression but then ran as a candidate for Congress. However, there are no records of Louis's candidacy in Chicago or anywhere else, and documents covering the family's property dealings have not survived. Those who knew the family during the 1940s and 1950s report that Louis Ellison was a rather modest accountant.

Larry might have done well in high school, but his mind-set was too unconventional for the place and the time. Complying with the accepted rules offered at least a chance of success, but Larry liked to do things his own way, and as a result he is remembered as a rebellious boy with few school friends. Those he did spend time with became good companions with shared enthusiasms, but he found it almost impossible to open up emotionally. When his adoptive mother died from kidney cancer, he did not even tell his friends; sharing his pain with them was, he later claimed, quite unthinkable.[22]

He had been close to his adoptive mother but hated his uncle Louis. He was constantly rebelling against the rules his adoptive father imposed, just as he did against the staid methods of his school, so his father constantly made him feel inadequate, prompting him to cite the age-old adage that his son "would amount to nothing." When Larry accidentally scored a basket for the opposing team in a school basketball game and the mistake made the local paper, Louis Ellison kept the clipping and took inordinate pleasure in bringing it out whenever the family had visitors.*

The stimulus that propels Bill Gates toward achieving massively and earning more money than he could ever possibly need is difficult to explain, but we need go no further than this tale of humiliation to see how a deep-rooted resentment manifested itself as Ellison's fanatical drive to succeed. It might also go some way to explaining why Ellison drapes himself in finery worthy of a Roman emperor.

But even with this impetus, Ellison was a very slow starter. He claims to have harbored dreams of becoming a doctor, and for a short time he attended the University of Chicago. It was here, in 1965, while enrolled in a physics course, that he was introduced to a computer for the first time and taught to program a 1401, the first IBM machine that used transistors instead of valves. He took to this new technology immediately and soon began to earn a few dollars from small programming jobs. Within a year he had dropped out of the university, but, unlike Gates, Ellison did not move straight into an entrepreneurial career as a software designer (the very concept of software was nothing more than an academic term in 1965); that prospect lay more than a decade in the future. Leaving Chicago and his despised father, in the summer of 1966 Ellison headed to California, to seek excitement and, he hoped, to make his fortune.

Arriving in San Francisco in an aquamarine Thunderbird at the start of the hippie era, Ellison was quite taken with the idea of free love but not greatly inspired by LSD, antiestablishment politics, or the fashions of the Haight-Ashbury crowd. Nevertheless, he spent the late 1960s and the first

---

*Incidentally, this story appears to be one of the more accurate tales from Ellison's childhood, verified by those who knew the family.

half of the 1970s, drifting from one job to another, enrolling in the odd college course, and scraping by. He earned a little from occasional programming work, which usually simply involved fixing coding problems for small companies and accepting a modest fee.

And so it continued. In those days Ellison, the mogul was so far from emerging it was almost as though the pin-striped executive who today heads one of the most valuable corporations on Earth and the undisciplined and unfocused Larry Ellison living in San Francisco were not merely aspects of the same individual but entirely different human beings.

In the summer of 1967, Ellison met a Berkeley student named Adda Quinn. Within a few months he had proposed, and before the year was out they were married. It was typical of the man's spontaneity, his way of thinking without planning, and, as with almost everything he tried to commit to at the time, the marriage was doomed.

Ellison was earning almost nothing, but he had already begun to cultivate expensive tastes. He found it difficult to meet rent payments, but he found a way to obtain a loan for a thousand-dollar mountain bike; he was frequently seen dining at expensive restaurants and even traveled to Beverly Hills to have plastic surgery on his nose. It was a lifestyle that could not be sustained. By 1974, as Bill Gates was entering his second year at Harvard, Ellison reached a critical juncture. Adda was about to leave him, they had accrued huge debts, and Larry was still drifting between part-time jobs. That summer he turned thirty.

Encouraged by his wife, Ellison agreed to undergo therapy, and although it came too late to save his marriage (Adda left that autumn) it appears to have propelled him toward some form of epiphany. He had been, he learned, yearning to prove his father wrong but had possessed no clear vision of how he could achieve this. Gradually he began to realize what his strengths and weaknesses were and how to manage both.

Within months Ellison had walked away from his corrosive lifestyle and found a promising position at the Silicon Valley audio and video equipment company Ampex. The job involved writing code for a data storage system Ampex was developing that would be capable of storing a terabit (one trillion bits) of data on videotape. Here, two years after the breakup with Adda, he met and married Nancy Wheeler, the daughter of a wealthy Kentucky businessman. A few months later he left Ampex when he was offered the position of vice president of systems development at a small technology firm called Precision Instrument Company (PIC). And it was from this career move that he stumbled upon the business opportunity that changed his life.

Precision Instrument needed software to improve one of its products, a data scanner called a PI180 that was used by large corporations to load information into their mainframe computers. However, estimates for producing this software began at $2 million, a sum a company the size of PIC could ill afford. Ellison quickly saw this as an opportunity to prove himself. He called two friends, both talented programmers he had met at Ampex, Bob Miner and Edward Oates, and proposed that they create the software PIC needed. Ellison knew that between them they could produce it for a fraction of the best estimate, and he soon convinced Miner and Oates. They put in a bid of $400,000, and a nervous PIC took it. Ellison, Miner, and Oates formed their first company, Software Development Laboratories, Inc., and immediately set to work producing the code to operate the PI180.

With this platform Ellison became convinced he had hit a rich seam of opportunity, and with his partners he began actively searching for new and profitable means to exploit what they were convinced was the start of a rush for innovative software. As Bill Gates was finding ways for the newly created Microsoft to link the muscle of IBM with the first wave of software users in small corporations, Ellison, Miner, and Oates were finding a different way into the software industry.

From the beginning Gates was almost exclusively committed to designing and selling software for the personal computer market, and he believed the greatest impact of the computer revolution would come from this sector. During Microsoft's formative years (the late 1970s and early 1980s), Gates expressed no interest in his company creating software for the large computers used in industry and government. If he had followed the path to designing for mainframes and multimillion-dollar machines, he and Microsoft may well have clashed with Ellison two decades before they did. However, Ellison, Miner, and Oates were as uninterested in talk of the PC as Gates was in mainframe and industrial networks. Ellison and his partners were concerned only with databases. As a consequence, for several years as Ellison's and Gates's companies expanded, they were hardly even aware of each other.

In a direct parallel with Gates and Allen, Ellison, Oates, and Miner found their entrée into what proved to be the remarkably lucrative industry of database software by being aware of the latest developments coming from the labs of IBM. Of particular interest to them were the theoretical ideas of Ted Codd and the System R group, then working on databases.

Although these developers were part of a gigantic international corporation, in some ways they (along with many other research groups within IBM) behaved like academics and published their findings as though they worked at a university. Because of this openness, Ellison and his partners were able to acquire all the information they needed to produce their own version of the database. Investing some of the capital from their first job writing software for the PI180, they went on to produce the first commercial database, a product they called Oracle.

Years later IBM's ethos of making public its research was considered by many to have been a huge mistake, and the way Oracle (the company) was founded in 1977 and Gates's Microsoft cashed in on this practice were used to demonstrate the foolhardiness of this approach. But others believed that a company the size of IBM was right to open up debate and communication in computer research (precisely in the way of pure science) and that in this direction lay the hope of more rapid advance that would lead to a broadening of markets rather than any weakening of the company.

The problem with this argument is that the link between pure research and commercial exploitation is so strong in the computer industry that companies like Microsoft and Oracle could not help but utilize the research funding of the industry giant of the time. In the pure sciences the results of physicists, biologists, and chemists are several stages removed from direct commercial exploitation.* It was a case not so much of Gates or Ellison stealing from IBM as of Big Blue offering them its research on a plate.

Many criticize Ellison for having taken IBM's idea for a database and further suggest he has done almost nothing to deserve his huge success. But this critique ignores the fact that although Codd's paper (and the follow-up from the System R group) was widely circulated in the industry, only Ellison and his partners succeeded in producing the first commercial database from this seminal research. He was certainly in the right place at the right

---

*Except perhaps in the pharmaceutical industry.

time, but he also understood that this piece of science could be exploited for financial gain. As Ellison's biographer, Mike Wilson, has highlighted: "Yes, IBM gave him the idea. But it did not give him six billion dollars. He made himself rich through ceaseless work, brilliant strategy, unrelenting optimism, and ruthless determination."[23]

In the same way Gates had stepped into the PC market at precisely the right moment, Ellison knew instinctively that the research IBM had produced would allow him and his partners an opening to exploit the database market. He, Miner, and Oates quickly dropped the name Software Development Laboratories and decided to rename the company after its most successful product, Oracle.

"To Ellison, life was a never-ending contest," Wilson has said, "every day, a new opportunity to prove himself. He would compete with anyone, any time, over anything."[24] And such competitiveness was an essential component in Oracle's success. In spite of having a commercial product in a wide-open market at exactly the right moment, it was an uphill struggle for a company with almost no initial capital investment. The handful of staff at Oracle had to work shifts because they could not afford to rent more office space, and sometimes salaries were paid late.

Indeed, many around him were convinced Larry Ellison and Oracle would fail even as their earliest software was being made ready for shipping out. All these critics were wrong, but none more so than Ellison's second wife, Nancy. The couple separated in 1978, and according to the terms of the divorce settlement, Mrs. Ellison was free to sell back shares in her husband's company. She asked her husband for a flat payment of five hundred dollars, and when he accepted she believed she had struck a good deal. Today those shares would be worth several billion dollars.

As well as struggling to balance the books, Oracle had competitors who were scrambling for the big contracts. But fighting off the competition was what Ellison did best, and he simply refused to be intimidated. During the early 1980s, Oracle's biggest rival was a software company called Ingres. Ingres had a turnover fifty times that of Oracle, but Ellison told everyone who would listen that within five years his company would overtake Ingres. His competitors laughed, but five years and three months later, Oracle had the greater annual turnover. When the sales figures were released, Ellison told his executives: "It isn't enough that we beat Ingres on a sale, Ingres had to go out of business. I want them on their knees. Begging for mercy. Pleading for their lives. Confessing their every sin." Then he began to chant: "Kill, kill, kill . . ."[25]

During the 1980s Oracle's sales figures more than doubled annually, and by the time the company was floated in March 1986, Ellison was worth $93 million on paper. But this was not enough. He looked back over the record: Oracle and Microsoft had been established in the same year 1977. Each year both companies had grown at almost the same rate (100 to 150 percent). By coincidence, they had even gone public within twenty-four hours of each other, Oracle on March 12, Microsoft on March 13. But even $93 million was not a powerful enough salve for Ellison's obsessively competitive zeal. He watched the Oracle shares open at $15, and by the end of the first day of trading they had risen to $20.75. Then he looked at Microsoft's flotation. Their shares had opened at $21, and in a single day they had rocketed to $28. Gates, Ellison calculated, was worth in excess of $300 million.[26]

That day the battle lines were drawn.

DURING THE EARLY 1960S, John F. Kennedy's government was referred to as Camelot; it had a fairy-tale aspect to it. During the second half of the 1990s, modern government and business in the United States bore some comparison with another, grittier medieval world, one both oligarchical and morally nebulous. Bill Clinton was king, and the media bosses, the old-style industrialists, and the billionaires of Silicon Valley and Seattle were his warlords and chieftains.

One of the most powerful warlords, Larry Ellison, had much in common with Bill Clinton. Close contemporaries, both had been raised, at least in part, by surrogate parents, both are exceptionally bright (although Clinton is academically more accomplished); both are obsessive characters and possess huge reserves of charm. Both Ellison and Clinton are supremely ambitious, and both have struggled and clawed their way to power. It is easy to imagine the two men sharing a beer. In 1993 Ellison began a four-year legal battle with a former female employee who had filed for wrongful dismissal and negligent mental distress. She ended the battle convicted of perjury and creating false evidence; Ellison's "monarch" is no stranger to legal fights with female associates.

Bill Gates and Larry Ellison are very different personalities, but in their approach to computer technology and business they have much in common. Each is utterly ruthless; each views business as war. They do not know each other well except as business adversaries at the cutting edge of technology. They are known to have had only one private meeting, during the 1980s, when Gates asked to meet at Ellison's house. The reason for the meeting remains unclear, even to Ellison, but he suspects it was simply for Gates to gain some intelligence on the activities of Oracle just as Ellison's company began to emerge as a powerful force in Silicon Valley.

Ellison claims the idea for the device at the center of his dispute with Gates, what he later called the network computer, or NC, came to him during a conversation with President Clinton. He is fond of drawing the analogy between this encounter and President Kennedy's decision to send an American to the Moon by the end of the 1960s. "The first time I used the term *network computer* and talked about a five-hundred-dollar computer was in a conversation with William Jefferson Clinton, the president of the United States, out here last August [1995]," Ellison told one journalist. "What you have to do," he explained, "—is attach human beings, students, kids, to the Internet. How do you do that? . . . I proposed that the president challenge our industry to build a machine for five hundred dollars that could be used in all the schools. That would be affordable and very, very easy to use. He could be like John Kennedy saying we're going to put a man on the Moon by 19——, choose any year. Bill Clinton could say we could put a computer on every student's desk by a certain year."[27]

Ellison could not have failed to notice the resonance of this statement with Bill Gates's much quoted comment that he wants to put a PC in every home, yet Ellison managed to sound purely philanthropic rather than like a man out to make even more money. However, this account is another example of Ellison's penchant for rewriting his personal history. He claimed he discussed the idea of the NC with President Clinton in August 1995, but the meeting occurred on September 21, two weeks after he had publicly announced his idea. The kernel of Ellison's concept was to produce a cheap computer that dispensed with much of the elaborate technology found in a PC. He believed most people did not want to have to pay in the region of two thousand dollars for a machine that had more power and gimmicks than they would ever need. He envisaged simple terminals, little more than screens and keyboards, that would retail for under five hundred dollars and would be linked up via the Internet to a big computer called a server. The NC would need only enough memory to run spreadsheets, word processors, and the hardware at the user's fingertips. Documents would then be stored by the server and retrieved with a password.

The idea was simple, elegant, provocative, and startlingly original and had evolved in Ellison's mind from three integrated elements. First, a set of cutting-edge changes in computer technology had made it feasible. Second, the dream of linking two existing communications industries (computers and TV) was foremost in Ellison's mind at the time; and third, was his lust for primacy, an almost pathological desire to beat Bill Gates.

The cutting-edge technology that made the idea possible was a new computer language called Java that had been created by Scott McNealy's Sun Microsystems. Java may be thought of as the Esperanto of the computing world, an easy-to-understand language that allows all forms of computers to communicate rapidly by chopping up large chunks of data into manageable pieces that are then reassembled in the user machine. Although it had not been created solely for the Internet, Java was the perfect language for the World Wide Web, and we may now see its arrival as one of the key factors in the emergence and staggering expansion of the Internet during the 1990s. One expert has referred to it breathlessly as "a fundamental break in the history of technology."[28]

Java provided the means by which Ellison's simple terminals, his network computers, could communicate with the "master" computer or server via a modem. The user of the NC might work on a document and then store it with the server by sending it down the line just like an e-mail or an attached document. Java would convert the document into a transportable form and the user could secure the files with passwords.

The second element in Ellison's thinking came from one of the most talked about concepts in Silicon Valley for some time, linking television with the Internet so that consumers could log on in their sitting rooms with a minikeyboard. The previous year, 1994, a small start-up company called WebTV Networks had begun to open a market for the relevant software and a stand-alone box that could turn a TV set into a computer.

The third element came from Ellison's desire to stymie Bill Gates. Angered by the global hysteria Microsoft PR had generated for the launch of Windows 95, Ellison needed to hit back, to strike at the heart of Microsoft's core business. How better to do this than create a globally popular computer that did not need Windows as an operating system? As one Oracle executive expressed it, Ellison wanted to "stick it to Bill."[29]

Ellison announced his idea to the public during a talk in Paris eleven days after the launch of Windows 95, and there was certainly no coincidence in the fact that Bill Gates was also booked to speak that day. Ellison stood beside a podium with a laptop linked via the Net to a powerful computer at Oracle headquarters. He began to talk about "a paradigm shift" in the world of the computer, of how the launch of Windows 95 was the high point of the PC and the personal computer was an unnecessary and overpriced extravagance. In the near future, the sort of graphics, video, and sound Ellison's audience was then seeing projected onto a screen from his laptop could be presented on a simple terminal with only a small memory. There would be no need for a hard drive, no need for a CD-ROM port, and of course, by implication, no need for an operating system like Windows 95. The terminal would be linked to the Web, and all memory-consuming material stored on a network. "This," and as Ellison spoke he grappled for the term, "this personal . . . this network computer" would retail for no more than five hundred dollars, and Oracle, experienced in transferring large chunks of information from computer to computer, could operate this new and revolutionary system.

Ellison's speech certainly had the initial desired effect; it created a frenzy in the computer media and beyond. His idea was discussed in *The New York Times* and *The Wall Street Journal,* and it would later find publicity on *Coast to Coast,* the *NBC Nightly News,* and, most bizarrely, *Oprah.* The idea of the network computer, a paradigm shift that would radically alter how we all used computers, and of course make more billions of dollars for the computing industry, immediately captured the imaginations of those who heard about it.

Bill Gates, taking the stage a few minutes after Ellison's announcement, did the only appropriate thing; he ignored the idea completely and went on with his prepared speech. But Gates must have known he could not leave the stage that afternoon without giving some response to Ellison. At the end of his speech, he was interviewed onstage by the Data Corporation analyst John Gantz, who asked the CEO of Microsoft what he thought of the idea of an Internet device that would threaten the PC. Gates chose his words carefully and did not mention Ellison or Oracle by name: "People who think we are going to have dumb terminals in the world of the Internet," he said and paused, searching for an appropriate rebuttal, "I just don't agree with that," he finished lamely.[30]

The media and the industry pundits were split in their reaction to Ellison's announcement. "Are you prepared to do without a floppy, hard, or CD-ROM drive?" wrote one astute pundit, "be unable to compute—or even access data—when the server goes down? Watch performance slow to a crawl during peak hours?"[31] Others went further. "In a few years, those NCs will make great doorstops," one expert announced.[32] Another called Ellison's computer "the $500 rip-of" and went on to say: "I don't think it can be done. Hey, Larry, the emperor has no clothes!"[33] But others were supportive. *Newsweek* carried a glowing report on the idea, declaring: "No one actually needs a complicated 'mini-mainframe' on the desktop anymore, the thinking goes. Instead people might prefer something easier and more friendly, a sort of information appliance that . . . is as nonthreatening as a television," and it went on to claim: "The PC is dead. It's the horse and carriage of the Information Revolution."[34]

Meanwhile, other computer technology companies in Silicon Valley were taking seriously Ellison's idea. By May 1996, some eight months after the announcement of the concept, four major players in Silicon Valley had joined Oracle to help produce the new device. Sun, IBM, Apple, and Netscape created Network Computer Inc. and allocated research funding for the production of the computer terminal, the software to run it, and the Internet links to an appropriate server. Within the industry this group, whose intentions had been apparent from the start, was dubbed "the anti-Microsoft coalition."

During 1996 the world seemed unusually rosy for Ellison but tinged with anxiety for Gates. To all outward appearances, Ellison and Oracle were working at full steam on a revolutionary computer system and had the support of many of the key figures in Silicon Valley. Gates was still suffering criticism of his book, supervising a hurried paperback reedit, and, after the shock of being caught ill prepared for cyberspace, he was working even longer hours than usual to turn his company into an Internet-based corporation.

Gates has professed an almost total lack of interest in popular culture or TV, so one can only wonder what he thought when Larry Ellison appeared on *Oprah* in September 1996.* Ellison had certainly pulled off a publicity coup, arranging for the program to feature him driving his Bentley convertible and wandering about his yacht, and sitting in the cockpit of his Italian jet fighter. He then announced that he would be giving a network computer to every pupil at the James Flood Science and Technology Magnet School close to Oracle Headquarters. Each of the 294 children at the school would have his or her own terminal. "The whole idea is to give each and every one of those students a computer to use at school," he told Oprah Winfrey. "The computers we currently make, personal computers, are rather expensive and very, very complex. So, we've introduced this new class of computer—I have a copy of it with me—for students, for normal human beings. It's very lightweight, very low-cost, and the kids can use it at school, and they can pick it up and unplug it and take it home with them and then use them at home."[35]

But as valuable as this publicity undoubtedly was, Ellison's statements were as superficial as most other items on talk shows, for they merely highlighted Oracle's thorniest problem with the NC; Ellison had made promises about the network computer he simply could not keep. First, he had price-tagged the NC at under five hundred dollars. This was not a realistic figure, and at around the same time, Intel-based PCs were appearing on the market for under a thousand dollars, diluting the impact of Ellison's price promise. Second, he had promised delivery of the first NCs by the autumn of 1996, but as that date passed the new delivery date became the end of the year, then the spring of 1997.

During the early 1990s Oracle had acquired something of a reputation for making promises it failed to keep; new software was invariably late, and

---

*This really is no exaggeration. Gates once told a gathering of twelve hundred media executives at a Radio and Television Society function that he did not like TV and that, having prepared for his talk by viewing tapes of the *Late Show with David Letterman*, he was assured his view had been correct and TV was "stupid" (James Wallace and Jim Erickson, *Hard Drives: Bill Gates and the Making of the Microsoft Empire* [New York: HarperCollins, 1993], p. 326). On another occasion, when the name Freddy Krueger came up in conversation, he asked: "Who's Freddy Kruger?" (ibid., p. 65). Ironically perhaps, by the end of the decade, Gates was joining the TV circus for the sake of PR. When Microsoft was faced with an antitrust ruling in 1999, he appeared on *20/20* to show his wedding video and sang "Twinkle, Twinkle Little Star" for Barbara Walters.

users had long learned to expect glitches and delays, but with such a high-profile, extensively publicized new product, these delays were becoming embarrassing. In February 1997 one of the most respected journalists in the field, John Foley of *Information Week*, declared: "For Larry Ellison, it's time to deliver the goods."[36] In response, Oracle senior executive Jeff Henley said: "If nothing happens we're going to look kind of stupid in a year or two. But that's Larry. Larry's got balls, and he'll take a risk."[37]

But Ellison's personal ambition and even the support of some of the giants of the industry, such as Sun and IBM, could not revolutionize computing, nor could it make Microsoft bow to pressure. Early in 1997 a widely read report noted that it cost corporations an average of $13,200 per year to run each PC they owned and that adoption of some form of network computer would make sound economic sense.[38] Unfortunately for Ellison and Oracle, this fact had not escaped Microsoft's CEO, and with their usual resourcefulness, the executives at Redmond had deflected what at first appeared a serious assault on their business and turned it to their advantage.

If we accept Ellison's idea of the four phases of any Microsoft shift into new territories, the reaction to the NC would be a classic example. At Microsoft the immediate response to Ellison's Paris speech was derision. Fifty million PCs had been sold that year globally, and the euphoria produced by the amazing sales of Windows 95 was still buzzing around the Redmond campus; how could Ellison be taken seriously?

But clearly Gates did take Ellison seriously, and soon Microsoft had entered Phase Two. It had become obvious to Microsoft that if it was to close the gap created by Ellison's innovation, it would have to acquire the research and development of an innovator in the field. The company did this by buying WebTV Networks, the start-up company that had pioneered linking PCs to television. The price, $325 million, shocked even an industry by then well used to excess.

With this sudden boost to its resources and research material, Microsoft was soon able to produce a rival to the NC, a machine called the Simply Interactive Personal Computer, or SIPC, which could be interfaced with TVs, VCRs, and other equipment. Gates claimed this was not the same as the NC because, although it would be linked to the Web, it was not what he insisted upon calling a dumb terminal. These devices would carry almost all the usual Windows applications.

Soon after the launch of the SIPC, Microsoft moved still closer to the concept of the NC with the NetPC, a simple terminal linked to a network that would soon find its way onto the desks of receptionists, airport check-in attendants, and bank tellers throughout the industrialized world. Crucially, all the Microsoft rivals to the NC carried Windows as a prominent feature, and, unfortunately for Ellison and Oracle, that appears to have been exactly what the public and business world wanted.

Ellison had intended his description of the way Microsoft entered new technology as an insult, a way of highlighting what he saw as industrial plagiarism, but he had also described a masterly application of Machiavellian rules in which Microsoft ended up winning even when the odds were stacked against it and it had been left standing on the starting block. Microsoft had seen, laughed, struggled to catch up, produced the goods on time and to order, and then others began to believe Microsoft had created the idea.

THE FIRST NETWORK computers were sold in the summer of 1997, nine months after they were promised. Because consumers wanted the option of using standard browsers on their terminals as well as the ability to access the server for the NC, the machines needed at least sixteen megabytes of memory. This meant they ended up far more expensive than originally imagined—just under a thousand dollars each—close to the price of a standard PC with CD-ROM, modem, and hard drive. Consequently, the launch of the NC failed miserably.

Ellison had claimed that when the NC was ready he would ship one million machines, but fewer than ten thousand were sold during the first year. Meanwhile, Microsoft was selling many times more of its competing product, prompting *The Wall Street Journal* to predict that within five years three out of four "dumb terminals" (as they are now generally known) would be Windows-based machines. "The irony," it wrote, "is that the anti-Microsoft alliance correctly identified some trends now sweeping the industry ... but it is Microsoft, which at first pooh-poohed the problem, that has most effectively used NC advocates' own marketing slogans."[39] *Forbes* reported: "Three years after Oracle chief Larry Ellison introduced the concept, the NC looks ready for a Smithsonian exhibition on promising technologies that went nowhere."[40]

However, Ellison's effort was not entirely wasted. Ironically, he had handed Microsoft another victory and had enabled it to earn even more money, but he had also contributed something truly innovative to the world of computers. Today many corporations are replacing their expensive conventional PCs with NCs and their successors, and the flat-screen dumb terminal used by corporate employees at the interface with the public—in banks, libraries, airports, hotels, restaurants, and retail outlets—is now ubiquitous.

But Ellison's idea did not initially reach as far as he had hoped or imagined it would. If we ignore the factors that drove him to the NC, we may see that he did genuinely believe the PC was set for obsolescence (as it inevitably is). He had a vision of a global computer network that gave users far greater power than they would have with any mass-market PC, however beefed up its memory.

Yet the most significant problem facing the NC was timing. Less than four years after Ellison's declaration that the days of the PC were numbered, the chairman of IBM, Lou Gerstner, was reported to say: "The PC era is over. The PC's reign at the center of gravity and the focus of investment and innovation in this industry is done. Full-blown PCs are going to be joined by a lot of other end-user appliances that bring computing and Net access to millions of people quickly and easily. In a matter of just a few years these devices will outnumber PCs. And I think it is reasonable to believe they'll have a greater lasting impact than the advent of the PC itself."[41]

Echoing precisely Ellison's dream from 1995 the computer pundit Garth Alexander has said recently: "As the Net's capabilities expand and the bandwidth provided by modems and transmission lines becomes ever broader, many of the storage and computing functions of today's PC will be provided by centralized servers communicating with ultra-simple home-based machines over the Net."[42] And, when asked to comment on this idea, Bill Gates responded: "[In the future] all your personal professional information will be stored in a 'cloud' on the Internet."[43]

Indeed, the NC is a much better idea than the PC. The problem for Ellison is that the concept came far too late; the PC system has become so entrenched, it will run its natural course. One day the Smithsonian will have a model of an NC and a PC.

Yet this is not the end of the story. For, as Ellison always knew, there is more to the computer industry than the PC. Oracle is now enjoying the latest victories in the war between the world's software giants. Ellison's idea of the NC was not workable, but it has led Oracle into a far greater commitment to Internet-related ideas and business development. After some five years in which the company was kept buoyant primarily by its core business of supplying database software, it has developed and launched a slew of exciting software packages that are driving e-commerce to new levels of innovation and profitability. The most important of these products include software that will allow companies to store huge amounts of information on the Web (just as Ellison imagined would be done by individual users working with the NC), along with packages big businesses need to create impressive Web sites and to communicate with each other through the Web.

Today not only does Oracle have 40 percent of the world's database business (compared with IBM's 18 percent) but it is poised, some believe, to overtake Microsoft as the world's number-one software company. Ellison's corporation and Ellison himself are constantly facing criticism; one journalist recently called their projects "vacuous,"[44] but they are leading the field in harnessing the full potential of the Web. This surge in development may be seen with hindsight as having its origins in the idea of the NC.

Most satisfying for Ellison is not the fact that Microsoft is facing the prospect of being split in two because of antitrust legislation, or even that Oracle is growing hugely, but that in April 2000 his personal wealth passed that of Bill Gates. Ellison is today worth $52.1 billion, while Bill Gates's shares in Microsoft are valued at $51.5 billion. Even so, Ellison cannot resist a twist of the knife. Referring to the claims of the U.S. Justice Department that Microsoft restricted the choice of computer users by packaging its own Internet server with Windows, he comments wryly: "They robbed the wrong bank."[45]

LARRY ELLISON AND BILL GATES, two men who between them are worth in excess of one trillion dollars, continue to struggle for supremacy as the king of the cyber-kings. But can we really consider them scientists or what they do as science? Can we legitimately speak of them in the same breath as Darwin or even Edison?

Any attempt to answer these questions is certain to generate controversy. Many believe that the term *scientist* describes only a very specific type of researcher. For example, many physical scientists (especially physicists) bristle at the suggestion that psychologists or archaeologists may be called *scientists*, and they are appalled by the idea of the words *social* and *sciences* being put together. Many scientists and science writers would claim that science is only science if it is rooted in mathematics, but what of the geologist or the biologist who uses mathematics less frequently than a civil engineer? If today money and power drive technology more than ever before, does this tale of innovation at the heart of big business illustrate that definitions of *science* and *scientist* have become eroded? And has any barrier that may once have divided science from technology been pulled down?

Of course, the theoreticians still juggle formulae and mold pure concepts, fueling innovation and technology, but perhaps we should accept the existence of a new species of "scientist." This beast has followed an evolutionary path that began with those who forged the Industrial Revolution, a species that gave us Thomas Edison and other great engineer-scientists of the nineteenth century and has led, in the twenty-first century, to men like Ellison and Gates. These figures do not simply enjoy money as a reward for their creativity; for them money unlocks imagination. They are a new breed, not mere technocrats but techno-moguls.

During some two hundred years, between the first flowering of the Scientific Renaissance and the Enlightenment, science (or natural philosophy, as it was then known) grew out of philosophic reasoning. Before the Enlightenment, the two disciplines were intermeshed, and no distinction was made between them. But now philosophy and science are confused only where pure science becomes so refined it is barely linked to the human experience at all (such as within some of the more exotic theories at the heart of quantum mechanics). Following the Industrial Revolution, technology and technologists emerged from the bedrock of the pure sciences. Whereas Roger Bacon and Aristotle were explorers who had no thought for the usefulness or practicalities of what they did, for Edison science was meaningful only if it could be applied. He was a supreme technologist whose work had nothing to do with philosophy.

Computer scientists, who dominate much of modern technology, are an extreme example of this evolution from philosophy to technology and beyond; they are a highly refined subspecies of what we might call *Homo scientificus*. They possess exaggerated characteristics that enable them not only to adapt to the environment in which they are flourishing but to mold their resources so skillfully they have become extremely powerful, dominant individuals.

These people work in a very different social framework from that of men like Newton or Priestley and have placed upon them demands that would have seemed altogether alien to any "scientist" before the mid-nineteenth century. But if, for a moment, we may allow ourselves to ignore the issue of money, it is possible to see that, like pure scientists, computer entrepreneurs are seeking both a form of Truth and recognition, just as Newton, Lavoisier, or Crick and Watson did. There is a certain, clear beauty in the lines of code that drive an operating system or interpret a database. There is a sublime symmetry within a microchip. The knowledge that technomoguls make vast fortunes from manipulating these beautiful things distracts us from the purity that is also there at the heart of the machines they manipulate and then develop to sell us.

Aside from these matters of definition, many people find it difficult to label the work of Gates or Ellison science because they work as part of teams and every move of everyone on those teams is overshadowed by corporate policy and corporate goals. But do we think any less of Caravaggio and Rembrandt because they used teams of assistants to help with their work? Are the innovations that came from Los Alamos or from the space program (atomic power and satellite communications, to name just two) any less profound because they derived from government-funded projects with definite objectives that had very little to do with scientific imperatives?

Almost half a century ago, Wernher von Braun wrote: "Take a group of people with the same general fund of knowledge, present them with a problem and you will get it done. The idea of a great genius sitting quietly in some corner dreaming up vast secrets is no longer real. So many things in modern science have become so wide in scope, so intricate, that more and more it takes groups of experts to do the work."[46]

Within the context of what von Braun was doing, this is almost certainly true. It took hundreds of thousands of people to send a rocket to the Moon, but I cannot agree entirely with his statement. Science is an evolutionary process; teams of engineers or scientists working on a vast project still need individual genius to link Nature's laws with the needs of humans, to tease out the deepest secrets. Without the ideas of Marie Curie or Niels Bohr, there could have been no development of the atomic bomb at Los Alamos; without the ideas of Lavoisier, Dalton, or Mendeleyev, there would be no plastics or superconducting materials; without Neumann or Turing, there would be no cyber-kings, no Oracle, no Microsoft.

In trying to decide whether the techno-moguls are also scientists, we have to come to terms with the fact that there are many types of scientists. There remain a few hardy souls who work as individuals: those in the theoretical research departments of universities, those who spend their entire careers contemplating the fine structure of the cosmos without ever looking through a telescope or dissecting the mathematics of the subatomic world, without a thought to future applications of such esoteric contemplation.

Complementary to these thinkers are the experimental scientists, the engineers who conduct research into all aspects of the material world. Then there are geneticists who work in large teams, their salaries paid by drug companies or corporations who plan to reap the potential rewards of gene patenting, and there are scientists who work for government projects to develop new weapons. All are scientists, all different in their motivations and the tools they use to prize open the secrets of the universe.

Technology grew at a staggering rate during the twentieth century, and it is set to develop exponentially faster this century. Such exaggerated and rapid change demands a redefining of what science means, what being a scientist entails. In the world of computing, science is finding public and commercial outlets ever more quickly. The path from idea to corporate bank statement is often astonishingly short, but it is surely wrong to believe that, by necessity, this sullies the intellectual purity of what the computer and software developers achieve.

At first reading, it may seem that the dispute between Larry Ellison and Bill Gates over the NC versus the PC was far less important than any of the rivalries in this book. Does it bear comparison with the Manhattan Project or the elucidation of the structure of DNA? In itself perhaps it does not, but it is included to illustrate how the competitiveness of companies that drive our computer-dominated world has an enormous influence upon modern life.

In the short term the PC versus NC battle will alter our lives in little more than a cosmetic sense. But in the future we are likely to see the TV and the computer merge; entertainment facilities, communications devices, and educational tools will be combined seamlessly. Our homes will contain many NCs, just as the average Western home now has several TVs and radios. Today in many industrialized nations, governments are sponsoring ways to link schools to the Internet using simple terminals just as Ellison predicted on *Oprah*.

As we have seen, Ellison's idea has also begun to be used to give large corporations the potential to store huge amounts of information in cyberspace. But, crucially, Ellison's dream of the NC did not simply make him the richest man on Earth; what multinational corporations are doing now with Oracle software will very soon become commonplace; the innovations driven by big business will be adapted for the home and the private office. Within a few years it is likely that a variety of computers will be found throughout our homes, offices, cars, even inside our bodies. Invisibly small computers will manage our lives; disposable wafer-thin computers will be used like paper, while larger machines capable of two-way voice communication will largely dispense with keyboards. All of these machines will be linked permanently and seamlessly with the Net. Just as we smile at old pictures of computers the size of rooms and code punched on cards, our children will think it quaint how we used big plastic boxes connected with cables and modems to send our words trickling across the screens of other big plastic boxes on other desks elsewhere in the world.

# CONCLUSION

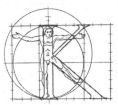

## Winded by Feather Punches

Do people seriously believe that we new artists do
not take our form from nature, do not wrest it from nature,
just like every artist that has ever lived? . . . Nature glows
in our paintings as it does in all art.
—*Wassily Kandinsky*, 1912

$\mathcal{I}$MAGINE A WORLD in which none of the rivalries discussed here had
ever happened. It's a parlor game, of course, but such things may sometimes
lead to small insights. We cannot quantify discovery, we cannot know its
precise effects, the details are invisible to us, but we may begin to imagine the
broad strokes, the grand sweep of progress. A discovery can often precipitate others; remove a piece of the mosaic and the image may be lost forever.

First we must consider the conflict between the Church and science,
which for two millennia dominated the intellectual world. Did this serve to
depress innovation or, perversely, to encourage it? Religion was certainly an

oppressive force. The Church not only attempted to restrain free thinking and to contain unorthodox ideology but presented a distorted version of reality to suit itself. The distinction between these rivals is clear: Science endeavors to explain what is really there; religion attempts to close eyes to the truth while presenting its own idiosyncratic vision of the universe.

However, this rivalry between innovators and suppressors had a positive aspect because some hardy souls were galvanized by the challenge and driven to try to show that the universe was not as the Church portrayed it. Most of these adventurers were forced into silence for fear of losing their lives, but some of their ideas survived, and a "resistance movement of the intellect" was always there, even through the darkest ages. Eventually the radicals could begin to emerge, word seeped out, and these words produced more words, each rebellious text encouraged further free thought, and the presses grew busy, with ink as their weapon. And so, eventually, enlightenment.

Newton and Leibniz were intellectual giants perhaps without peer. If Newton had not conceived the three things for which he is most famous— the law of gravitation, the laws of motion, and the foundations of optical theory—these changes may have been a long time in coming, for there was no one with his depth or breadth of vision alive at the time. Descartes had died when Newton was seven and Christiaan Huygens was fading as Newton's star rose. On mathematical ground, Leibniz could meet Newton as an equal, but he offered few insights into mechanics or optics.

So Newton had no rival in his development of the laws of motion or his other wonders, and, as we have seen, his hatred for Hooke led to the effects of "negative rivalry" in that the irascible Lucasian professor withheld his own descriptions of optical experiments. No, Newton would have been Newton without rivalry, but, crucially, if the dispute over the calculus had not been fought, his wonderful ideas would not have been applied by others with such relish and imagination.

As we know, in his early days at least, Newton was often loath to publish, so without the need to support his priority against Leibniz he may not have published his ideas at all. But beyond this, the fight with Leibniz spurred on the allies and intellectual descendants of the German mathematician and led to great advances in Europe because Leibniz's notation was successfully blended with Newton's physics. This combination gave posterity both the tools (calculus) and the theoretical foundation (the law of gravitation, laws of motion, and, eventually, optical laws) to make Newtonian mechanics practical.

Until Lavoisier, chemistry was something of a sleeping beauty. Priestley, the experimenter, produced the seemingly insignificant result that a particular gas, oxygen, could be made from an oxide and is responsible for respiration and combustion. Lavoisier took these observations and constructed from them an entire framework for chemistry. Lavoisier did not despise Priestley in the way Newton detested Leibniz: His own accounts, merely indicate a belief that Priestley was misguided. But Lavoisier was more than keen to do battle with the phlogistonists, a group whose ideologies he did despise. The phlogiston theory had held chemistry back for almost a century, but, ironically, it made up for this by becoming the catalyst for Lavoisier's endeavors. The phlogistonists fueled Lavoisier's sense of purpose and provided him with an enemy to kick out against.

And what of Darwin, gentle Darwin, whose only enemy it seems was the institution of the irrational? Richard Owen set himself up as representative for this institution and pulled no punches. Having only just emerged from another dispute with Gideon Mantell and Rev. William Buckland over the nature of the fossil record, Owen spent much of the 1860s and 1870s flag-waving for the anti-Darwinians. The resultant clash fueled by Owen's acid pride and Huxley's almost anarchic rhetoric generated public interest in the theory of evolution like no other scientific debate before it.

But what if there had been no Darwin? Would the history of science have been very different? Alfred Wallace had created a theory to compare with Darwin's. Formulated during a malaria-induced fever some two decades late, it nevertheless contained the same elements, almost the same language as the early drafts of *Origin*. But there was one crucial, fundamental difference between Darwin's theory of evolution via natural selection and Wallace's—their interpretations of how natural law operates.

According to Darwin, there is no external force, no God, no guide to control the evolutionary process: Evolution works purely by chance and selection. But Wallace insisted upon adding an indefinable spiritual element to the mix. His "natural selection" was not natural in the sense Darwin intended. In Wallace's worldview natural selection was not entirely self-supporting or self-contained; instead God played a key role steering species toward their destiny.

So, if evolution had been brought to us by Wallace, the bigots of the southern states may be slightly happier people today, but it would have been a theory with far less philosophical or sociological impact, one that would have failed to open debate as Darwin's pronouncements did. Most important, it would have been quite wrong.

And so, from bones and blood and philosophy to generators, wires, and electricity, from the internal world, the contemplation of what it means to be human, to those things that make technology and civilization what they are. Mirroring the dispute between Priestley and Lavoisier, the clash between Edison, the ultimate inventor, and Tesla, the visionary theoretician, came in part from a chafing of approaches. Each man was pushed to more fervent expression of his characteristic genius; each wanted to win at any cost. If Tesla had never lived, the arrival of domestic electricity might have been delayed for years, perhaps until World War I. Maybe if Edison had listened to others with theoretical knowledge and teamed up with them to crack the problems of creating an AC system, he would have triumphed. If Edison had never lived, there were others with practical skills (for the theoretical talents of Tesla were certainly the rarer).

But we must not forget George Westinghouse, for his role was instrumental in the story. He hated Edison and fought him at every turn. In stumbling upon Tesla, he found a way to strike out at the Wizard of Menlo Park and, at the same time, create a system far superior to anything Edison's company could offer. The Battle of the Currents fused a scientific and an industrial rivalry and from it forged a significant technological advance.

And so we come to the dawn of the twentieth century, the start of one hundred years of change greater than the sum of scientific advance in all previous times. In 1900, at the threshold, the idea of the "professional natural philosopher" or "professional scientist" was perhaps only seventy years old; indeed, the name *scientist* had been coined only halfway through the century just passed.*

And here, in this new age, the spirit of rivalry never faltered. How could it? Science was in a state of frenzy. As Schönberg struggled with his revolu-

---

*During Newton's time, two centuries earlier, there were really only two "professional natural philosophers" or "salaried scientists" in Britain. One was the astronomer royal, John Flamsteed, the other was the curator of experiments at the Royal Society, Robert Hooke.

tionary atonal music, Picasso, Schiele, and Munch perceived new vistas, and James Joyce deconstructed literature, Einstein, de Broglie, Bohr, Schrödinger, and many others forged new sciences every bit as radical as the new arts.

Egon Schiele fought persecution from the conservative society in which he lived, and Arnold Schönberg enraged and horrified almost everyone as he entered into vicious battles with traditionalist rivals, prompting him to declare: "I am a conservative who was forced to become a revolutionary."[1] The same may be said for some of the early quantum mechanists, for they too were reevaluating traditional theories and meeting huge resistance because of it. Quantum mechanics and relativity rewrote the score of science, they covered the canvas with strange oils; but these too were born out of conflict. Einstein loathed some of the later developments of quantum mechanics, but, within its early forms at least, he saw the universe reflected and knew the old models would not do.

The young rebels of the early twentieth century fought hard against their elders: The grandees of late-nineteenth-century science did not like the coming of the quantum revolution. But just as Lavoisier resisted anachronism, just as Darwin confronted the accepted, the new age of physics—quantum theory and relativity; the dawning of modern biology—an awakening to the impetus of genetics; and another rebirth for chemistry—the radicalism of the new breed of organic chemist and the biochemist; all these were fought for against the resistance of the old guard.

And then war, and more war. From the nascent sciences of the early twentieth century came practical applications: nerve gas for the trenches, ammonia to make explosives, radio, the internal combustion engine, the telephone, and of course the atomic bomb. Big change may come from little science, but during the twentieth century, big change more usually came from big science. And few projects have been bigger than the Manhattan Project, few changes as dramatic as the dawning of the Atomic Age. For the first time, science was used as an instrument for the implementation of national interest on a grand scale. All wars from the dawn of time have been driven by scientific innovation, from the sharpened stone to the automatic rifle, but the building of the atomic bomb represented an application of intellect to violent effect taken to a degree without precedent.

Fortunately, much more came from it than vaporized bodies. The entire science of nuclear physics and the second half century of quantum mechanics derived from the work on the Hill. From the effort to build the bomb came nuclear power, modern electronics, and much more. And this remains a growing, evolving science, one that had its roots in revolution, was weaned on military application, and then blossomed into the maturity of the present.

And as the physicists wrote the future in chalk, labyrinthine equations on a blackboard, the biologists were running a parallel race to enlightenment. Mendel's research was rediscovered in 1900 by three botanists— Hugo De Vries, Karl Correns, and Erich Tschermak von Seysenegg—and their find initiated another feud between the old school, believers in continuous generation—and the radical biologists who saw that hereditary traits were communicated through genes. This led to arguments spreading across the pages of learned journals and into the magisterial halls of academic institutions, fights reminiscent of the struggles between Huxley and his rivals. Once more generations clashed, the young and adventurous challenging the status quo. This time, the row came to a head with a fierce debate at a meeting of the British Association in 1904, when the geneticists carried the day.

But as the ideas of genetics became refined, there were still dissenting voices. In the Soviet Union, the biologist and political extremist Trofim Lysenko, who was a close personal friend of Stalin, branded Mendelian genetics "bourgeois" and "capitalist" science. Described as "a man possessed of invincible ignorance, but a sure instinct for power and self-advancement," Lysenko argued against the idea of inheritance of discrete entities or genes.[2] "All this DNA, DNA!" he once declared. "Everybody speaks about it, nobody has seen it!"[3]

Elevated to head of the Institute of Genetics of the Soviet Academy of Sciences in Moscow and empowered by the Kremlin, Lysenko proceeded to implement a retrogressive agricultural policy based upon a Lamarckian conception of heredity (named after the French biologist Jean-Baptiste Lamarck), in which it was believed that acquired characteristics are passed from one generation to the next.

With Stalin's support, Lysenko trampled any opposition. He had his scientific opponents incarcerated or executed and forced Soviet science into a cul-de-sac in which truth and reality were overruled by his misguided perception. But Stalin died in March 1953, and a few weeks later news of Crick and Watson's discovery had reached as far as the cold halls of the Moscow Academy and the farms of the Ukraine, where, thanks to Lysenko, crops were failing and people were dying of starvation. Lysenko was exposed: Scientific truth had removed one crutch, the death of Stalin, the other.

Lysenko knew well the power of science supported by political will, and he lived to see the apotheosis of such a blend. During the 1940s governments had unleashed the power of science and turned it into the ultimate destructive force; two decades later they started to use science subversively, acting out a shadow play in which rockets and satellites became the props, cosmonauts and astronauts took the mantel of actors, and the world was their stage.

The script for the space race was an exciting one, the production big-budget. All of human life was there—courage, deception, death, defeat, and triumph. And during the early 1970s, when the final lines fizzled out, the cameras shut down, and the journalists turned to other stories, the effects of a rivalry that had enthralled the world for more than a decade continued quietly offering one technological advance after another. This is how you have come to read these words on this page.

Little more than three hundred years had taken us from a man on a Vatican floor forced by mere cardinals to deny a heliocentric worldview to an astronaut viewing the world over the lunar horizon.

So FINALLY WE must consider the question: Is there any pattern to the spirit of rivalry? Are there threads that tie together the disputes and battles that tell this tale of scientific progress?

Do the personalities of the individuals themselves hold a key? When asked recently about the feud between the astronomers Fred Hoyle and Martin Ryle, the astrophysicist Virginia Trimble commented: "I knew them both and neither person was overwhelmingly easy to get along with." But then she added: "Nobody who does something earthshaking is likely to be."[4]

This seems to be the case in most examples of scientific rivalry, yet, intriguingly, some of the scientists who have fought the most fiercely over priority or to assert their theory over others have also shown extreme naïveté in the way they viewed the world. Consider Tesla, who was convinced his electrical system was superior to Edison's but was almost childlike in his dealings with society, so easily impressed by wealth and show, a man who understood nothing of money and commerce. Think of Heisenberg and Einstein, who argued vigorously against classical physics and openly challenged the views of their illustrious teachers, yet believed they could change politics by creating a union of scientists to guide the world's leaders. And Szilard, a physicist who struggled for years to push the Allies into creating the atomic bomb but who also believed in a nebulous *Bund*.

It is tempting to try to find links between the personalities of scientists who have engaged in feuds, but such connections are at best vague. Might we find some element common to the childhoods of these people? Newton's mother left him when he was a small child; Larry Ellison's mother deserted him. Flamsteed's mother died young, as did Lavoisier's, Priestley's, and Darwin's. But Tesla's mother lived until her son was a grown man; Bill Gates's mother saw her son reach almost forty.

What then of some other childhood trauma? Here again, there is no apparent connection to later aggression. Bill Gates comes from a very wealthy family, so did Oppenheimer, von Braun, and Darwin; but Thomas Huxley was born into poverty, his father an alcoholic; Hooke's father was penniless, and Korolev's father deserted his family. Between these extremes, Crick, Watson, Franklin, and Heisenberg came from, broadly speaking, middle-class families. It is true that some of the great figures of science have been compelled by disadvantage, left to chase recognition, but others have felt overendowed with good fortune.

There is also no sex bias within rivalry. The few women who have climbed to scientific peaks have met conflict head-on, unflinching. Rosalind Franklin behaved as aggressively as her male counterparts; Marie Curie, struggled to assert her convictions; and the Nobel laureate Dorothy Hodgkin, a pioneering biochemist who found the chemical structure for insulin and penicillin, often had to resist strong opposition to her ideas.

Indeed, all of this shows that peculiarities of circumstance are largely irrelevant, for what is truly significant as a catalyst for rivalry is not superficial idiosyncrasy but the nature of science itself and the character of "the scientist."

History shows us that three factors motivate scientists: They must satisfy the demands of their society, they must struggle to understand, and they must seek recognition. The first of these changes constantly, but the need to unravel Truth and the desire for acceptance never do. And for the scientist to succeed and to create meaningful work, work that lasts and inspires, work that reveals and explains, these three factors must, over time, achieve some form of balance.

Newton, alone in his room at Trinity College, worked within a society in which science (natural philosophy) was invisible, its connection to everyday life threadlike and bare: The invention of technology lay at least a century in the future. Newton's fierceness and joy in isolation was abnormal, but so was his devotion to Truth, so even he, the most extreme example of the defensive scientist, achieved a form of balance.

The physicists who worked at Los Alamos faced huge pressures from society, but they could not have worked on the bomb if they were not also searching for deeper truths within the nuclear inferno; clearly they would not have been stimulated by a project to construct a conventional bomb. Yet, to make progress in what they had chosen to do, they needed to maintain a balance among the same three factors that guided Newton: They needed to satisfy the generals, feed the spirit of discovery, and satisfy their egos. Indeed, the story of the Manhattan Project illustrates how delicate the balance may be. Teller's virtual defection came about because his desire to find a different aspect of Truth obsessed him and he possessed an ego that was barely containable.

Taken further, obsession may lead to a complete loss of balance, and then true science founders. For example, Lysenko cannot be considered a genuine scientist because he allowed political fervor to eliminate rationality. For the scientist, there is no greater truth than reality itself, no greater crime than to allow Truth to be subsumed.

As I suggested in the introduction to this book, and as I hope this narrative has shown, science is a process with argument at its core. The Truthseeker impels Nature to reveal a little of itself and approaches his work with blood and thunder as well as the gossamer touch. The scientist is an inveterate quarreler, but an individual winded by feather punches, as passionate as any political zealot or poet burning up with creative energy. Just as the artist gorges on reality and is consumed by beauty, so is the scientist.

And so, whither rivalry? This tale began in the amphitheaters of

Greece, moved on to the hallowed halls of great societies and academic institutions, and finished in the boardrooms of multinational corporations. What next?

Of course it is impossible to judge precisely where rivalry may spur innovation. Rivalry permeates all areas of science, but it is probably reasonable to assume that the next great disputes will come from those sectors of science that have already become or soon will become most intimately linked with commerce. During the twenty-first century two of the most fertile areas for the marriage of mind and money are already gaining high media profiles: the development of biotechnology, in particular genetic science, and the exploitation of space travel.

In spite of the fact that innovation within the computer industry appears to be advancing technology at a dizzying pace, we now merely stand at the foothills of what is possible. The Internet is changing every aspect of our lives at a rate never before witnessed in any field of science or technology, and the repercussions of this change are still barely imaginable. However, as progress continues, humans are also looking inward and outward simultaneously—inward to the realm of the gene and outward to the planets and stars. In both these arenas wonderful new opportunities are already presenting themselves for the scientist. And at the same time, for those more interested in money than in molecules, there have never been so many opportunities to exploit Nature and the human intellect.

As we have seen, throughout human history science itself has been enough to stir the imagination of the creative soul, at least to initiate a desire to investigate and explore. But commercial and political factors have played an increasingly powerful role and have succeeded in driving innovation while drawing upon the energy of scientific rivalry to complete its work. And it is from here, within this grand synergy, that the future of technological advance will spring, to dominate our future as it has our past.

# NOTES

## INTRODUCTION: The Long Road to Reason

1. Lucretius, "The Persistence of Atoms," *On the Nature of Things* (c. 60 B.C.).

2. W. C. Dampier, *A History of Science: And Its Relations with Philosophy and Religion* (Cambridge: Cambridge University Press, 1984), p. 25.

3. Leonardo da Vinci, *Codex Atlanticus*, 182v c., Ambrosiana Library, Milan.

4. Leonardo da Vinci, *Trattato della pittura (Treatise on Painting)*, A. P. McMahon, ed. (Princeton: Princeton University Press, 1956), p. 686.

5. Nicolaus Copernicus, *De revolutionibus orbium coelestium (On the Revolutions of the Heavenly Spheres)*, bk. 1, ch. 10.

6. Quoted in Jacob Bronowski, *The Ascent of Man* (London: BBC Books, 1973), p. 209.

7. See John L. Heilbron, *The Sun in the Church* (Cambridge, Mass.: Harvard University Press, 2000).

8. Quoted in Melvyn Bragg with Ruth Gardiner, *On Giant's Shoulders: Great Scientists and Their Discoveries from Archimedes to DNA* (London: Hodder and Stoughton, 1997), p. 205.

9. James Glanz, "What Fuels Progress in Science? Sometimes, a Feud," *New York Times*, September 14, 1999, p. F2.

10. John Wallis, *Philosophical Transactions of the Royal Society*, August 6, 1666, from the abridged collection (1665–1800), vol. 1, p. 108, Royal Society, London.

11. Quoted in Rachel H. Westbrook, "John Turbeville Needham and His Impact on the French Enlightenment" (Ph.D. diss., Columbia University, 1972), p. 28.

12. Karl Popper, *The Open Society and Its Enemies* (New York: Routledge, 1966).

13. Arthur Meyer, *The Rise of Embryology* (Palo Alto: Stanford University Press, 1939), p. 60.

## ONE: Second Inventors Count for Nothing

1. John Flamsteed, *Self Inspections of J. F.* vol. 33, fols. 104–106, Archives of the Royal Greenwich Observatory at Herstmonceaux.

2. R. S. Westfall, "Short-Writing and the State of Newton's Conscience, 1662 (1)," *Notes and Records of the Royal Society of London* 18 (1963): 10–11.

3. See Michael White, *The Last Sorcerer* (London: Fourth Estate, 1997).

4. H. W. Turnbull et al., eds., *The Correspondence of Isaac Newton*, 7 vols. (Cambridge: Cambridge University Press, 1959–1977), vol. 3, pp. 241–242.

5. Ibid., vol. 1, pp. 3–4.

6. Ibid., pp. 110–111.

7. Ibid., p. 416.

8. Frank E. Manuel, *A Portrait of Isaac Newton* (New York: Da Capo, 1968), p. 305.

9. Francis Baily, *An Account of the Rev^d. John Flamsteed, the First Astronomer Royal* (London, 1835), p. 7.

10. Ibid., p. 60.

11. *Correspondence of Newton*, vol. 4, p. 134.

12. Ibid., pp. 54–55.

13. Ibid., p. 58.

14. Ibid., p. 134. (Flamsteed's comments added to Newton's letter of July 9, 1695).

15. Baily, *Flamsteed*, p. 92.

16. Flamsteed, *Self-Inspections*, vol. 32c, fols. 78–82.

17. G. MacDonald Ross, *Leibniz* (Oxford: Oxford University Press, 1984), p. 4.

18. David Millar et al., eds., *Chambers Concise Dictionary of Scientists* (Cambridge: Chambers, Cambridge University Press, 1989), p. 242.

19. Manuel, *Newton*, p. 323.

20. See Joseph E. Hoffman, *Leibniz in Paris, 1672–1676: His Growth to Mathematical Maturity* (Cambridge: Cambridge University Press, 1974), p. 291.

21. *Correspondence of Newton*, vol. 1, p. 15.

22. Ibid., p. 356.

23. Ibid., vol. 2, p. 134.

24. Ibid., p. 110.

25. Ibid., p. 163.

26. Hoffman, *Leibniz in Paris*, ch. 20.

27. Isaac Newton, *Principia Mathematica*, trans. and ed. Florian Cajori (Berkeley and Los Angeles: University of California Press, 1934), pp. 655–656.

28. Gottfried von Leibniz, *Sämtliche Schriften und Briefe*, ser. 1, vol. 4 (1950), pp. 475–476.

29. Ibid., p. 477.

30. University Library, Cambridge, (ULC), Add MS. 3968.41, f.85^r.

31. Manuel, *Newton*, p. 345.

32. W. G. Hiscock, ed., *David Gregory, Isaac Newton and Their Circle: Extracts from David Gregory's Memoranda, 1677–1708* (Oxford: Oxford University Press, 1937), p. 40.

33. *Correspondence of Newton*, vol. 4, p. 100.

34. Bernoulli to Leibniz, quoted in D. T. Whiteside, ed., *The Mathematical Papers of Isaac Newton*, 8 vols. (Cambridge: Cambridge University Press, 1967–1980), vol. 7, p. 181.

35. A. Rupert Hall, *Philosophers at War: The Quarrel Between Newton and Leibniz* (Cambridge: Cambridge University Press, 1980), p. 118.

36. Nicholas Fatio du Duillier, *Lineae brevissimi descensus investigatio geometrica duplex (A Two-fold Geometrical Investigation of the Line of Briefest Descent)* (London, 1699), p. 18.

37. Quoted in L. T. More, *The Life and Works of the Honourable Robert Boyle* (Oxford: Oxford University Press, 1944), p. 98.

38. *Correspondence of Newton*, vol. 3, p. 286.

39. Voltaire to Mme. Denis, Berlin, December 18, 1752, quoted in Ben Ray Redman, ed., *The Portable Voltaire* (New York: Viking, 1949), pp. 487–488.

40. John Keill, "Epistolia . . . de Legibus Virum Centripetarum," *Philosophical Transactions of the Royal Society* 26 (1708): 185.

41. "Isaaci Newtoni tractatus duo, de speciebus & magnitudine figurarum curvilinearum," *Acta Eruditorum* ( January 1705), p. 35 (italics Leibniz's).

42. Pierre Des Maizeaux, *Recueil des pièces sur la philosophie, la religion naturelle, l'historie, les mathèmatiques, etc. Par Mrs. Leibniz, Clarke, Newton, & autres autheurs célèbres* (Amsterdam, 1720), vol. 2, p. 49.

43. *Correspondence of Newton*, vol. 5, p. 142.

44. Ibid., p. 207.

45. Joseph Raphson, *The History of Fluxions* (London, 1715), p. 100.

46. *Journal Book of the Royal Society*, March 11, 1712.

47. "On a point connected with the dispute between Keill and Leibniz about the invention of Fluxions," *Philosophical Transactions of the Royal Society* 136 (1846): 107.

48. Journal Book (Copy) of the Royal Society, London, vol. 10, p. 391.

49. ULC, Add MS.4007B, fol. 617.

50. Journal Book (Copy) of the Royal Society, London, vol. 10, p. 391.

51. ULC, Add MS.3968.37, f.539ʳ.

52. Isaac Newton, "Account of the Commercium epistolicum," *Philosophical Transactions of the Royal Society* 29 (1714–16): 211.

53. ULC, Add MS.3968.37, f.539ʳ.

54. William Whiston, *Historical Memoirs of the Life of Dr. Samuel Clarke* (London, 1730), p. 132.

55. *Correspondence of Newton*, vol. 6, p. 3.

56. Ibid., pp. 8–9.

57. Ibid., pp. 18–19.

58. First quoted in a letter from John Chamberlayne to the Royal Society, the Royal Mint Library, Newton MSS., II, fol., 334. Chamberlayne to Newton, November 25, 1713.

59. Gottfried von Leibniz, *Die Philosophischen Schriften*, ed. C. J. Gerhardt (Berlin, 1875–1890; Hildesheim, 1960–61), vol. 3, p. 589.

60. *Correspondence of Newton*, vol. 6, p. 288.

61. Quoted in John Theodore Merz, *Leibniz* (New York: Lippincott, 1884), p. 126.

62. Keynes MS.133, p. 10. King's College Library, Cambridge,

63. Quoted in Merz, *Leibniz* p. 89.

64. Morris Kline, *Mathematics and the Physical World* (New York: Thomas Y. Crowell, 1959), p. 419.

65. Robert Fox. "Laplacian Physics," in R. C. Olby et al., eds., *Companion to the History of Modern Science* (London: Routledge, 1990), p. 291.

TWO: The Fanatic and the Tax Collector

1. Lewis Pyenson and Susan Sheets-Pyenson, *Servants of Nature: A History of Scientific Institutions. Enterprises and Sensibilities* (London: HarperCollins, 1999), p. 87.

2. Benjamin Franklin to Antoine Lavoisier, October 23, 1788 (Library of Congress).

3. Joseph Priestley, *Memoirs of Dr. Joseph Priestley to the Year 1795 with a Continuation by His Son*, 2 vols. (London, 1806).

4. C. G. Jung, *Man and His Symbols* (London: Aldus, 1964), p. 10.

5. Lavoisier's private laboratory notebook, 1772, as quoted in W. R. Aykroyd, *Three Philosophers: Lavoisier, Priestley and Cavendish* (London: Heinemann, 1935).

6. Memorandum by Antoine Lavoisier, November 1, 1771, Library of the Royal Academy of Sciences, Paris.

7. *Ibid.*, February 20, 1772.

8. Letter from Pierre Bayen, *Observations*, January 1775, Royal Academy of Sciences, Paris.

9. Jean Rey, *Essays* (Bazas, 1630).

10. Joseph Priestley, *Experiments and Observations on Different Kinds of Air* (London: Royal Society, 1775).

11. Priestley, *Memoirs*.

12. Ibid.

13. Ibid.

14. Ibid.

15. Ibid.

16. *Philosophical Transactions* (Royal Society, London), March 1775.

17. As quoted in Aykroyd, *Three Philosophers*, p. 117.

18. Antoine Lavoisier, *Memoirs* (Paris: Royal Academy of Sciences, 1786).

19. Joseph Priestley to Thomas Henry, December 31, 1775 (Royal Society, London).

20. Joseph Priestley to James Keir, February 4, 1778 (Royal Society, London).

21. Ibid.

22. Joseph Priestley to Josiah Wedgwood, May 26, 1781 (Royal Society, London).

23. Joseph Priestley to Sir Joseph Banks, June 23, 1783 (Royal Society, London).

24. Joseph Priestley to Josiah Wedgwood, March 21, 1782 (Royal Society, London).

25. Joseph Priestley, in *Philosophical Transactions* (Royal Society, London), March 1775.

26. Diary of Arthur Young, October 1787, as quoted in Douglas McKie, *Antoine Lavoisier* (New York: Da Capo, 1952).

27. Jean Marat, *L'Ami du peuple* (Paris, August 1792).

28. Aykroyd, *Three Philosophers*.

29. Priestley, *Memoirs*.

30. Priestley, *Experiments and Observations*.

31. Joseph Priestley, *History of Electricity*, 1st ed. (London: Royal Society, 1767), p. 442.

32. Ibid., p. 444.

33. Quoted in Sir Philip Hartog, "Newer Views of Priestley and Lavoisier," *Annals of Science* (London: Taylor Francis, August 1941), p. 7.

34. Antoine-Laurent Lavoisier, *The Elements of Chemistry* (Paris, 1789), preface.

35. Edmund Blair Bolles, ed., *Galileo's Commandment: An Anthology of Great Science Writing* (London: Little, Brown, 1997), p. 379.

THREE:  Of Monkeys and Men

1. Thomas Huxley to F. Dysater, September 9, 1860, Thomas Huxley Papers, Imperial College, University of London, 15.115.

2. John Hooker to Charles Darwin, July 2, 1860, Darwin Archive, Cambridge University Library, 100: 141–142.

3. Francis Darwin, ed., *Life and Letters of Charles Darwin*, 3 vols. (1887), vol. 2, pp. 323–324.

4. F. Burkhardt and S. Smith, eds., *Correspondence of Charles Darwin*, 8 vols. (Cambridge: Cambridge University Press, 1985–1993), vol. 1, p. 133.

5. R. Keynes, ed., *Charles Darwin's Beagle Diary* (Cambridge: Cambridge University Press, 1988), p. 4.

6. Burkhardt and Smith, *Correspondence of Darwin*, vol. 1, p. 368.

7. Fabienne Smith, "Charles Darwin's Ill Health" and "Charles Darwin's Health Problems: The Allergy Hypothesis," *Journal of the History of Biology* 23, no. 3 (Fall 1990): 443–459, and 25, no. 2 (Summer 1992): 285–306.

8. *Life and Letters of Darwin*, vol. 1, p. 31.

9. Ibid., p. 55.

10. Ronald Clark, *The Survival of Charles Darwin* (London: Weidenfeld and Nicolson, 1984).

11. Quoted by, for example, Antony Flew, *Malthus* (London: Pelican, 1970).

12. See Darwin's "Notebook on Transmutation of Species," *Bulletin of the British Museum (Natural History), Historical Series* 2 (1960).

13. Charles Darwin, *On the Origin of Species by Means of Natural Selection, or the Preservation of Favoured Races in the Struggle for Life* (1859; reprint, London: Penguin, 1985).

14. Ibid.

15. Burkhardt and Smith, *Correspondence of Darwin*, vol. 3, p. 103.

16. Ibid., p. 184.

17. Quoted by John Bowlby, *Charles Darwin: A New Biography* (London: Hutchinson, 1990).

18. *Life and Letters of Darwin*, vol. 1, p. 33.

19. Clark, *Survival of Darwin*, p. 66.

20. *Life and Letters of Darwin*, vol. 2, p. 116.

21. Richard Owen, *Anatomy of Vertebrates* 3 (1866–1868): 798.

22. Gavin de Beer, ed., *Charles Darwin and T. H. Huxley, Autobiographies* (Oxford: Oxford University Press 1983), p. 61.

23. In a letter from Richard Owen to William Whewell, February 24, 1845, in "Richard Owen, William Whewell and the Vestiges," *British Journal for the History of Science* 10 (1977): 142.

24. W. H. McMenemey, "Education and the Medical Reform Movement," in F. N. L. Poynter, ed., *The Evolution of Medical Education in Britain* (London: Pitman, 1966), p. 45.

25. Thomas Hobbes, *Leviathan* (1651; rept., New York: Penguin, 1986), p. 186.

26. Thomas Huxley, "Science and the Church," *Reader* 4 (1864): 804.

27. Adrian Desmond, *Huxley: The Devil's Disciple* (London: Michael Joseph, 1994), p. xiii.

28. Quoted in Clark, *Survival of Darwin*, p. 137.

29. John Fiske, "Reminiscences of Huxley" *Annales Reptilia Smithsonian Inst.*, 1901, in *The Personal Letters of John Fiske* (Cedar Rapids, Iowa: Torch, 1939), pp. 121–122.

30. H. F. Jones, ed., *Samuel Butler: A Memoir*, 2 vols. (London: Macmillan, 1920), vol. 1, p. 385.

31. Henrietta Anne Heathorn, *Reminiscences*, in T. H. Huxley Papers, Imperial College, London, Huxley Archives, 62. 1.

32. Burkhardt and Smith, *Correspondence of Darwin*, vol. 5, p. 133, as quoted in Adrian Desmond, *Darwin*, (London: Michael Joseph, 1991), p. 433.

33. As quoted in Desmond, *Darwin*, p. 432.

34. Darwin, *Origin*, p. 255.

35. *Times* [London], December 26, 1859.

36. Darwin to T. H. Huxley, December 28, 1859, T. H. Huxley Papers, 5.92.

37. Richard Owen, "Darwin on the Origin of Species," *Edinburgh Review* 3 (1860): 503–504.

38. Samuel Wilberforce, *Quarterly Review* 108 (1860): 247.

39. William Irvine, *Apes, Angels and Victorians: A Joint Biography of Darwin and Huxley* (London: Jonathan Cape, 1956), p. 5.

40. *Life and Letters of Darwin*, vol. 2, p. 300.

41. Ibid., pp. 300–301.

42. Francis Darwin and A. C. Seward, eds., *More Letters of Charles Darwin*, 2 vols. (London: John Murray, 1903), vol. 1, p. 149.

43. Ibid., p. 185.

44. Ibid., p. 203.

45. Ibid., p. 232.

46. Darwin, *Origin*, p. 420.

47. L. Huxley, *Life and Letters of Thomas Henry Huxley*, 2 vols. (London: Collins, 1900), vol. 1, p. 152.

48. As quoted in Melvyn Bragg, with Ruth Gardiner, *On Giant's Shoulders: Great Scientists and Their Discoveries from Archimedes to DNA* (London: Hodder and Stoughton, 1997), p. 172.

49. Thomas Huxley to W. Sharpley, October 13, 1862, Sharpley Correspondence MSS Ad.227, University College London Library.

50. Adrian Desmond, *Archetypes and Ancestors: Palaeontology in Victorian London, 1850–1875* (London: Blond and Briggs, 1982), p. 143.

51. Darwin to John Hooker, March 4, 1874, Darwin Archive, Cambridge University Library, 93: 313–316.

52. Huxley to John Tyndall, July 30, 1873, T. H. Huxley Papers, 9.73.

53. "Monkeyana," *Punch* (May 18, 1861).

54. Ernst Mayr, *One Long Argument: Charles Darwin and the Genesis of Modern Evolutionary Thought* (Cambridge, Mass., Harvard University Press, 1991), p. 25.

55. One of the best examples of this is to be found in Lee Smolin, *The Life of the Cosmos* (London: Weidenfeld and Nicolson, 1997).

FOUR: The Battle of the Currents

1. *Buffalo Express*, August 7, 1890.

2. Carlos F. MacDonald, *Report of Carlos F. MacDonald, M.D., on the Execution by Electricity of William Kremmler, alias John Hart, Albany Argus*, 1890, p. 6.

3. Ibid.

4. *New York Times*, August 7, 1890, p. 16.

5. "Mrs. Edison As the Wife of a Genius Has a Full-time Job," *Delineator* 3 (1927): 77.

6. Caroline Farrand Ballantine, "The True Story of Edison's Childhood and Boyhood," *Michigan History Magazine* 4 (1927): 177.

7. Paul Israel, *Edison: A Life of Invention* (New York: John Wiley, 1998), p. 6.

8. T. H. Metzger, *Blood and Volts: Edison, Tesla and the Electric Chair* (New York: Autonomedia, 1996), p. 52.

9. Waldo P. Warren, "Edison on Invention and Inventors," *Century* 82 (1911): 415–416.

10. Thomas Edison to Pupils of the Grammar Schools of New Jersey, April 30, 1912, Document File, Edison National Historical Site, New York.

11. Matthew Josephson, *Edison: A Biography* (New York: McGraw-Hill, 1959), p. 23.

12. Sidney Warren, *American Freethought, 1860–1914* (London: P. S. King and Staples, 1943), p. 227.

13. Draft of a letter from Thomas Edison to S. H. Norton, February 21, 1911, Document File, Edison National Historical Site.

14. James Symington to Thomas Edison, February 23, 1891, in *Thomas A. Edison Papers Microfilm Edition* (Frederick, Md.: University Publications of America, 1985–), reel 130, frame 988.

15. *Christian Herald and Sign of Our Times*, July 25, 1888, in *Thomas A. Edison Papers Microfilm Edition*, reel 146, frame 286.

16. "A Visit to Edison," *Philadelphia Weekly Times*, April 29, 1878, in *Thomas A. Edison Papers Microfilm Edition*, reel 25, frame 189.

17. G. M. Shaw, "Sketch of Thomas Alva Edison," *Popular Science Monthly* (August 13, 1878), p. 489–490.

18. Quoted in Josephson, *Edison*, p. 180.

19. Metzger, *Blood and Volts*, p. 65.

20. Josephson, *Edison*, p. 180–181.

21. Ibid., p. 180.

22. Garret P. Serviss, "Edison's Conquest of Mars," *Journal*, December 15, 1897–January 1898.

23. *New York Herald*, December 29, 1879, p. 17.

24. Nikola Tesla, *My Inventions: The Autobiography of Nikola Tesla* (Austin, Tex.: Hart Brothers, 1982), p. 28.

25. Ibid., p. 56.

26. Quoted in John O'Neill, *Prodigal Genius* (1944; reprint, New York: Ives Washburn, 1968), p. 256.

27. Tesla, *My Inventions*, p. 57.

28. Ibid., p. 60,

29. Ibid., p. 59.

30. Ibid., p. 61.

31. Ibid., p. 66.

32. Ibid.

33. Robert Lomas, *The Man Who Invented the Twentieth Century: Nikola Tesla, Forgotten Genius of Electricity* (London: Headline, 1999), p. 54.

34. Margaret Cheney, *Tesla: Man Out of Time* (1968; reprint, New York: Dorset, 1989), p. 32.

35. Quoted in O'Neill, *Prodigal Genius*, p. 64.

36. Quoted in Lomas, *Man Who Invented the Twentieth Century*, p. 69.

37. Quoted in O'Neill, *Prodigal Genius*, pp. 95–96.

38. Paraphrased from the dialogue reported in the *New York Times*, July 31, 1888, p. 17.

39. Josephson, *Edison*, p. 347 n. 11.

40. *New York Times*, December 13, 1888, p. 13.

41. Ibid., December 18, 1888, p. 11.

42. Harold Passer, *The Electrical Manufacturers, 1875–1900* (New York: Arno, 1972), p. 174.

43. Metzger, *Blood and Volts*, p. 105.

44. Quoted in *Electrical Engineer* (February 1889), p. 4.

45. Harold Brown, "The New Instrument of Execution," *North American Review* (November 1889), p. 41.

46. *New York Times*, May 8, 1889, p. 5.

47. Quoted in Cheney, *Tesla*, p. 49 n. 16.

48. Thomas Alva Edison, *The Diary and Sundry Observations* (New York: Philosophical Society, 1948), pp. 212–213.

49. B. C. Forbes, "Edison Working on How to Communicate with the Next World," *American Magazine* (October 1920), p. 85.

FIVE:  Of Atom Bombs and Human Beings

1. G. Conn and H. Turner, *The Evolution of the Nuclear Atom* (Boston: American Elsevier, 1965), pp. 136 ff.

2. Spencer Weart and Gertrud Szilard, eds., *Leo Szilard: His Version of the Facts* (Cambridge, Mass.: MIT Press, 1978).

3. Frédéric Joliot, "Number of Neutrons Liberated in the Nuclear Explosion of Uranium 143" *Nature* (April 22, 1939): 680.

4. David Irving, *The German Atomic Bomb* (New York: Simon and Schuster, 1967), p. 34.

5. *New York Times*, April 30, 1939, p. 35.

6. Original in the Hebrew University, Jerusalem, Israel.

7. Records of the U.S. Senate, 1945, p. 7.

8. Report from Lt. Col. S. V. Constant (U.S. Military Intelligence), August 13, 1940, quoted in Max F. Perutz, "An Intelligent Bumblebee," *New York Review of Books*, October 7, 1993.

9. Werner Heisenberg, *Physics and Beyond* (New York: Harper and Row, 1971), pp. 169 ff.

10. Interview with Victor Weisskopf, June 5, 1990, quoted in Thomas Powers, *Heisenberg's War: The Secret History of the German Bomb* (London: Jonathan Cape, 1993), p. 125.

11. Letter from Werner Heisenberg to Robert Jungk, January 18, 1957, quoted in Robert Jungk, *Brighter Than a Thousand Suns* (New York: Harcourt, Brace, 1958), p. 103.

12. Powers, *Heisenberg's War*, p. 126.

13. Ibid., p. 107.

14. Peter Goodchild, *J. Robert Oppenheimer: Shatterer of Worlds* (Boston: Houghton Mifflin, 1980), pp. 56 ff.

15. Off-the-record remark to a journalist, March 8, 1946, Leo Szilard Papers, University of California at San Diego.

16. Letter to Francis Fergusson, January 23, 1926, Oppenheimer Collection, Library of Congress, Washington, D.C.

17. Interview with Jeffries Wyman by Charles Weiner, May 28, 1975, pp. 21–22, quoted in Alice Kimball Smith and Charles Weiner, eds., *Robert Oppenheimer: Letters and Recollections* (Palo Alto: Stanford University Press, 1980), p. 93.

18. Interview with John. T. Edsall, July 16, 1975, ibid., p. 94.

19. Hans Bethe, "J. Robert Oppenheimer," *Biographical Memoirs FRS* 14, no. 391, p. 396.

20. Interview of March 8, 1946, Leo Szilard Papers.

21. Ronald W. Clark, *Einstein* (New York: Avon, 1971), p. 685.

22. Nuel Pharr Davis, *Lawrence and Oppenheimer* (New York: Simon and Schuster, 1968), p. 216.

23. Ibid.

24. Lawrence Badas et al., *Reminiscences of Los Alamos* (New York: D. Reidel, 1980), p. 49.

25. Hans Bethe, "Comments on the History of the H-bomb," *Los Alamos Science* (Fall 1982): 2.

26. J. Robert Oppenheimer to Leslie R. Groves, May 1, 1944, Manhattan Engineer District Records (Record Group 77), National Archives, Peierls. R.

27. Jo Ann Shroyer, *Secret Mesa: Inside Los Alamos National Laboratory* (New York: John Wiley, 1998), p. 28.

28. Ibid., p. 211.

29. Robert Couglan, "The Tangled Drama and Private Hells of Two Famous Scientists," *Life*, December 13, 1963.

30. Smith and Weiner, *Oppenheimer: Letters and Recollections*, p. 273.

31. Quoted in Paul Strathern, *Oppenheimer and the Bomb* (London: Arrow, 1998), p. 84.

32. *In the Matter of L. Robert Oppenheimer, Transcript of Hearings* (Washington, D.C.: Government Printing Office, 1954), p. 665.

33. As quoted in Davis, *Lawrence and Oppenheimer*, p. 302.

34. Edward Teller, *The Legacy of Hiroshima* (Garden City, N.Y.: Doubleday, 1962), p. 17.

35. Studs Terkel, *The Good War* (New York: Pantheon, 1984), pp. 512 ff.

36. *Los Alamos—Beginning of an Era, 1943–1945* (New Mexico: Los Alamos Scientific Laboratory, n.d.), p. 53.

37. Farm Hall Reports, vol. 14, National Archives, London.

38. See Powers, *Heisenberg's War*.

39. Interview with Victor Weisskopf, October 23, 1991, quoted in Powers, *Heisenberg's War*, p. xi.

40. William Lanouette with Bela Szilard, *Genius in the Shadows: A Biography of Leo Szilard, the Man Behind the Bomb* (New York: Scribner's, 1993), p. 134.

41. Report from the Directorate of Tube Alloys to the British Government, January 1944, quoted in Max Perutz, *I Wish I Had Made You Angry Earlier: Essays on Science and Scientists* (Oxford: Oxford University Press, 1998), p. 51.

42. Leslie R. Groves, *Now It Can Be Told* (New York: Harper and Row, 1962), p. 186.

43. Weart and Szilard, eds., *Leo Szilard*, p. 184.

44. Martin Sherwin, *A World Destroyed* (New York: Alfred A. Knopf, 1975), p. 136.

45. James Conant to Vannevar Bush, May 18, 1945, Bush-Conant File, National Archives, Washington, D.C., folder 5.

## SIX: The Race for the Prize

1. James Watson, *The Double Helix: A Personal Account of the Discovery of the Structure of DNA* (London: Weidenfeld and Nicolson, 1968), p. 26.

2. Note Complementaire in Robert Stoops, *Les Protennes: Rapports et Discussions* (Brussels: Institut International de Chimie Solvay Neuvième Conseil de Chimie, 1953), p. 113.

3. Watson, *Double Helix*, p. 28.

4. Ibid., p. 29.

5. Ibid., p. 36.

6. Ibid., p. 21, 24.

7. Ibid., p. 21.

8. Francis Crick, *What Mad Pursuit: A Personal View of Scientific Discovery* (London: Penguin, 1990), p. 64.

9. Ibid.

10. Max Perutz, *I Wish I Had Made You Angry Earlier: Essays on Science and Scientists* (Oxford: Oxford University Press, 1998), p. 88.

11. Crick, *What Mad Pursuit*, p. 69.

12. Watson, *Double Helix*, p. 46.

13. Crick, *What Mad Pursuit*, p. 75.

14. Quoted in Horace Freeland Judson, *The Eighth Day of Creation: Makers of the Revolution in Biology* (New York: Cold Spring Harbor Laboratory Press, 1996), p. 126.

15. Watson, *Double Helix*, p. 27.

16. Stoops, *Protennes*, p. 626.

17. Ibid., p. 627.

18. Ibid., p. 127.

19. Ibid., p. 626.

20. Crick, *What Mad Pursuit*, p. 69.

21. Meyer Friedman and Gerald Friedland, *Medicine's Ten Greatest Discoveries* (London: Yale University Press, 1998), p. 206.

22. Ibid.

23. Arthur Meyer, *The Rise of Embryology* (Palo Alto: Stanford University Press, 1939), p. 80.

24. Watson, *Double Helix*, p. 123.

25. Quoted in Judson, *Eighth Day of Creation*, p. 133.

26. Ibid., p. 135.

27. Watson, *Double Helix*, p. 131.

28. Ibid., p. 132.

29. Linus Pauling to Peter Pauling, February 18, 1953, quoted in Judson, *Eighth Day of Creation*, p. 126.

30. James Watson to Max Delbrück, February 20, 1953.

31. Maurice Wilkins to Francis Crick, March 7, 1953, much quoted, in Watson's *Double Helix* and elsewhere.

32. J. D. Watson and F. H. Crick, "Molecular Structure of Nucleic Acid: A Structure for Deoxyribose Nucleic Acid," *Nature* 171 (April 25, 1953): 737.

33. Watson, *Double Helix*, p. 171.

34. Ibid., p. 17. Watson was obviously moved by this comment and for a while even considered using the title "Honest Jim" for his account of the discovery.

35. Friedman and Friedland, *Medicine's Ten Greatest Discoveries*, p. 192.

36. Ibid., p. 200.

37. Candace B. Pert, *Molecules of Emotion: Why You Feel the Way You Feel* (New York: Simon and Schuster, 1997), p. 109. It is also worth noting that on page 109 she refers to Watson as John Watson.

38. James Watson's inaugural address before the students and staff of the Harvard Center for Genomics Research, Boston, September 3, 1999.

39. Crick, *What Mad Pursuit*, p. 76.

40. Aaron Klug, Nobel Prize Acceptance Lecture, Stockholm, 1982.

41. Friedman and Friedland, *Medicine's Ten Greatest Discoveries*, p. 225.

<br>

SEVEN: Reaching for the Moon

1. *New York Times*, October 5, 1957.

2. *Pravda*, October 5, 1957, p. 1.

3. *Le Figaro*, Paris, October 7, 1957, pp. 4–5.

4. Martin Summerfield, "Problems of Launching an Earth Satellite," *Astronautics* (November 1957): 8–21.

5. Project Rand Report SM-11827, May 2, 1946, from *Rand Twenty-fifth Anniversary Volume* (Santa Monica, Calif.: Rand Corp., 1973), p. 3.

6. Wernher von Braun quoted in *Life*, March 6, 1952, p. 33.

7. F. J. Krieger, *Behind the Sputniks* (Washington, D.C.: Public Affairs Press, 1958), p. 330.

8. *New York Times*, August 4, 1955, p. 4.

9. Keith T. Glennan, *The Birth of NASA* (Washington, D.C.: NASA History Office, 1993), p. 13.

10. Quoted in James Shefter, *The Race* (New York: Doubleday, 1999), p. 7.

11. Ibid., p. 112.

12. Edwin Diamond, *Newsweek* (October 6, 1959): 11.

13. Quoted in Shefter, *The Race*, p. 23.

14. Ibid., p. 25.

15. Ibid., p. 18.

16. K. E. Tsiolkovsky, *Selected Works* (Moscow: Mir Publishers, 1968), p. 14.

17. Milton Lehman, *This High Man: The Life of Robert H. Goddard* (New York: Farrar, Straus, 1963), p. 144.

18. As quoted in James Harford, *Korolev: How One Man Masterminded the Soviet Drive to Beat America to the Moon* (New York: John Wiley, 1997), p. 1.

19. Ibid., p. 3.

20. Frederick I. Ordway and Mitchell R. Sharpe, *The Rocket Team* (New York: Thomas Y. Crowell, 1979), p. 106.

21. Walter Dornberger, *V-2* (New York: Viking, 1954), p. 27.

22. P. T. Atashenkov, *Akademik S. P. Korolev* (Moscow: Mashinostroyenniye, 1969), p. 10.

23. Ibid.

24. A. Y. Ishlinsky, *Akademik S. P. Korolev, Uchyonii, Inzhenyer, Chelovek* (Moscow: Nauka, 1986), pp. 38–39.

25. Ibid.

26. Yaroslav Golovanov, *Sergei Korolev: The Apprenticeship of a Space Pioneer,* (Moscow: Mir, 1975), pp. 230–231.

27. M. Rebrov, "Notes on Career of Designer V. P. Glushko," *Journal of Propulsion and Rocket Science* 90-001 (March 15, 1990): p. 57.

28. Quoted in Harford, *Korolev,* p. 48.

29. Editorial, *Manchester Guardian,* October 7, 1957.

30. George Kennan, *George F. Kennan Memoirs (1950–1963)* (New York: Pantheon, 1972), p. 140.

31. Von Braun in an interview with Richard B. Stolley, *Life,* November 18, 1957, p. 136.

32. Von Braun, "The Big Adventure," *Colliers,* March 22, 1952, p. 39.

33. Lyndon B. Johnson., speech before the Senate Democratic Caucus, quoted in Doris Kearns, *Lyndon Johnson and the American Dream* (New York: Harper and Row, 1976), p. 145.

34. Ordway and Sharpe, *Rocket Team,* pp. 453–454.

35. In "Vanguard" from *The Coming of the Space Age,* Arthur C. Clarke, ed. (New York: Meredith, 1968), p. 19.

36. *London Daily Herald,* December 7, 1957, p. 1.

37. Clayton Koppes, *JPL and the American Space Program* (New Haven: Yale University Press, 1982), p. 83.

38. William J. Broad, "'U.S. Planned Nuclear Blast on the Moon,' Physicist Says," *New York Times,* May 16, 2000, p. A17.

39. As quoted in Harford, *Korolev,* p. 163.

40. *The Apollo Spacecraft: A Chronology,* NASA SP4009, vol. 1, p. 8.

41. Alan Shepard and Deke Slayton, *Moon Shot* (Atlanta, Ga.: Turner, 1994), p. 91.

42. *New York Times,* April 12, 1961.

43. *Pravda,* April 13, 1961.

44. Kennedy Press Conference, April 12, 1961, quoted in Harford, *Korolev,* p. 176.

45. John M. Logsdon, *The Decision to Go to the Moon* (Cambridge, Mass.: MIT Press, 1970), p. 109.

46. Memorandum from John F. Kennedy to Lyndon B. Johnson, April 20, 1961, John F. Kennedy Library, Boston, Massachusetts.

47. Wernher von Braun memo to Vice President Johnson, April 29, 1961, NASA Historical Archives.

48. Shepard and Slayton, *Moon Shot,* p. 109.

49. Quoted in Shefter, *The Race,* p. 143.

50. Quoted in Logsdon, *Decision to Go to the Moon,* p. 128.

51. Sergei Kryukov, "The Brilliance and the Eclipse of the Lunar Program," in *Nauka I Zhizn,* no. 4 (Moscow, 1994): 81.

52. Harford, *Korolev,* p. 258.

53. Ibid., p. 14.

54. Ibid., p. 15.

55. Gyorgi Vetrov, "The Difficult Fate of Rocket N-1," in *Nauka I Zhizn,* no. 4 (Moscow, 1994): 79–80.

56. Quoted in Kryukov, "Brilliance and Eclipse of the Lunar Program," p. 84.

57. Sergei Khrushchev, *Khrushchev on Khrushchev* (Boston: Little, Brown, 1990), p. 192.

58. *The Times* [London], June 17, 1963, p. 1.

59. As quoted in Shefter, *The Race*, p. 181.

60. John F. Kennedy, speech prepared for delivery to Dallas Citizens Council, November 22, 1963, John F. Kennedy Library.

61. Quoted in Shefter, *The Race*, p. 177.

62. Quoted in Harford, *Korolev*, p. 271.

63. Andrei Sakharov, *Andrei Sakharov Memoirs* (New York: Alfred A. Knopf 1990), p. 178.

64. James E. Oberg, *Red Star in Orbit* (New York: Random House, 1981), p. 87.

65. Harford, *Korolev*, p. 15.

66. Sergei Kryukov, "Brilliance and Eclipse of the Lunar Program," in *Nauka I Zhizn*, no. 4 (Moscow, 1994): 85.

67. Vassily Mishin, "Why Didn't We Fly to the Moon?" *Novoye v Zhizni*, Teknike, no. 12 (1990), translated in *Journal of Propulsion and Rocket Science* USP-91-006 (November 12, 1991): 5.

68. Ibid., p. 16.

EIGHT: The Battle of the Cyber-Kings

1. Larry Ellison in *New York Times*, February 28, 1988.

2. James Wallace, *Overdrive: Bill Gates and the Race to Control Cyberspace* (New York: John Wiley, 1997), p. 294.

3. Gary Rivlin, *The Plot to Get Bill Gates: An Irreverent Investigation of the World's Richest Man . . . and the People Who Hate Him* (London Quartet Books, 1999), p. 88.

4. Author's interview with Bill Gates, November 21, 1995.

5. Ibid.

6. Stephen Manes and Scott Andrews, *Gates: How Microsoft's Mogul Reinvented an Industry and Made Himself the Richest Man in America* (New York: Doubleday 1993), p. 16.

7. Quoted in James Wallace and Jim Erickson, *Hard Drive: Bill Gates and the Making of the Microsoft Empire* (New York: HarperCollins, 1993), p. 37.

8. Quoted in Rivlin, *Plot to Get Bill Gates*, p. 61.

9. Manes and Andrews, *Gates*, p. 51.

10. Quoted in *Triumph of the Nerds*, a PBS-TV program produced by Oregon Public Broadcasting and John Gau Productions.

11. *People* (December 26–January 2, 1984): 37.

12. Ibid.

13. Rivlin, *Plot to Get Bill Gates*, p. 249.

14. Ibid., p. 145.

15. Charles Ferguson and Charles Morris, *Computer Wars: The Fall of IBM and the Future of Global Technology* (New York: Times Books, 1993), p. 156.

16. "Software Hardball," *Wall Street Journal*, September 25, 1987, p. 1.

17. *Fortune* (August 26, 1991): 43.

18. *Fortune* (November 29, 1993).

19. Quoted in Mike Wilson, *The Difference Between God and Larry Ellison* (New York: William Morrow, 1998), p. 8.

20. Ibid., p. 9.

21.  Rivlin, *Plot to Get Bill Gates*, p. 326.

22.  Ferguson and Morris, *Computer Wars*, p. 27.

23.  Wilson, *Difference Between God and Larry Ellison*, p. 71.

24.  Ibid., p. 87.

25.  Wallace and Erickson, *Hard Drive*, p. 224.

26.  The information about Oracle's float comes from a press release by Oracle's chief financial officer, Jeffrey O. Henley, March 1996. Information about Microsoft's float comes from Manes and Andrews, *Gates*.

27.  Richard Brandt, "Bill Gates—At the Gates," *Upside* magazine ( January 1996).

28.  George Gilder, "The Coming Software Shift," *Forbes ASAP* (August 1995).

29.  Raymond Lane quoted in *PC Week*, February 26, 1996.

30.  As quoted in Wilson, *Difference Between God and Larry Ellison*, p. 305.

31.  Eric Knorr, *Multimedia World*, September 29, 1995.

32.  Christopher Barr, *c/net* (Internet news service), October 4, 1995.

33.  Woody Leonhard [publication unknown], December 8, 1995.

34.  Michael Meyer, "Is Your PC Too Complex? Get Ready for the NC," *Newsweek* (October 12, 1995).

35.  *Oprah*, NBC, September 15, 1996.

36.  John Foley, "Time to Deliver," *Information Week* (February 24, 1997).

37.  Quoted in Wilson, *Difference Between God and Larry Ellison*, p. 349.

38.  Gartner Group Report published in "Weighing the Case for the Network Computer," *Economist* ( January 18, 1997).

39.  Don Clark and David Bank, "Network Computers Fall Short in Contest Against Cheap PCs," *Wall Street Journal*, May 23, 1998.

40.  Julie Pitta, "Compu-money," *Forbes* (September 1998).

41.  Garth Alexander, "Giants Count Cost of Great PC Give-away," *Sunday Times*, April 25, 1999, (Business), p. 6.

42.  Ibid.

43.  Ibid.

44.  Steve Hamm, Jay Greene, and David Rocks, "Oracle: Why It's Cool Again," *Business Week* (May 8, 2000): 42–47.

45.  Ibid.

46.  Quoted by George Barrett, *New York Times Magazine*, October 20, 1957, p. 86.

CONCLUSION:  Winded by Feather Punches

1.  Quoted in Norman Lebrecht, *The Companion to Twentieth-Century Music* (London: Simon and Schuster, 1992), p. 307.

2.  Walter Gratzer, ed., *The Longman Literary, Companion to Science* (London: Longman, 1989), p. 454.

3.  Horace Freeland Judson, *The Eighth Day of Creation: Makers of the Revolution in Biology* (New York: Cold Spring Harbor Laboratory Press, 1996), p. 468.

4.  James Glanz, "What Fuels Progress in Science? Sometimes, a Feud," *New York Times*, September 14, 1999, p. F3.

# BIBLIOGRAPHY

Aczel, Amir D. *Fermat's Last Theorem: Unlocking the Secret of an Ancient Mathematical Problem.* London: Viking, 1997.

Aykroyd, W. R. *Three Philosophers: Lavoisier, Priestley and Cavendish.* London: Heinemann, 1935.

Baily, Francis. *An Account of the Revd. John Flamsteed, the First Astronomer Royal.* London, 1835.

Barbour, Ian G. *Issues in Science and Religion.* New York: Harper and Row, 1966.

Birch, Beverley. *Fleming.* Exley, 1990.

Bolles, Edmund Blair, ed. *Galileo's Commandment: An Anthology of Great Science Writing.* London: Little, Brown, 1997.

Bowlby, John. *Charles Darwin: A New Biography.* London: Hutchinson, 1990.

Bragg, Melvyn, with Ruth Gardiner. *On Giant's Shoulders: Great Scientists and Their Discoveries from Archimedes to DNA.* London: Hodder and Stoughton, 1997.

Brennan, Richard P. *Heisenberg Probably Slept Here: The Lives and Ideas of the Great Physicists of the Twentieth Century.* New York: John Wiley, 1997.

Bronowski Jacob. *The Ascent of Man.* London: BBC Books, 1973.

———. *Science and Human Values.* New York: Harper and Row, 1956.

Brown, Andrew. *The Darwin Wars: How Stupid Genes Became Selfish Gods.* London: Simon and Schuster, 1999.

Burkhardt, Frederick, ed. *Charles Darwin's Letters: A Selection, 1825–1859.* Cambridge: Cambridge University Press, 1996.

Carey, John, ed. *The Faber Book of Science.* London: Faber and Faber, 1995.

*Chambers Concise Dictionary of Scientists.* Cambridge: Cambridge University Press, Chambers, 1989.

Cheney, Margaret. *Tesla: Man out of Time.* New York: Dorset, 1989.

Clark, Ronald W. *Einstein.* New York: Avon, 1971.

———. *The Survival of Charles Darwin.* London: Weidenfeld and Nicolson, 1984.

Clarke, Arthur C., ed. *The Coming of the Space Age.* New York: Meredith, 1968.

Collins, Michael. *Flying to the Moon and Other Strange Places.* London: Piccolo, 1979.

Conn, Gordon, and Henry Turner. *The Evolution of the Nuclear Atom.* Boston: American Elsevier, 1965.

Crick, Francis. *What Mad Pursuit: A Personal View of Scientific Discovery.* London: Penguin, 1990.

Dampier, W. C. *A History of Science: And Its Relations with Philosophy and Religion.* Cambridge: Cambridge University Press, 1984.

Darwin, Charles. *Correspondence of Charles Darwin.* 8 vols. Edited by Burkhart and Smith. Cambridge: Cambridge University Press, 1985–1993.

———. *On the Origin of Species by Means of Natural Selection, or the Preservation of Favoured Races in the Struggle for Life.* 1859. Reprint, London: Penguin, 1985.

Darwin, Francis, ed., *Life and Letters of Charles Darwin.* 3 vols. London, 1887.

Darwin, Francis, and A. C. Seward, eds. *More Letters of Charles Darwin.* 2 vols. London: John Murray, 1903.

Davies, Kevin, and Michael White. *Breakthrough: The Quest to Isolate the Gene for Hereditary Breast Cancer.* London: Macmillan, 1995.

Davis, Nuel Pharr. *Lawrence and Oppenheimer.* New York: Simon and Schuster, 1968.

de Beer, Gavin, ed. *Charles Darwin and T. H. Huxley, Autobiographies.* Oxford: Oxford University Press, 1983.

Desmond, Adrian. *Archetypes and Ancestors: Palaeontology in Victorian London, 1850–1875.* London: Blond and Briggs, 1982.

———. *Huxley: The Devil's Disciple.* London: Michael Joseph, 1994.

———. *Huxley: Evolution's High Priest.* London: Michael Joseph, 1997.

Desmond, Adrian, and James Moore. *Darwin.* London: Michael Joseph, 1992.

Dornberger, Walter. *V-2.* New York: Viking, 1954.

Dyson, Esther. *Release 2.0: A Design for Living in the Digital Age.* London: Viking, 1997.

Eamon, William. *Science and the Secrets of Nature: Books of Secrets in Medieval and Early Modern Culture.* Princeton: Princeton University Press, 1994.

Einstein, Albert. *Ideas and Opinions.* New York: Bonanza, 1954.

Emsley, John. *The Shocking History of Phosphorus: A Biography of the Devil's Element.* London: Macmillan, 2000.

Ferguson, Charles, and Charles Morris. *Computer Wars: The Fall of IBM and the Future of Global Technology.* New York: Times Books, 1993.

Flew, Antony. *Malthus.* London: Pelican, 1970.

Friedman, Meyer, and Gerald Friedland. *Medicine's Ten Greatest Discoveries.* New Haven: Yale University Press, 1998.

Fromm, Erich. *The Anatomy of Human Destructiveness.* London: Pimlico, 1973.

Gartman, Heinz. *Science as History: The Story of Man's Technological Progress from Steam Engine to Satellite.* London: Hodder and Stoughton, 1961.

Gates, Bill. *Business @ The Speed of Thought: Using a Digital Nervous System.* London: Penguin, 1999.

———. *The Road Ahead.* New York: Viking, 1995.

Gjertsen, Derek. *Science and Philosophy: Past and Present.* London: Penguin, 1989.

Goodchild, Peter. *J. Robert Oppenheimer: Shatterer of Worlds.* Boston: Houghton Mifflin, 1980.

Gratzer, Walter, ed. *The Longman Literary Companion to Science.* London: Longman, 1989.

Gray, Mike. *Angle of Attack: Harrison Storms and the Race to the Moon.* New York: W. W. Norton, 1992.

Greenstein, George. *Portraits of Discovery: Profiles in Scientific Genius.* New York: John Wiley, 1998.

Gribbin, John. *Companion to the Cosmos.* London: Weidenfeld and Nicolson, 1996.

Groves, Leslie R. *Now It Can Be Told.* New York: Harper and Row, 1962.

Hager, Thomas. *Force of Nature: The Life of Linus Pauling.* New York: Simon and Schuster, 1995.

Hall, A. Rupert. *From Galileo to Newton.* New York: Dover, 1981.

———. *Philosophers at War: The Quarrel Between Newton and Leibniz.* Cambridge: Cambridge University Press, 1980.

Harford, James. *Korolev: How One Man Masterminded the Soviet Drive to Beat America to the Moon.* New York: John Wiley, 1997.

Harré, Rom. *Great Scientific Experiments: Twenty Experiments That Changed Our View of the World.* Oxford: Phaidon, 1981.

Heilbron, John L. *The Sun in the Church.* Cambridge, Mass.: Harvard University Press, 2000.

Heisenberg, Werner. *Physics and Beyond.* New York: Harper and Row, 1971.

Henshall, Philip. *The Nuclear Axis: Germany, Japan and the Atom Bomb, 1939–1945.* London: Sutton, 2000.

Hersey, John. *Hiroshima.* London: Penguin, 1946.

Hibbert, Christopher. *The French Revolution.* London: Penguin, 1980.

Hiscock, W. G., ed. *David Gregory, Isaac Newton and Their Circle: Extracts from David Gregory's Memoranda, 1677–1708.* Oxford: Oxford University Press, 1937.

Hobbes, Thomas. *Leviathan.* New York: Penguin, 1986.

Hodges, Andrew. *Turing.* London: Phoenix, 1997.

Hoffman, Joseph E. *Leibniz in Paris, 1672–1676. His Growth to Mathematical Maturity.* Cambridge: Cambridge University Press, 1974.

Hunter, Michael, ed. *Robert Boyle Reconsidered.* Cambridge: Cambridge University Press, 1994.

*Hutchinson Dictionary of Scientists.* London: Helicon, 1996.

Huxley, L. *Life and Letters of Thomas Henry Huxley.* 2 vols. London: Collins, 1900.

Irvine, William. *Apes, Angels and Victorians: A Joint Biography of Darwin and Huxley.* London, 1956.

Irving, David. *The German Atomic Bomb.* New York: Simon and Schuster, 1967.

Israel, Paul. *Edison: A Life of Invention.* New York: John Wiley, 1998.

Jones, H. F., ed. *Samuel Butler: A Memoir.* 2 vols. London: Macmillan, 1920.

Josephson, Matthew. *Edison: A Biography.* New York: McGraw-Hill, 1959.

Judson, Horace Freeland. *The Eighth Day of Creation: Makers of the Revolution in Biology.* New York: Cold Spring Harbor Laboratory Press, 1996.

Jung, C. G. *Man and His Symbols.* London: Aldus, 1964.

Kennan, George. *George F. Kennan Memoirs (1950–1963).* New York: Pantheon, 1972.

Kevles, Daniel J. *The Physicists: The History of a Scientific Community in Modern America.* Cambridge, Mass.: Harvard University Press, 1987.

Keynes, R. *Charles Darwin's Beagle Diary.* Cambridge: Cambridge University Press, 1988.

Khrushchev, Sergei. *Khrushchev on Khrushchev.* Boston: Little, Brown, 1990.

Kimball Smith, Alice, and Charles Weiner, eds. *Robert Oppenheimer: Letters and Recollections.* Palo Alto: Stanford University Press, 1980.

Kline, Morris. *Mathematics and the Physical World.* New York: Thomas Y. Crowell 1959.

Koestler, Arthur. *The Sleepwalkers: A History of Man's Changing Vision of the Universe.* London: Penguin, 1964.

Koppes, Clayton. *JPL and the American Space Program.* New Haven: Yale University Press, 1982.

Krige, John, and Dominique Pestre, eds. *Science in the Twentieth Century.* Amsterdam: Harwood Academic, 1997.

Lanouette, William, with Bela Szilard. *Genius in the Shadows: A Biography of Leo Szilard, the Man Behind the Bomb.* New York: Scribner's, 1993.

Lattimer, Dick, ed. *All We Did Was Fly to the Moon.* Gainesville, Fla.: Whispering Eagle Press, 1992.

Lavoisier, Antoine-Laurent. *The Elements of Chemistry.* Preface. Paris, 1789.

Lebrecht, Norman. *The Companion to Twentieth-Century Music.* London: Simon and Schuster, 1992.

Lehman, Milton. *This High Man: The Life of Robert H. Goddard.* New York: Farrar, Straus, 1963.

Leonardo da Vinci. *Trattato della pittura (Treatise on Painting)*. Edited by A. P. McMahon. Princeton: Princeton University Press, 1956.

Lewontin, R. C. *The Doctrine of DNA: Biology as Ideology*. London: Penguin, 1993.

Logsdon, John M. *The Decision to Go to the Moon*. Cambridge Mass.: MIT Press, 1970.

Lomas, Robert. *The Man Who Invented the Twentieth Century: Nikola Tesla, Forgotten Genius of Electricity*. London: Headline, 1999.

Lorimer, David, ed. *The Spirit of Science: From Experiment to Experience*. London: Floris Books, 1998.

MacDonald Ross, Gordon *Leibniz*. Oxford: Oxford University Press, 1984.

Magee, Bryan. *Popper*. London: Fontana, 1992.

McKie, Douglas. *Antoine Lavoisier*. New York: Da Capo, 1952.

Manchester, William. *A World Lit Only by Fire: The Medieval Mind and the Renaissance*. London: Macmillan, 1996.

Manes, Stephen, and Scott Andrews. *Gates: How Microsoft's Mogul Reinvented an Industry and Made Himself the Richest Man in America*. Garden City, New York: Doubleday, 1993.

Manuel, Frank E. *A Portrait of Isaac Newton*. New York: Da Capo, 1968.

Marinacci, Barbara, ed. *Linus Pauling: In His Own Words: Selections from His Writings, Speeches, and Interviews*. New York: Touchstone, 1995.

Mayr, Ernst. *One Long Argument: Charles Darwin and the Genesis of Modern Evolutionary Thought*. Cambridge, Mass.: Harvard University Press, 1991.

Meadows, Jack, ed. *The History of Scientific Discovery: The Story of Science Told Through the Lives of Twelve Great Scientists*. Oxford: Phaidon, 1987.

Merz, John Theodore. *Leibniz*. New York: Lippincott, 1884.

Metzger, T. H. *Blood and Volts: Edison, Tesla and the Electric Chair*. New York: Autonomedia, 1996.

Meyer, Arthur. *The Rise of Embryology*. Palo Alto: Stanford University Press, 1939.

Millar, David, et al., eds. *Chambers Concise Dictionary of Scientists*. Chambers, Cambridge: Cambridge University Press, 1989.

More, L. T. *The Life and Works of the Honourable Robert Boyle*. Oxford: Oxford University Press, 1944.

Moszkowski, Alexander. *Conversations with Einstein*. London: Sidgwick and Jackson, 1972.

Naughton, John. *A Brief History of the Future: The Origins of the Internet*. London: Weidenfeld and Nicolson, 1999.

Newton, Isaac. *Principia Mathematica*. Translated and edited by Florian Cajori. Berkeley and Los Angeles: University of California Press, 1934.

Oberg, James E. *Red Star in Orbit*. New York: Random House, 1981.

Olby, R. C., et al., eds. *Companion to the History of Modern Science*. London: Routledge, 1990.

O'Neill, John. *Prodigal Genius*. New York: Ives Washburn, 1944; reprint 1968.

Ordway, Frederick I., and Mitchell R, Sharpe. *The Rocket Team*. New York: Thomas Y. Crowell, 1979.

Pais, Abraham. *The Genius of Science: A Portrait Gallery of Twentieth-Century Physicists*. Oxford: Oxford University Press, 2000.

Passer, Harold. *The Electrical Manufacturers, 1875–1900*. New York: Arno, 1972.

Peat, David. *In Search of Nikola Tesla*. Bath: Ashgrove, 1983.

Pert, Candace B. *Molecules of Emotion: Why You Feel the Way You Feel*. New York: Simon and Schuster, 1997.

Perutz, Max. *I Wish I Had Made You Angry Earlier: Essays on Science and Scientists*. Oxford: Oxford University Press, 1998.

Popper, Karl. *The Open Society and Its Enemies*. New York: Routledge, 1966.

Porter, Roy. *The Greatest Benefit to Mankind: A Medical History of Humanity from Antiquity to the Present.* London: HarperCollins, 1997.

———. *Man Masters Nature: Twenty-five Centuries of Science.* London: BBC Books, 1987.

Powers, Thomas. *Heisenberg's War: The Secret History of the German Bomb.* London: Jonathan Cape, 1993.

Poynter, F. N. L., ed. *The Evolution of Medical Education in Britain.* London: Pitman, 1966.

Priestley, Joseph. *Experiments and Observations on Different Kinds of Air.* London: Royal Society, 1775.

———. *History of Electricity.* 1st ed. London: Royal Society, 1767.

———. *Memoirs of Dr. Joseph Priestley to the year 1795 with a Continuation by his Son.* 2 vols. London, 1806.

Pyenson, Lewis, and Susan Sheets-Pyenson. *Servants of Nature: A History of Scientific Institutions, Enterprises and Sensibilities.* London: HarperCollins, 1999.

Ralling, Christopher, ed. *The Voyage of Charles Darwin: His Autobiographical Writings Selected and Arranged.* London: BBC Books, 1979.

Redman, Ben Ray, ed. *The Portable Voltaire.* New York: Viking, 1949.

Regis, Ed. *Who Got Einstein's Office?: Eccentricity and Genius at the Princeton Institute for Advanced Study.* London: Penguin, 1989.

Rhodes, Richard. *Dark Sun: The Making of the Hydrogen Bomb.* New York: Simon and Schuster, 1995.

———. *The Making of the Atomic Bomb.* London: Penguin, 1986.

Richter, Jean Paul, ed. *The Notebooks of Leonardo da Vinci: Compiled and Edited from the Original Manuscripts, vols. 1 and 2.* New York: Dover, 1970.

Rivlin, Gary. *The Plot to Get Bill Gates: An Irreverent Investigation of the World's Richest Man . . . and the People Who Hate Him.* London: Quartet Books, 1999.

Rohm, Wendy Goldman. *The Microsoft File: The Secret Case Against Bill Gates.* New York: Times Business, Random House, 1998.

Rupke, Nicolaas, A. *Richard Owen: Victorian Naturalist.* New Haven: Yale University Press, 1994.

Schrödinger, Erwin. *What Is Life?* Cambridge: Cambridge University Press, 1967.

Shefter, James. *The Race.* New York: Doubleday 1999.

Shepard, Alan, and Deke Slayton. *Moon Shot.* Atlanta, Ga.: Turner, 1994.

Sherwin, Martin. *A World Destroyed.* New York: Alfred A. Knopf, 1975.

Shroyer, Jo Ann. *Secret Mesa: Inside Los Alamos National Laboratory.* New York: John Wiley, 1998.

Silver, Brian L. *The Ascent of Science.* Oxford: Oxford University Press, 1998.

Simmons, John. *The 100 Most Influential Scientists: A Ranking of the 100 Greatest Scientists Past and Present.* London: Robinson, 1997.

Smith, Alice Kimball, and Charles Weiner, eds. *Robert Oppenheimer: Letters and Recollections* (Palo Alto: Stanford University Press, 1980).

Smith, Arthur. *Planetary Exploration: Thirty Years of Unmanned Space Probes.* Wellingborough: Patrick Stephens, 1988.

Smolin, Lee. *The Life of the Cosmos.* London: Weidenfeld and Nicolson, 1997.

Spangenburg, Ray, and Diane K. Moser. *Wernber von Braun: Space Visionary and Rocket Engineer.* New York: Facts on File, 1995.

Storr, Anthony. *Music and the Mind.* London: HarperCollins, 1992.

Strathern, Paul. *Oppenheimer and the Bomb.* London: Arrow, 1998.

Teller, Edward. *The Legacy of Hiroshima.* Garden City, New York: Doubleday, 1962.

Terkel, Studs. *The Good War.* New York: Pantheon, 1984.

Tesla, Nikola. *My Inventions: The Autobiography of Nikola Tesla.* Austin, Tex.: Hart Brothers, 1982.

Turnbull, H. W., et al., eds. *The Correspondence of Isaac Newton.* 7 vols. Cambridge: Cambridge University Press, 1959–1977.

Van Doren, Charles. *A History of Knowledge: The Pivotal Events, People, and Achievements of World History.* New York: Ballantine, 1991.

Wallace, James. *Overdrive: Bill Gates and the Race to Control Cyberspace.* New York: John Wiley, 1997.

Wallace, James, and Jim Erickson. *Hard Drive: Bill Gates and the Making of the Microsoft Empire.* New York: HarperCollins, 1993.

Watson, James. *The Double Helix: A Personal Account of the Discovery of the Structure of DNA.* London: Weidenfeld and Nicolson, 1968.

———. *A Passion for DNA: Genes, Genomes and Society.* Oxford: Oxford University Press, 2000.

Weart, Spencer, and Gertrude Szilard, eds. *Leo Szilard: His Version of the Facts.* Cambridge, Mass.: MIT Press, 1978.

Weiner, Jonathan. *The Beak of the Finch: An Extraordinary Scientific Adventure Story About Birds, Biology and the Origin of Species.* London: Jonathan Cape, 1994.

White, Michael, *The Last Sorcerer.* London: Fourth Estate, 1997.

———. *Leonardo: The First Scientist.* London: Little, Brown, 2000.

White, Michael, and John Gribbin. *Darwin: A Life in Science.* London: Simon and Schuster, 1995.

———. *Einstein: A Life in Science.* London: Simon and Schuster, 1993.

Wilson, Mike. *The Difference Between God and Larry Ellison.* New York: William Morrow, 1998.

Wolpert, Lewis, and Alison Richards. *A Passion for Science: Renowned Scientists Offer Vivid Personal Portraits of Their Lives in Science.* Oxford: Oxford University Press, 1988.

Wyndham, John, and Lucas Parkes. *The Outward Urge.* London: Penguin, 1967.

# INDEX